Environment and the Developing World

Principles, Policies and Management

AVIJIT GUPTA
School of Geography, University of Leeds

and

MUKUL G. ASHER
Department of Economics and the Public Policy
Programme, National University of Singapore

JOHN WILEY & SONS
Chichester • New York • Weinheim • Brisbane • Singapore • Toronto

Baffins Lane, Chichester,
West Sussex PO19 1UD, England

National 01243 779777
International (+44) 1243 779777
e-mail (for orders and customer service enquiries): cs-books@wiley.co.uk
Visit our Home Page on http://www.wiley.co.uk
or http://www.wiley.com

John Wiley & Sons, Inc., 605 Third Avenue,
New York, NY 10158-0012, USA

WILEY-VCH Verlag GmbH, Pappelallee 3,
D-69469 Weinheim, Germany

Jacaranda Wiley Ltd, 33 Park Road, Milton,
Queensland 4064, Australia

John Wiley & Sons (Asia) Pte Ltd, 2 Clementi Loop #02-01,
Jin Xing Distripark, Singapore 129809

John Wiley & Sons (Canada) Ltd, 22 Worcester Road,
Rexdale, Ontario M9W 1L1, Canada

Library of Congress Cataloging-in-Publication Data

Gupta, Avijit.
Environment and the developing world : principles, policies, and management / Avijit Gupta and Mukul G. Asher.
p. cm.
Includes bibliographical references and index.
ISBN 0-471-96604-5 (cloth : alk. paper). — ISBN 0-471-98338-1 (pbk. : alk. paper)
1. Environmental policy—Developing countries. 2. Environmental policy—Tropics. 3. Developing countries—Economic policy. 4. Tropics—Economic policy. 5. Tropics—Environmental conditions. I. Asher, Mukul G. II. Title.
GE190.D44G86 1998
363.7'009172'4—dc21 97–51765
CIP

British Library Cataloguing in Publication Data

A catalogue record for this book is available from the British Library

ISBN 0-471-96604-5 (hardback)
0-471-98338-1 (paperback)

Typeset in 10/12pt Times from the authors' disks by Dorwyn Ltd, Rowlands Castle, Hants
Printed and bound in Great Britain by Bookcraft (Bath) Ltd, Midsomer Norton
This book is printed on acid-free paper responsibly manufactured from sustainable forestry, for which at least two trees are planted for each one used for paper production.

Environment and the Developing World

To Ella and Radha

Contents

PART II. MANAGING THE ENVIRONMENT

Preface

We wrote this book because we needed a text which specifically and simultaneously discusses environmental management, economic development and the tropical environment. Environmental issues require a multifaceted and multidisciplinary approach. Environmental managers, however, are usually trained in a single discipline, such as earth or natural sciences, economics or engineering. The demand of the job then requires them to cultivate a multidisciplinary approach. *Environment and the Developing World* is a textbook which provides a broad-based multidisciplinary exposure at the advanced undergraduate level.

The book is in three parts. The first part is an introduction which describes the physical environment of the tropics (where most of the developing countries are located), the nature of global population, and the principles of environmental economics. This provides a theoretical base prior to discussing environmental issues. A reader who is familiar with one or more of these topics may, however, continue to the next.

The second part of the book deals with specific environmental issues such as land, water and air, and techniques of environmental impact identification. In the final section, the nature and mechanisms of environmental governance are discussed at various levels: global, national and local. This part also includes discussions on the history of the environmental movement, sustainable development, the role of international organizations such as the World Bank and the United Nations in environmental management, and the increasing involvement of NGOs and industrial establishments.

The text has been strengthened by a large number of examples from a wide range of the developing countries and from different types of environment. The more important examples have been highlighted as boxed case studies. Most chapters end with an exercise, the function of which is to make readers think and also to expose them to real-life situations, where the required information is not always available for decision-making. This is particularly useful in the context of the developing world. All monetary values are given in US dollars.

We would like to thank Stephen Lintner and Susan Becker for providing us with the extremely useful background information regarding environmental management practices in the World Bank and the United Nations Development Programme respectively. We are indebted to A. K. Saha, Dipankar Chakraborty and

Goutam Ghosh for discussions and information regarding the environment of Calcutta. We acknowledge the technical advice and help received from John Boland, John Costa and Helen Fox-Moody. Figure 5.5 is an outcome of the National University of Singapore Research Project RP3890040. We thank the United Nations Publications Board for sending us a copy of their forthcoming book on world urbanization prospects. Anthea Fraser Gupta has critically read the entire text. We have also benefited from M. G. Wolman and Neil Buhne's reading of Chapters 13, 15 and 16 which deal primarily with environmental governance; Ian Turner's comments on Chapter 5 on vegetation; and G. Shantakumar's review of Chapter 3 on population. All the diagrams were drawn by Lee Li Kheng.

We thank the individuals and organizations who gave us permission to reproduce their texts, tables and diagrams. We have acknowledged them in the appropriate places.

Part I
Basic Tools and Concepts

1

Introduction

1.1 Environment and Development

The word *environment* has been around for a long time but the importance and usage of the word has increased dramatically over the last forty years. Traditionally it has been used to describe the surroundings in which an organism lives. The term *ecology* is also used as a surrogate for 'environment', although, strictly speaking, ecology is the study of the relationship between living organisms and their environment. The relationship is a two-way street: environment and organisms affect each other. For people, environment is the set of physical and social conditions that determines the existing living conditions. Here the relationship is highly reciprocal. People modify their environmental conditions to such an extent that the altered environment in turn tends to have an impact on their lives. The impact could be either positive or negative. Providing a city with clean drinking water is a positive impact; a city discharging polluted water to a stream is a negative one.

Development refers to the better utilization of factors of production, including natural resources, leading to improvement of the income and the quality of life for a set of people. It is thus a wider concept than that of higher income or economic growth. Economic policy-making and political and media attention tend to centre on economic growth instead of development, as quality of life indicators are difficult to quantify with sufficient vigour and regularity. An attempt has been made to incorporate both growth and quality of life indicators together in an index, called the Human Development Index (United Nations Development Programme, 1995). Unfortunately, this index does not contain any component directly relating to environment.

Both growth and development were earlier perceived as being in conflict with the environment. A water power project which requires the construction of a dam and a reservoir will alter some of the characteristics of the physical environment of the river basin, and certainly that of the river. Such alterations may have striking negative impacts on the physical environment. Furthermore, they may force a number of the residents of the valley to relocate themselves as their settlements are submerged. On the other hand, it is difficult to exist without electricity. Environment is for the people. So is development.

Over time this perception of development being a degrader to environment has changed. We now attempt to integrate environmental conservation with economic development. It is perceived more and more that development which is sustainable comes only when environmental conditions (both physical and social) are respected. Environmental management is involved with the principles and techniques for achieving this.

It should be noted, however, that development may come at a price. It may be impossible to extract mining resources, supply water to agricultural fields or generate power without causing some degradation to the environment. Environmental management keeps this price as low as possible, and also considers modifying the project to bring in some positive impacts. This, however, is not an easy task and, besides strong motivation on the part of the manager, requires expertise in many areas. An environmental manager often needs to deal with experts from four disciplines beginning with an 'e': earth sciences, ecology, economics and engineering. The environmental manager may also need to consult experts in areas other than these, such as organic chemistry or law, depending on the development project and the specific nature of environment involved. The manager, however, should be in a position to communicate properly with the experts, and be able to implement their recommendations. A good environmental manager is not only exposed to a wide range of disciplines but is also a specialist in at least one of them. This book provides a broad foundation of environmental knowledge and as such helps in fulfilment of the first requirement.

Box 1.1

The Planned Projects on the Narmada River, India and the Possible Impacts

The Narmada (Figure 1.1) is the biggest west-flowing river of India, 1300 km long and draining a long and narrow basin, nearly 99 000 km^2 in area. The river flows through three rock gorges with alluvial plains in between. It is entirely rain fed, with nearly 90 percent arriving in the wet monsoon between June and September. Occasional tropical storms by bringing in very high rainfall in the middle of the wet season contribute to large floods on the Narmada. Although a large proportion of the basin is under forest or is thinly populated, there is a considerable demand for water in the region.

Plans for using the river water mainly for irrigation and power generation had been around for a number of years but only in 1979 was a final decision taken to share the water between four neighbouring states: Madhya Pradesh, Gujarat, Rajasthan and Maharashtra. Almost all the benefits, however, are planned to go to Madhya Pradesh and Gujarat, the states through which the river flows.

The proposal has raised considerable controversy and has involved a number of non-governmental organizations. Arguments both for and against the projects are summarized here without attempting to take sides. The Narmada projects illustrate the necessity of and difficulty in environmental management when development projects (especially large-scale ones) are put into operation. Failure to do so properly leads to serious environmental degradation as many examples presented later in this book will testify.

Figure 1.1 *The Narmada River at Dardi Falls, near the site of the proposed Indira Sagar Dam and reservoir*

The projects will take decades to complete, and in their final form a number of dams and reservoirs will be constructed across the basin. Of these, the two biggest reservoirs will be Sardar Sarovar in Gujarat and Indira Sagar in Madhya Pradesh (Figure 1.2). Both dams are designed mainly for irrigation and power generation in co-ordination with other dams and reservoirs, such as the planned Omkareshwar multipurpose project below Indira Sagar, and the Maheswar power project further downstream between Omkareswar and Sardar Sarovar. Flood control is not a major feature of the projects.

There is a need for water in this region for various reasons. Taken together, the two large projects are expected to irrigate about 30 000 km^2 of land. There is also a small demand for drinking water and industrial water which is expected to rise in the future. The demand for power at present is not fully met and more power is needed for domestic use, industries, and modernized agriculture which uses electric pumps to raise water from irrigation canals to the fields.

The environmental consequences of the projects could be serious. A large number of people need to be settled out of the area designated for inundation, and the livelihood of an even larger number will be affected. With estimates varying between about 160 000 to nearly one million, the real number probably lies somewhere in between. Controversies also exist regarding the amount of forested or cultivated land that will be submerged. A loss of flora and fauna is also possible. The river carries a large sediment load, especially in floods, and concern has been expressed regarding rapid filling in of the reservoirs. The Narmada Valley is a seismically active area and the weight of the huge volume of water stored in the reservoirs may cause earthquakes. An earthquake, reaching a magnitude of 6

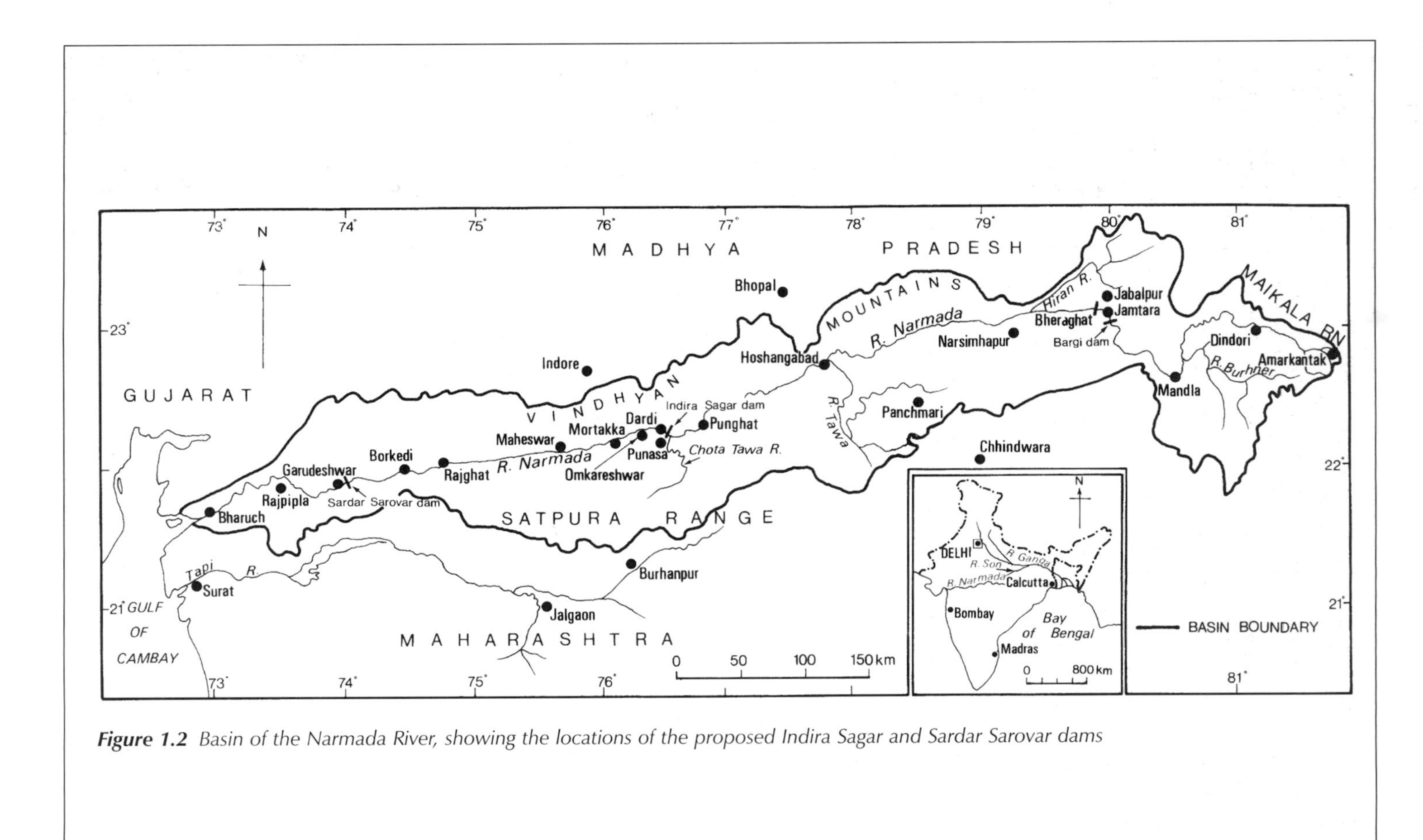

Figure 1.2 Basin of the Narmada River, showing the locations of the proposed Indira Sagar and Sardar Sarovar dams

on the Richter scale, occurred naturally on 22 May 1997. The water from the reservoirs may seep into the subsurface and raise the groundwater table, leading to waterlogging. Irrigation in dry areas, as in parts of Gujarat, may lead to deposition of a crust of salt on agricultural fields, a process known as salinization. Waterborne diseases, particularly malaria, may spread after the reservoirs are filled. Lastly, the Narmada is an extremely holy river to the Hindus and its banks are lined with temples, some of which are of great antiquity. A substantial number of these may be inundated.

All these concerns are rational, and some of these could be taken care of by careful planning. For example, afforestation elsewhere could compensate for the forest loss. Lined irrigation canals and regular draining of agricultural fields are techniques against salinization. But some of the environmental degradation on the list will certainly occur. As experience of past large water-related projects indicates, at times unforeseen types of environmental degradation also take place. On the other hand, at least a part of the planned benefits should arrive to help the local inhabitants in the future. But given such a long list on both sides it is not surprising that considerable controversy has arisen around the set of Narmada projects, and such controversy is likely to continue until some time after completion when hindsight will reveal what the correct decision should have been.

1.2 Environment at Different Scales: Local, National and Global

Two points should be kept in mind while managing the environment. First, environment is both degraded and managed at various levels: local, national and global. Second, all levels are hierarchically linked. Extraction of timber from the rain forest is a common example of environmental degradation. Removal of a number of large trees from a hillslope creates a *local* condition of intense erosion, gully development, sediment accumulation in stream channels, and increased flood potential. Deforestation of many hillslopes creates loss of *national* biodiversity, *regional* erosion and floods, and unrecoverable loss of a *national* resource (forests). In extreme cases, the country concerned may end up needing to import timber several years later. The current removal of the rain forest in many tropical countries has raised the degradation to the *global* level where the problems are extensive biodiversity losses and global warming (Setzer and Pereira, 1991).

Environmental management also may need to operate at the same three hierarchical levels. At the global scale, a world arrangement between countries is needed to control rain forest degradation which requires international agreements and the supervision by a world agency such as an agency of the United Nations. Some of the arrangements attempted at the Rio de Janeiro summit on the environment in June 1992 were of this nature. Proper forest management, however, is possible only as a part of the policy at the national level. This requires planning, regulations and monitoring by the national government. The regulations, however, only become effective if the forest is supervised by local officials or residents. Proper environmental management happens when management and planning are carried out at each of the three levels and acts of management at different scales are hierarchically co-ordinated. This is difficult but necessary.

Environment is not necessarily a state prerogative. It has always attracted public participation and pressure groups. Such pressure groups or non-governmental organizations (NGOs) continue to perform two important roles in environmental management. First, local NGOs often identify an environmental problem before it receives governmental recognition. Second, they also pressurize the authorities into taking action. The impacts of a project are properly understood only when two classes of information are available: an account of the environment prior to project implementation (baseline data), and an understanding of the operating local environmental processes such as rainfall, river systems, vegetation dynamics, income sources, asset distributions, health hazards, etc. Baseline research is the first step to environmental management and one should not wait until an act of environmental degradation actually occurs before getting involved in the acquisition of baseline information.

1.3 Environment and Sustainable National Income

Construction of *national income accounts*, which quantify a country's overall economic performance and its major economic components, has been a most important innovation. These accounts enable policy-makers and other participants in a country's economy to take informed decisions.

The most widely used measure of the conventional national income is the Gross Domestic Product (GDP). GDP provides market-based valuation of a country's total output of all goods and services during a given period. Environment, of course, is affected in the process of production and generation of GDP. Conventional National Income Accounts do not take into consideration the environmental factor. To address this deficiency, the accounts should be adjusted for the value of environmental degradation and resource depletion. If this is not carried out, then policy decisions are taken on the basis of inappropriate indicators. When such adjustments are made, a measure of sustainable national income is available.

The amount which the people of a country can consume without impoverishing themselves of future resources is known as Sustainable National Income (SNI). Estimating SNI requires making adjustments in monetary terms to conventional national income estimates. Such adjustments are done in two areas: environmental degradation and resource depletion. Environmental degradation includes items such as soil erosion or pollution of air or water. Resource depletion involves the extraction of non-renewable natural resources, e.g. timber or coal, which leaves less available for future use.

The difference such adjustments cause to a country's conventional accounts was discussed by Steer and Lutz (1994) who used the data from an earlier case study for Mexico. Net Domestic Product (NDP), not GDP, is used in the study. NDP is GDP less depreciation of human-made capital such as machine tools and computers. For Mexico in 1985, resource depletion reduced NDP by 5.7 percent. Such resource depletion involved oil (3.5 percent), land use changes (1.8 percent) and timber (0.4

percent). Furthermore, environmental degradation reduced it by another 7.6 percent. Such degradation included air pollution (3.9 percent), water pollution (1.6 percent), soil erosion (1.1 percent), solid wastes (0.5 percent) and groundwater (0.5 percent). The total reduction was 13.3 percent. The environmentally adjusted 1985 NDP for Mexico was 86.7 percent of the original amount. The adjusted national accounts therefore better indicate the reality by providing the necessary database which represents growth and environment simultaneously. To ascertain whether a country's current production and consumption levels are sustainable in future, it is essential to make two more adjustments which are seldom undertaken. The first is to account for the environmental damage done by a country to the commons. For example, the United States is responsible for a very high share of the damage to the global commons such as the ozone layer and the lower atmosphere. Second, depreciation of the national capital embodied in international trade should also be included. As an example, a country may maintain its lifestyle through more pollution or natural resource intensive import, thus degrading the environment of the exporting countries.

Productive capacity of the economy depends on natural (physical), social and human capital as well as human-made capital. Conventional National Income Accounts consider changes in stock of human-made capital but not of other types. Exclusion of natural resource depletion is clearly inappropriate. Common (1995) has defined such costs as 'the reduction over the period in the value of the total stock of environmental assets, to be measured by multiplying the physical change in the size of each environmental asset by a corresponding unit price and then summing the values arising across the different assets.' It is a major task requiring enormous amounts of information, resources, and technical expertise. It is also difficult to assign monetary values to specific environmental impacts such as air pollution.

Many high-income countries (e.g. USA, Japan, the Netherlands, Denmark, Sweden and Norway) are actively undertaking the task of constructing accounts which consider the environment. In the developing countries, however, this task has barely begun. The United Nations, in collaboration with the World Bank, is developing a system of Integrated Environmental and Economic Accounts (United Nations, 1993a). Some progress has been made, particularly regarding the linking of production and consumption activities with the use of natural resources, but a lot more work in this area needs to be done. The current trend seems to be in favour of keeping the environmental accounts separate from the conventional national accounts until further progress is made (Bloem and Weisman, 1996).

1.4 Environment and the Developing World

The nature of the developing world emphasizes certain environmental and project characteristics. The developing countries are located almost entirely within the tropics and the subtropics. Their physical environment is therefore controlled to some extent by their climate, although a considerable variation in the physical environment does occur within the tropics. An environmental manager has to be especially careful regarding certain sensitive areas of the tropical environment. A few examples of such sensitive areas are as follows:

- given the nature of tropical rainfall, any destruction of a vegetation cover in the humid tropics leads to very high erosion and associated problems;
- soil fertility is generally low, apart from the soil on floodplains and volcanic slopes;
- the plant and the animal worlds are extremely rich in species.

On the other hand, certain types of development projects are common in these countries. These include projects related to water, agriculture, power generation, and population distribution and control. Attempts are sometimes made to achieve a more equitable distribution of land and income. Currently the cities of the developing world are expanding rapidly and require careful planning. For example, the air pollution in many cities is becoming a serious hazard.

It is therefore necessary to have a basic knowledge of the tropical environment, the socio-economic conditions of these countries, their limited political clout on the world scene, and also the standard problems that may arise from the common types of development projects. The baseline information is growing, but in comparison to the developed nations, it is still rather limited. This hampers proper environmental management. Again, the low-income status of most of the citizens of the developing world necessitates rapid implementation of a large number of development projects. A large part of environmental management in the developing countries is therefore in designing development projects in order to maximize their benefits while at the same time reducing environmental degradation as far as possible. This is not an easy task, given the general shortage of funds and technology, and in many areas, lack of environmental information and political will. It has been possible, however, to circumvent some of these difficulties by sheer innovation and commitment but such examples are not as common as they should be.

1.5 The Basic Structure of the Book

In this book we chart the successes and failures of such projects to illustrate the principles of environmental management in the developing countries. Basic concepts of the physical and socio-economic environment are introduced in Chapters 2–4 which form Part I of this book. The objective is to provide all readers with a conceptual and information base for the rest of the book. A reader familiar with the content of any of these chapters can proceed to the next one. In Part II, specific aspects of environment and projects are discussed, mostly at local and regional scales. Part III describes the global scale of environmental problems and management, and presents the environmental status of the developing countries against such a background. A number of the problems that affect the world, including the developing countries, are best managed at the global scale which requires world agreement and co-operation. There are some success stories but also a lot of failures.

Exercise

Collect information on the Narmada projects or any other large-scale river development project such as the Aswan Dam or the Three Gorges project. List both the benefits and negative impacts of the chosen project. Do you approve of the project? Would you want to modify it? Would you want to abandon it altogether?

2

Introduction to the Physical Environment

2.1 The Planet Earth

Our planet is a closed system. It neither loses nor gains matter apart from the arrival of meteorites and the departure of spacecrafts which are not programmed to return. Energy behaves differently to meteorites and spacecrafts. The earth receives short-wave solar radiation on its surface, and in exchange radiates back enough long-wave terrestrial energy to space to maintain the heat balance. Otherwise the earth would get progressively hotter or cooler.

Many constituents of the earth, e.g. water or carbon, occur naturally in different parts of the planet and in different forms. Water may exist underground, on land, in oceans, and in the air. It exists in solid, liquid and gaseous states. Carbon occurs inside the earth in rocks, on the land surface in plants and animals, in water, and in the air as a constituent of certain gases, e.g. carbon dioxide (CO_2). Burning of vegetation increases the amount of CO_2 in the air. Plants absorb CO_2 from the air to reduce the amount of carbon in the atmosphere. The earth's constituents are thus recycled, although the total amount remains constant.

Anthropogenic activities tend to alter the natural location of the earth's constituents which in turn may create environmental problems. Increasing the amount of CO_2 in the air, for example, brings in global warming. Rivers carry less sediment downstream from dams after most of the sediment load is deposited in the reservoirs. This alters the characteristics of rivers downstream, and deltas at river mouths retreat from the sea as the dwindled sediment supply cannot keep pace with coastal erosion. Such problems have become extremely serious world-wide, and the bulk of the book deals with these. However, in order to understand the current state of our environment, we first need to know the natural conditions of our planet. In order to measure the degradation we have caused, we need to know what we started with. Knowledge of the earth material and of the various processes that transfer and store such material in different locations on our planet are necessary before we can plan any recovery from most types of environmental degradation.

This chapter is a brief presentation of such information. Further details are available in standard introductory accounts, such as Strahler and Strahler (1992).

2.2 The Material and Structure of Our Planet

The earth has a layered structure (Figure. 2.1). Three compositionally different units form its interior. At its centre lies a dense spherical mass of iron and nickel which is known as the *core*. The core is surrounded by a shell of different material which is called the *mantle*. The mantle is less dense, mostly in a solid state, and contains mainly silicon, magnesium, iron and oxygen. The thinnest outermost shell enveloping the mantle is the *crust*. Chemically, the crust is mainly composed of oxygen, silicon, aluminium, calcium, sodium, potassium and magnesium. Our environment mostly involves the top of the crust where its rocky shell (lithosphere) interfaces with water on the surface of the planet (hydrosphere) and the shell of air (atmosphere) on top. Building materials (e.g. carbon, water, sulphur etc.) move from one sphere to another across this interface.

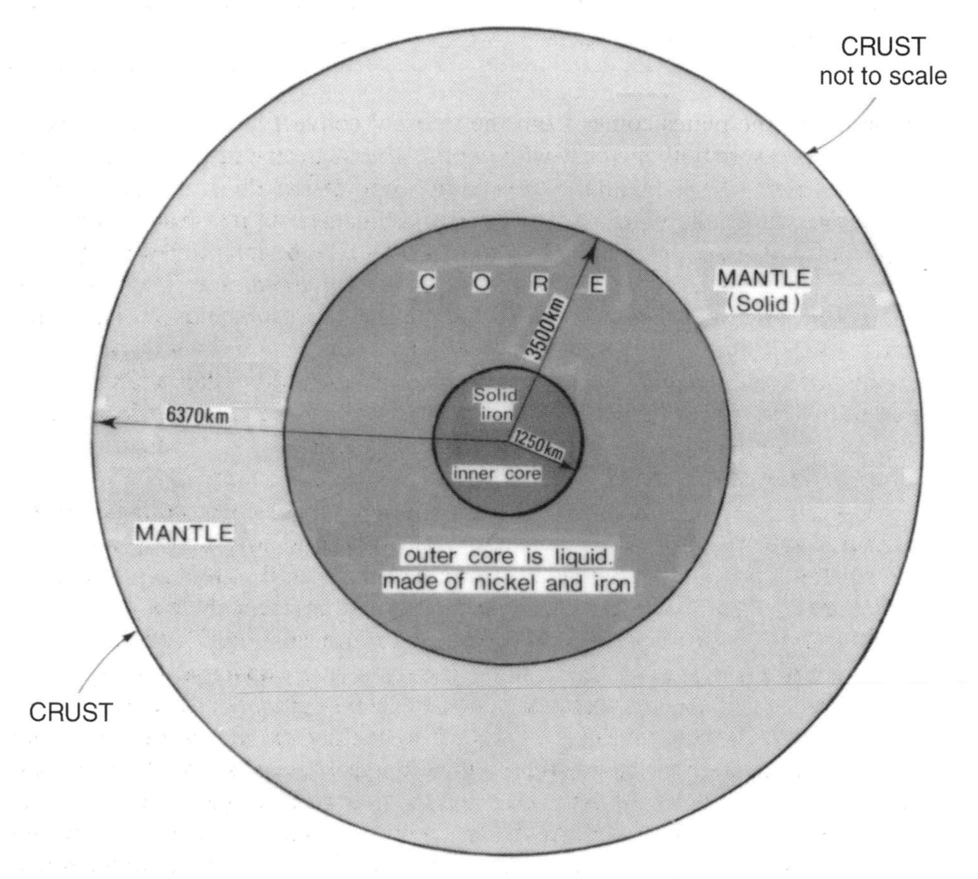

Figure 2.1 *The interior of the earth*

The thickness of the crust is variable, being about 8 km under the oceans (the oceanic crust) and on average 45 km under the continents (the continental crust). The thickness of the continental crust ranges from 30 to 70 km. Its compositional difference makes the oceanic crust denser than the continental one. The surface of the crust is divided into several units which are called plates (Figure 2.2). A plate can be entirely oceanic, entirely continental, or composed of both types. The plates are mobile, although they tend to move only at a pace of several centimetres per year. The time available for their movement, however, is vast and counted in hundreds of thousands of years. The present pattern of plates, for example, started to evolve from a single continent and a single ocean about 200 million years ago. They have taken all this time to move into their present positions to give us the map of the world we are familiar with (Figure 2.2). If we can find a time machine and travel 100 million years in the past, not only will we see dinosaurs but the map of the world's continents and oceans will also look very different. Similarly, if we were to travel 10 million years into the future, there would not be much point in carrying a current atlas with us.

2.3 Plate Tectonics and the Surface Configurations

The mobility of the plates comes from the internal convection of the earth. Temperature and pressure both increase with depth, and rocks melt into a fluidized state in the upper part of the mantle. Convectional circulation then carries hot fluid material upwards to below the crust. Where such material arrives from below, the crust splits into two separate plates (Figure 2.3), which then move slowly away from each other. This separation between the two plates is known as a divergent plate boundary. It is also known as a divergent margin or a spreading centre. New rock (i.e. crust) is created at this margin by material moving into the void from below and solidifying. This is happening now, for example, below the Atlantic Ocean. The submerged volcanic Mid-Atlantic Ridge, parts of which are exposed in volcanic islands (e.g. Iceland) are being built in this fashion. The Rift Valley of East Africa is a location where the African Plate is breaking up in a similar way.

As the surface of the earth is finite, creation of new crust has to be balanced by destruction of old crust. This happens when two plates come together and collide. If we assume that one of them is made of oceanic crust and that the other is continental in nature, the denser oceanic crust tends to be forced below the lighter continental variety after collision. The process is known as subduction and the contact is known as a convergent plate boundary or a convergent margin. As subduction continues, the advanced edge of the diving plate reaches a zone where temperature and pressure are high enough to melt the rock into a fluidized state and thereby return the material to the mantle. The process of subduction or crust destruction therefore balances crust creation at divergent margins. However, the pushing down of one solid plate below another solid plate creates friction and stress. The stress builds up until the rocks snap back, giving rise to vibrations which we know as earthquakes. The frictional heat and the high temperature and pressure

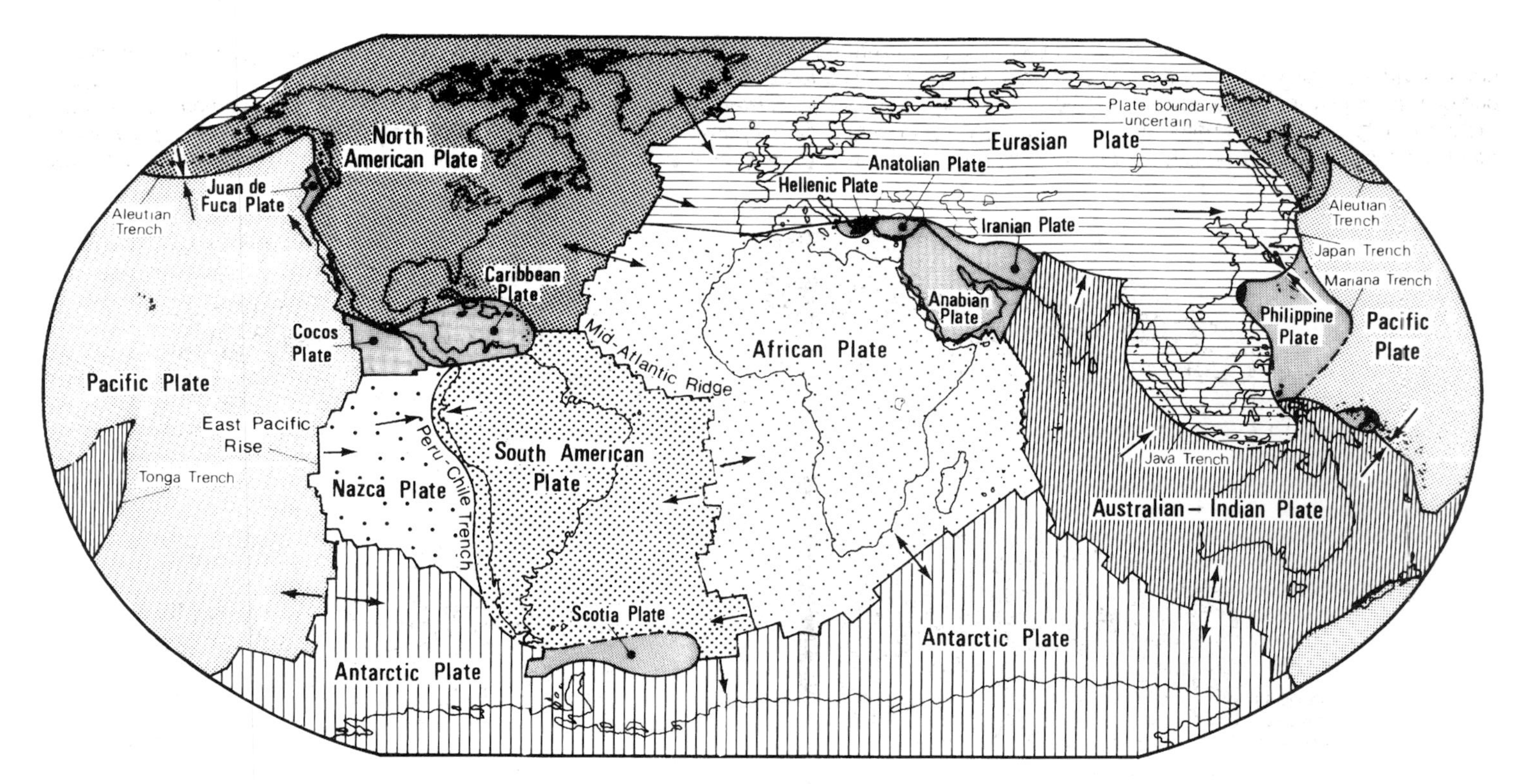

Figure 2.2 *Six large and a number of small lithospheric plates cover the surface of the earth, moving at the rate of several centimetres per year in the directions shown by the arrows. Volcanic activities, earthquakes and mountain-building occur at the plate boundaries. From Skinner, B.J. and Porter, S.C. (1995)* The Dynamic Earth: An Introduction to Physical Geology. *Copyright © 1995 John Wiley & Sons, Inc. Reprinted by permission of John Wiley & Sons, Inc.*

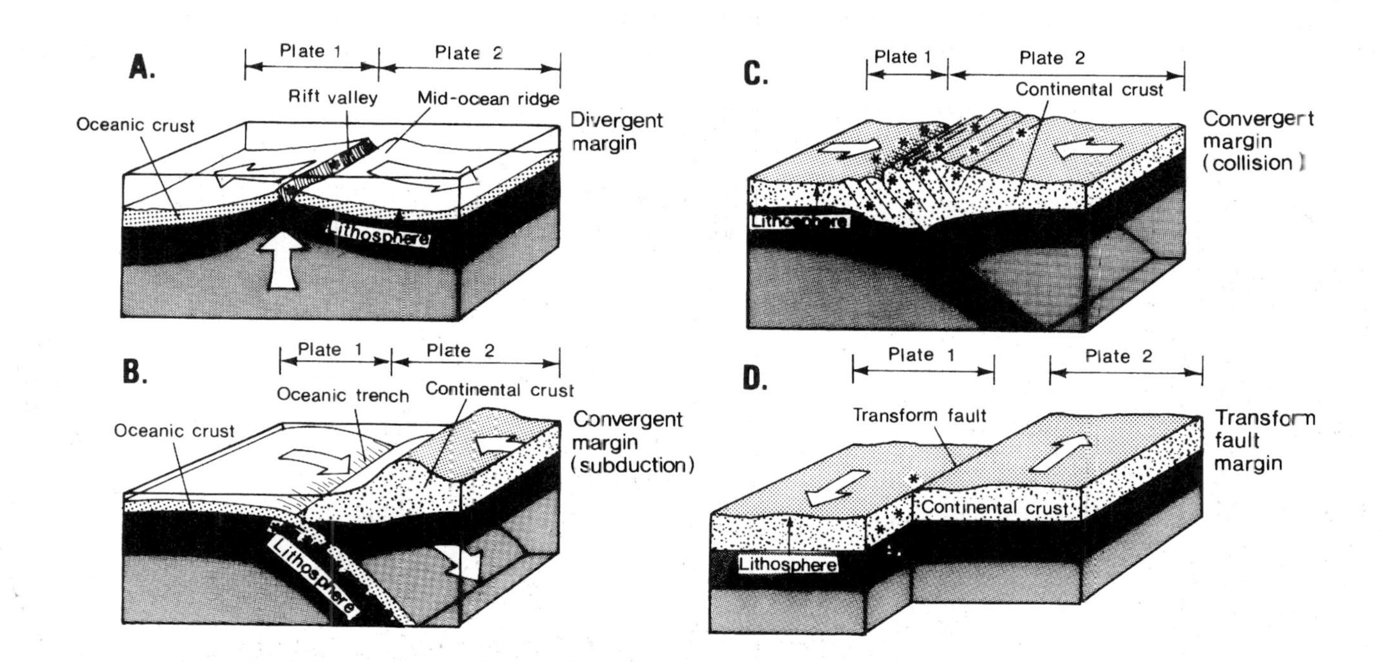

Figure 2.3 *Schematic sketches of the types of plate boundaries (margins). (A) Divergent boundary building a mid-oceanic ridge. (B) Convergent boundary between an oceanic and a continental plate indicating subduction, a seafloor trench, and a volcanic mountain range. (C) Convergent boundary between two continental plates, where collision and squeezing of the rocks between the plates leads to the formation of a folded mountain range. (D) Transform boundary associated with earthquakes. From Skinner, B.J. and Porter, S.C. (1995)* The Dynamic Earth: An Introduction to Physical Geology, *Copyright © 1995 John Wiley & Sons, Inc. Reprinted by permission of John Wiley & Sons, Inc.*

at greater depths melt rocks. Parts of this melt then move up as fluids which are less dense and come out at the surface to build volcanoes. Convergent margins where subduction takes place are therefore areas prone to earthquakes and have a range of steep volcanoes which erupt from time to time next to an oceanic trench created by subduction (Figure 2.3). Examples of such areas cover Indonesia, the Philippines, Japan, and the Andes Mountains of South America.

A variation of the convergent margin occurs when two continental plates collide with each other. As both plates are of comparable density, subduction is unimportant, and crustal material is piled up as folds between the two plates at the collision zone to form mountains. Earthquakes are common if the plates continue to converge. The Himalaya Mountains were built in this way, with the Indian plate colliding with the Eurasian plate. Areas at or near convergent margins should be developed with caution. Apart from natural disasters such as earthquakes and volcanic eruptions, the resultant steep mountains make these areas fragile and prone to landslides.

The third type of plate margin is the transform fault margin. Here two plates slip past each other. In this process, their ragged edges sometimes lock and stress builds up to be released in earthquakes. Such a boundary runs roughly north–south near the coast of western North America, where the Pacific plate to the west moves past the North American plate to the east. This particular plate boundary is responsible for many of the disastrous earthquakes of California, including those associated with the well-publicized San Andreas Fault.

The term *plate tectonics* describes the occurrence of plates and their movements. Plate tectonics determines the distribution of many physical features: mountains, volcanoes, steep slopes, submerged ridges, deep sea trenches and shallow seas, as well as stable land areas with a very low probability of earthquakes. The hazardousness of a place and its sensitivity for environmental degradation therefore depend to a large extent on its location with reference to plate margins and plate interior.

2.4 The Climate of the Earth

Short-wave solar radiation reaches the surface of the earth after some of it is lost by scattering in space, absorbed by a layer of ozone high in the atmosphere, and reflected to space by clouds. After absorbing the short-wave solar radiation that reaches its surface, the earth sends out long-wave radiation, part of which is lost to space but the rest heats the atmosphere from below. The temperature in the lower atmosphere therefore falls with rising altitude. This is the lapse rate (6.4 °C per 1000 m) which changes sharply at an altitude of about 14 km, beyond which the temperature of the atmosphere increases. The lowest part of the atmosphere is called the *stratosphere*, the altitude at which the abrupt temperature change occurs is known as the *tropopause*, and the part of the atmosphere immediately above it is the *troposphere*. There are other changes in the temperature–altitude relationship higher up in the upper atmosphere, but details of such changes are not necessary at this stage. Warmer air tends to rise upwards from the earth's surface. Under certain conditions, this pattern of decreasing temperature upwards is changed when a very cold layer of air temporarily finds itself next to the surface of the earth. The phenomenon is known as temperature inversion. This may happen in several ways:

- on a very cold night the lowest part of the atmosphere may be chilled in contact with the cold surface of the earth;
- in mountainous areas, dense cold air from high up in the mountains may sink to over nearby plains or into mountain-girt basins;
- air cooled over a cold sea may move on to coastal areas.

Temperature inversion prevents air from rising, and until it is destroyed (usually by the heat of the sun later in the day) it acts as a lid in the lower atmosphere, trapping any rising matter such as smoke underneath it. Thus polluted air is prevented from dissipation as long as inversion continues.

The variability in the amount of solar radiation reaching the earth has given rise to the decreasing temperature pattern from the equatorial areas to the polar zones. This general pattern is modified by the distribution of continents and oceans and by the presence of chains of high mountains. The centres of continents, for example, show a wide annual temperature range between cold winters and hot summers. The distribution of temperature controls the atmospheric pressure system at the surface of the earth, which in turn controls the major wind systems (Figure 2.4). The circulation of the wind systems occurs as a three-dimensional phenomenon. Winds move across the surface of the earth and vertically rise and sink to complete the circulation. The wind pattern also regulates the amount of precipitation received across the world, and temperature and rainfall together control the vegetation cover of the land masses.

This general model is useful for explaining diffusion of large-scale air pollution and also diffusion of material erupted by volcanoes at convergent plate margins. These volcanoes release quantities of fine ash into the air. Pinatubo in the Philippines, for example, did this repeatedly in 1991 and 1992. Certain other climatic phenomena also tend to accelerate the effects of environmental degradation. The El Niño Southern Oscillation (ENSO) is a striking example (Box 2.1).

Box 2.1
El Niño Southern Oscillation

In most years, winds blow across the Pacific Ocean from the high pressure area of the west coast of tropical South America towards the low pressure area of the Asian coast. As it blows over the water, this wind generates surface currents in the ocean which travel in the same direction. Off the coast of Peru, therefore, surface water moves westward and water comes up from the deep to fill in the gap. This process is known as upwelling. Water that upwells from the depths is colder and rich in plankton. This has two results: Peru has a cool and dry coastal climate and its offshore waters are rich in fish which arrive for the plankton.

There are certain years when the atmospheric pressure difference between East Pacific and West Pacific disappears. Then the wind across the ocean weakens and the surface water which was being piled up against the Asian coasts by the prevailing wind tends to flow back across the Pacific. When fully developed, this brings warm water to the Peruvian coast. Not only does this alter the weather of the Peruvian coast, but it also lowers the fish catch drastically. This change in the usual pattern is locally known as El Niño, the name given to this phenomenon by the fishers of Peru after the infant Jesus, as it usually develops towards the end of December near Christmas. It has also been demonstrated that El Niño years strongly correlate with years of drought in Australia, parts of Southeast and South Asia, and Africa. In contrast, these could be years of heavy rain and floods on the western coasts of the Americas. Such long-range connections are called teleconnections. This teleconnection indicates that parts of the world will suffer from drought conditions which could be extreme. Development projects in these areas which carry large demands for water therefore exacerbate the effect of El Niño years on land, people and animals. Such droughts are especially disastrous near the edge of deserts, due to this pattern of periodic shift in the weather system. El Niños tend to recur after several years but they vary in strength. The El Niño of 1982–83 was an especially strong one which brought drought, floods, landslides, forest fires, and death and destruction across the globe. Since then El Niños have occurred in the early 1990s and again in 1997–98. The most recent could end up being as disastrous as the one of 1982–83.

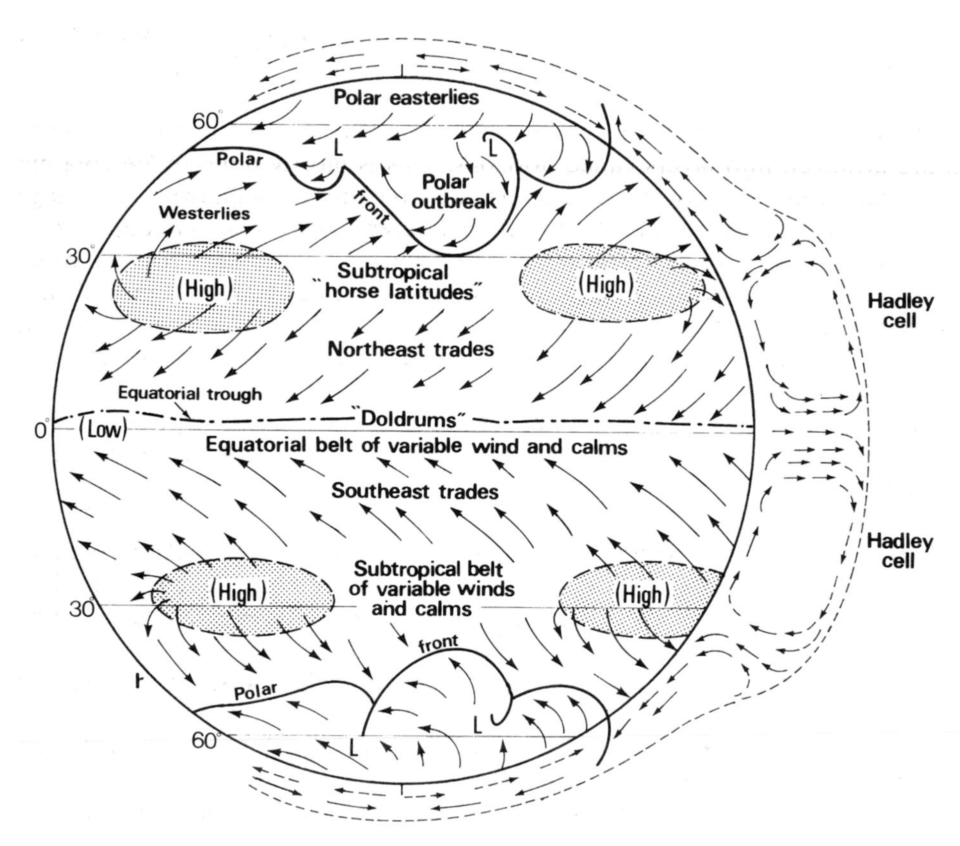

Figure 2.4 *Major pressure belts (high or low) and wind systems of the earth. The arrows to the right of the circle show the vertical movement of air at different latitudes to indicate the three-dimensional circulation of air. From Strahler, A.N. (1975)* Physical Geography, *John Wiley, New York. Copyright by Arthur N. Strahler*

2.5 Erosion of the Land Surface

The rocks at or very near the surface of the earth are weakened by the action of various climatic agencies such as temperature, wind and rainfall. Exposed to this, rocks break down into smaller components, and chemical reaction takes place on the surface of the rocks. The constituents of rocks are thus altered mostly by rainwater and chemicals dissolved in the rainwater. The process is speeded up at high temperature. As this type of rock alteration or decomposition is related to the prevailing weather conditions, the process of rock decomposition is known as weathering. Weathering makes rocks soft, and under such conditions two further developments may take place:

- weathered rocks may be further altered by a combination of dissolution by circulating water, action of burrowing organisms, and further chemical changes to form a fine-grained layered structure immediately below the surface called soil;
- soft weathered rocks are prone to be removed by rivers, glaciers, waves, etc. – a process known as erosion.

Soil is formed when the constituents of the top section of the weathered material are arranged into layers called horizons. This is mostly done by percolating water which removes clay and soluble material (the process is known as leaching) from the top and deposits them below. Thus different horizons are created. Figure 2.5 shows the final products in a generalized fashion. Rivers, landslides, glaciers, etc., which are known as geomorphic processes, remove the softer soil and weathered rocks, break them into small grains, and deposit the small grains on flatter grounds at lower elevation or where the velocity of flowing water or moving glacier slackens.

We are now in a position to list a number of key statements which provide useful background information for understanding both environmental degradation and environmental management.

1. Certain areas of the world, because of (i) their location in a plate tectonics context or (ii) the type of rocks they are on, are more sensitive to erosion than other places. Development of such sensitive areas requires enlightened planning followed by careful execution and monitoring of such plans.
2. Areas susceptible to periodic large storms should also be considered as sensitive to changes.
3. A linkage occurs between various aspects of climate such as temperature and rainfall and natural vegetation cover.
4. High temperature and heavy rainfall lead to rapid rock weathering and soil formation.
5. Elsewhere, weathering and soil formation happen but at a slower speed.

The following types of physical environment are particularly sensitive to change:

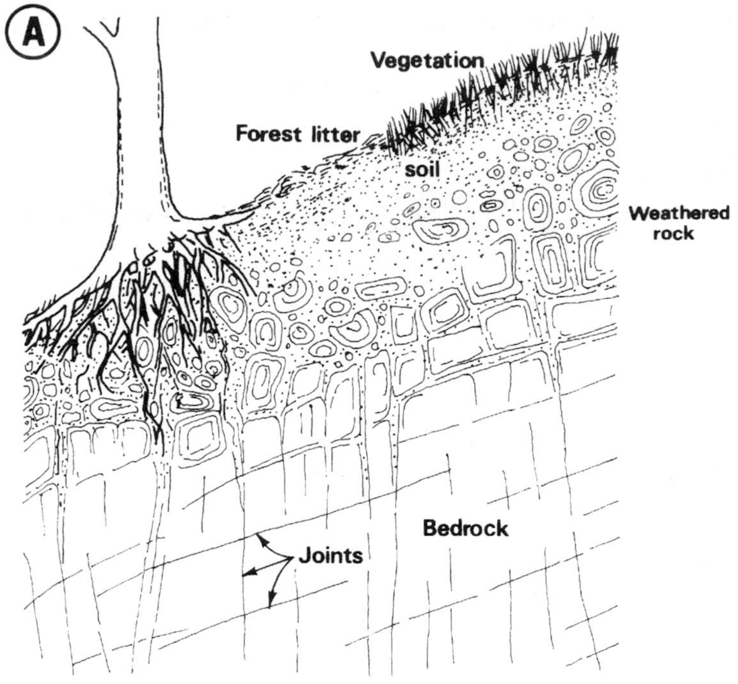
A
Vegetation
Forest litter
soil
Weathered rock
Bedrock
Joints

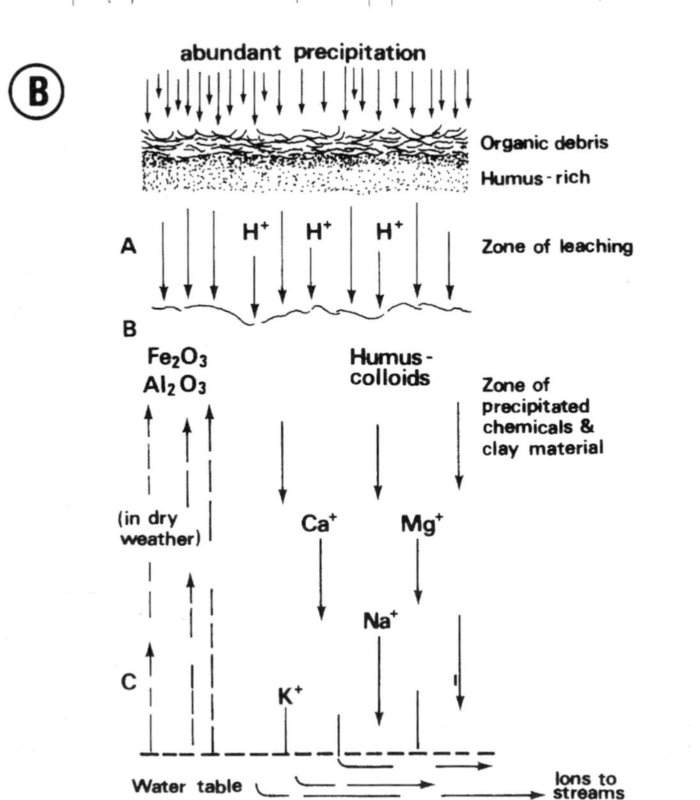
B
abundant precipitation
Organic debris
Humus-rich
A
H+
H+
H+
Zone of leaching
B
Fe2O3
Al2O3
Humus-colloids
Zone of precipitated chemicals & clay material
(in dry weather)
Ca+
Mg+
Na+
C
K+
Water table
Ions to streams

- rocks with a chemical composition that is prone to weathering
- fractured rocks
- rocks which are at a high angle to the horizontal
- steep slopes
- areas with heavy rainfall
- areas with low vegetation cover
- areas subject to large-scale but temporary changes in climatic conditions from time to time
- areas at or near plate margins
- coastal areas

Any development in these areas needs careful planning.

2.6 Cyclic Movements of the Constituents of the Earth

The constituents of the earth are finite in amount, and a particular constituent may exist in various locations and in different states of matter. Water, for example, exists in a solid state in glacier ice, in a fluid state in oceans, and as vapour in the atmosphere. There is exchange between these storage places, often requiring a changing of state. This is a natural cyclical system which keeps the surface of the earth habitable by controlling the amount of various constituents in each of these storage places. For example, we have an amount of water vapour and carbon dioxide in the air. These gases absorb terrestrial radiation as it is diffused from the planet and radiate it back to the earth. This raises the average surface temperature of our planet to 15 °C. If such gases were absent, the temperature would have plunged to –18 °C. The neighbouring planet Mars has a much less dense atmosphere and little water vapour. Its average temperature therefore is much below freezing. One of the characteristic signals of environmental degradation is a change in the amount of these constituents in various places of storage. We have increased the amount of CO_2 in the lower atmosphere by activities such as deforestation, biomass burning, generation of electricity by coal burning, and automobile emission. As a result, it is opined that the surface temperature of the earth is increasing slowly; a process termed global warming. Global warming is discussed later in the book (Chapter 14), but it is sufficient to say at this stage that this phenomenon is one of the most worrying of all types of environmental degradation. An understanding of these cycles is crucial for environmental managers.

Figure 2.5 *(A) Layered structure at the ground surface. Hard bedrock is superseded by soft weathered rock and then by soil under vegetation. Note the process of weathering attacking the bedrock along openings called joints. Blocks of rock between joints are then separated from each other and decomposed from all sides. (B) In a humid area, soil is formed by transportation of chemical elements and compounds by water moving up and down through the source weathered rock – a process known as leaching. Soils have three horizons: A is the zone of leaching; B is the zone where most of the chemicals and clays are precipitated; and C is the original weathered rock. Copyright by Arthur N. Strahler*

We describe three of these cycles, starting with the one dealing with water, the hydrological cycle. This is followed by two members (carbon and nitrogen) of a group collectively known as the biogeochemical cycles.

The hydrological cycle

Figure 2.6 is a graphical presentation of the hydrological cycle. The energy which drives the cycle comes from the sun. Under solar radiation, evaporation of water takes place mostly from the surface of the oceans, with a small amount evaporating from the land surface. The water vapour then becomes a constituent of the lower atmosphere and travels in the prevailing wind system. The vapour on reaching a certain height or a cold location condenses to form clouds, a collection of small water droplets. Nearer the earth such condensation may give rise to fog and mist. Further cooling creates bigger water droplets that fall as rain, or in colder places form solids which fall as snow. All water that returns to earth is collectively called precipitation. Precipitated water drains off the land surface or moves through soil and rocks in the subsurface to fill rivers and lakes. The rivers carry the water into the seas to complete the cycle. Rainwater that enters the ground takes various paths. It can flow out into a river, or stay stored in pore spaces in soil for plants to extract via their root systems. Plants release the water to the atmosphere after use, via pores in their leaves called stomata; a process known as transpiration. The transfer of water to the atmosphere by evaporation from water surfaces and by transpiration by plants, is collectively termed evapotranspiration.

Deeper penetration into the earth material brings water to a level below which nearly all the pores are filled with water. This is the groundwater which may occur at varying depths, even to several thousand metres. Such deep groundwater is seldom recycled and could stay in storage for thousands (even several millions in some cases) of years. The upper level of the groundwater (the groundwater table) fluctuates with the addition and extraction of water. Water is extracted from this level in wells.

Precipitation as snow may fall on polar areas, on the tops of high mountains, and (in winter) at high latitudes. Snow from the third example eventually melts and joins the circulatory pattern. Snow in the polar areas or on the tops of mountains, however, may turn into solid ice and remain in that state, out of active circulation, for a very long time. For example, part of the ice over Antarctica and Greenland has been dated to be as old as several hundred thousand years. Figure 2.6 also indicates that the vast majority of the earth's water remains in the oceans. Of the rest, most is stored as polar ice or deep groundwater. Water that sustains us, and is actively cycled, constitutes only a minute part of the total water content of the world.

The carbon cycle

The carbon cycle delineates the movement of carbon between several reservoirs. Carbon at or near the earth's surface has four important reservoirs:

- the atmosphere, where it exists primarily as CO_2;

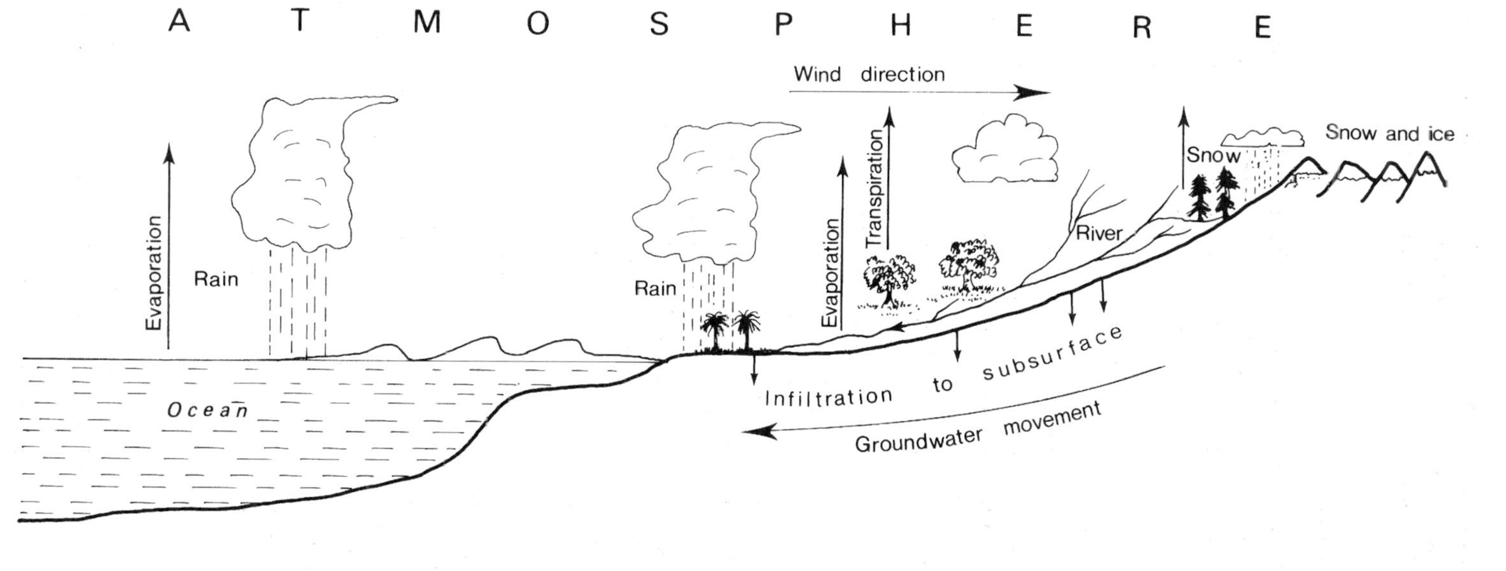

Figure 2.6 The hydrological cycle. Note that 97 percent of all water is stored in the oceans. Of the rest, 75 percent is locked up in ice sheets and glaciers, and about 25 per cent remains as groundwater. Only a small fraction remains in the atmosphere, rivers, lakes and soil

- the crust, where carbon is bound as $CaCO_3$ and $MgCO_3$ in limestone and marble, and also occurs in buried organic matter such as peat, coal and petroleum;
- the hydrosphere, where carbon occurs as dissolved CO_2 gas;
- the biosphere, where it occurs in organic compounds in plants and animals.

Figure 2.7 shows the reservoirs of the carbon cycle and the fluxes between them. In the biosphere, plants extract carbon from the atmosphere to form organic compounds in the presence of sunlight (photosynthesis) and water. The organic compounds are stored in the plants. After their death, the plant material decays on the ground; carbon is oxidized by combining with atmospheric oxygen; and CO_2 is returned to the atmosphere. The process of such recycling between the biosphere and atmosphere is extremely rapid, calculated in years. A fraction of the decaying plants may be buried as sediment in suitable places, e.g. coastal plains or river deltas. Over time this buried organic layer becomes richer in carbon by expelling volatile matter which was originally buried with it. Gradually such organic matter alters into peat which is a low-energy fuel. Given a lot more time (on a scale of hundreds of thousands of years) organic material changes into fossil carbon reservoirs such as coal or petroleum.

Carbon dioxide from the atmosphere is dissolved in waters of the ocean. This dissolved CO_2 is extracted by both aquatic plants and a number of aquatic fauna to form shells or coral reefs. Both shells and coral reefs over time are compacted and bound together by $CaCO_3$ acting as cement, being precipitated from water that circulates through an accumulation of shells or parts of coral reefs. Carbon is held in these cemented rocks which for obvious reasons are known as limestone. When limestone becomes exposed to the atmosphere and is eroded and decomposed, carbon returns to the atmosphere. The carbon cycle is extremely dynamic, large changes are possible, and the content of the main reservoirs may show short-term variations.

The carbon cycle is being studied in detail due to the large-scale anthropogenic alteration of carbon content between three of its main reservoirs. Carbon in the biosphere is depleted by tropical deforestation and urban expansion. The rising demand for energy has led to accelerated use of fossil fuel and biomass burning. In turn, all such processes increase the amount of CO_2 in the atmosphere. As the amount of CO_2 in the atmosphere is directly related to planetary temperature, the temperature of the land–atmosphere interface is slowly but steadily increasing. This is known as global warming. Its possible occurrence and associated consequences have prompted detailed investigation into the reservoirs and fluxes of carbon (Schimel, 1995).

The nitrogen cycle

Nitrogen is stored in five reservoirs (Figure 2.8): the atmosphere, biomass, soils, sediments and oceans. Most of the nitrogen is stored in the atmosphere, 78 percent of which is nitrogen by volume. Atmospheric nitrogen is assimilated by a process known as nitrogen fixation which can be carried out only by certain microorganisms. The reverse step, returning nitrogen from the biosphere to the atmosphere, happens via the process of denitrification.

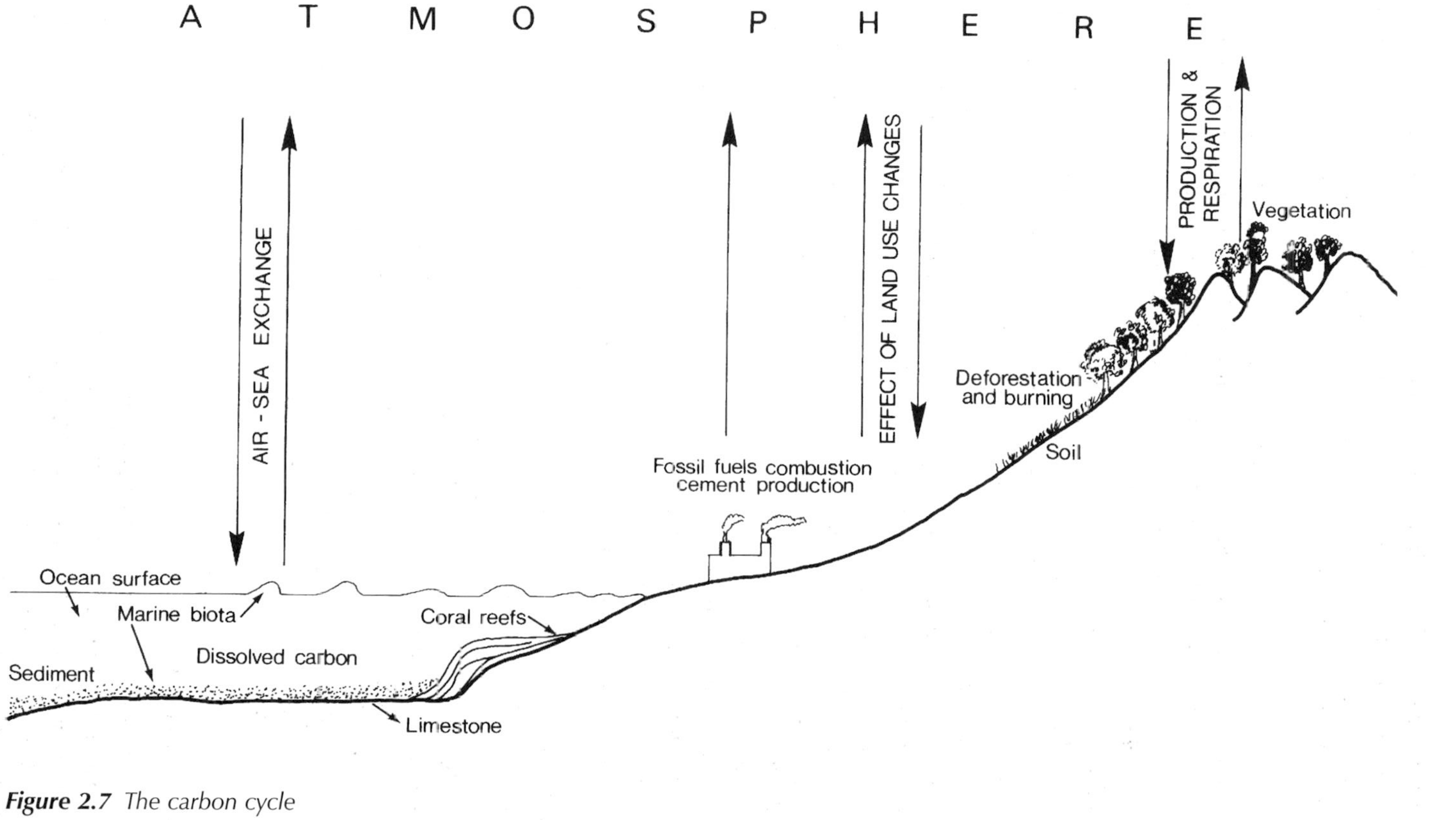

Figure 2.7 The carbon cycle

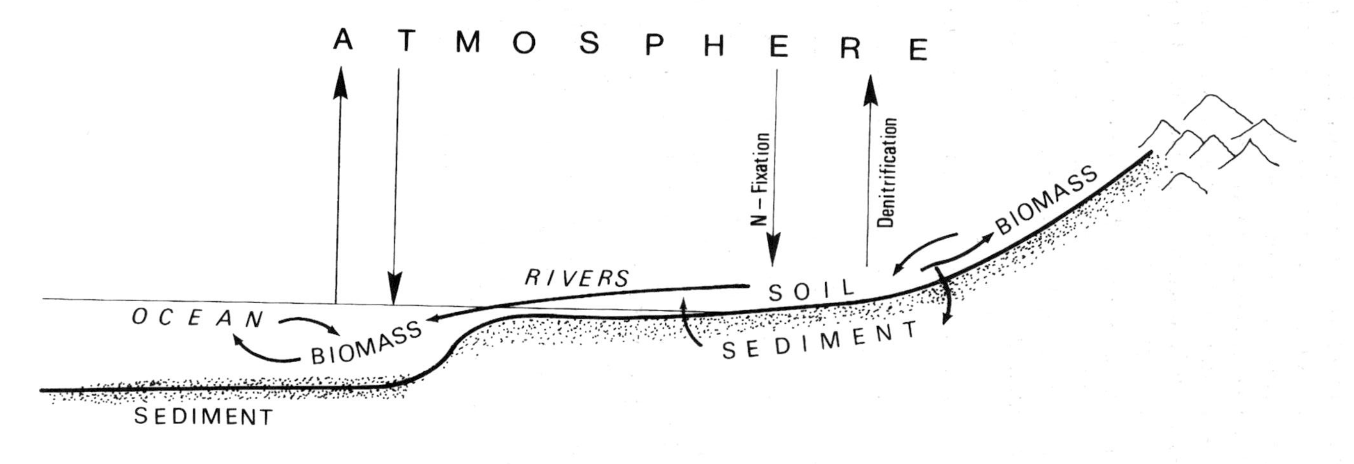

Figure 2.8 The nitrogen cycle

Nitrogen-fixation can be carried out by a selected species of soil bacteria, certain types of blue-green algae, and bacteria of the genus *Rhizobium* which are found in a number of floristic species, especially those which belong to the family of legumes. Legumes such as alfalfa, clover, beans, peas, soybeans and peanuts are used to fix nitrogen in the soil. This is achieved by rotating legumes with other crops such as cereals. One takes nitrogen out of the soil; the other retrieves it.

Currently, nitrogen-fixation is much more active than denitrification due to the use of manufactured fertilizers which contain nitrogen, increased farming of legumes, and combustion of fossil fuels. This is potentially degrading to the environment. Natural processes of erosion and transportation may carry nitrogen with soil grains into rivers, lakes and shallow coastal seas. This is likely to increase the growth of algae and phytoplanktons in the water which could have a detrimental effect on varieties of aquatic life. This new development of increased storage of nitrogen in water and sediment therefore needs to be examined and monitored.

2.7 Conclusions

Apart from providing a thumbnail sketch of the materials that constitute our planet and the processes that move such material around, this chapter introduces two important concepts. First, certain locations on earth are particularly vulnerable to environmental impact due to their location with reference to underlying geology, surface configurations, and prevailing climatic conditions. The intensity of environmental impacts such as those arising out of deforestation therefore varies according to location. Second, many constituents such as water, carbon or sediment have several main reservoirs and some of the material moves between them in fluxes. This is a natural process but anthropogenic activities tend to disrupt this type of distribution by moving a considerable amount of material to one or two reservoirs at the expense of the others. This, as we will see many times in this book, usually foretells disaster.

3

World Population: Distribution and Trends

3.1 Introduction

Both environment and development are for the people. It is therefore necessary to acquire an understanding of the world population, its distribution, and its economic and environmental impacts. The relationship between population and environment and development, particularly regarding the developing countries, has given rise to views that vary across a broad range. High population growth has been perceived as the main reason for poverty and environmental degradation. On the other hand, it has also been argued that it is poverty that leads to high population growth and environmental degradation. Population growth, poverty, and the state of the local environment are perceived to be interconnected (Dasgupta, 1995). Each of these factors influences the others and, in turn, is affected by them. The number of people that can be supported by the earth is also a controversial figure.

3.2 World Population and Its Distribution

For most of human history, the population of the world grew very slowly. The current pattern of rapid increase only started to appear in the second half of the eighteenth century, following the Industrial Revolution. From about 900 million in 1800, the population of the world had increased to 1600 million by 1900, and to 2500 million by 1950. The number reached 5600 million in 1994.

Of the 1994 world population, 56.7 percent lived in the low-income countries, mostly in Africa and Asia (Table 3.1). These countries had a per capita Gross National Product (GNP) which was only 8.5 percent of the world average, and generated only 4.9 percent of the total GNP of the world. At the other extreme, the high-income countries, with an average per capita GNP of $24 170, contained only 15.2 percent of the world's population but nearly 80 percent of the world's GNP.

Table 3.1 *Distribution of the world population by per capita GNP, 1994*

GNP per capita income group	Number of countries	GNP ($ billion)	Population (million)	GNP per capita ($)
World	209 (100)	25 793 (100)	5603 (100)	4600 (100)
Low income	64 (30.6)	1251 (4.9)	3178 (56.7)	390 (8.5)
Lower-middle income	66 (31.6)	1818 (7.0)	1100 (19.6)	1650 (35.9)
Upper-middle income	35 (16.7)	2207 (8.6)	476 (8.5)	4640 (100.9)
High income	44 (21.1)	20 517 (79.5)	849 (15.2)	24 170 (525.4)

Figures within parentheses are percentages.
Source: calculated from World Bank (1996a) The World Bank Atlas, *1996, p. 20.*
The World Bank classifies countries according to their 1994 per capita GNP as: low – $725 or less; lower-middle – $726–2895; upper-middle – $2896–8955; high – $8956 or more.

Most of these countries are in Europe, North America and Australasia. Elsewhere, only Japan, Singapore, Hong Kong, Israel and some of the oil-rich countries of Western Asia are in the high-income category.

Thus, a gross imbalance exists between the responsibility for supporting the world's population in a sustainable manner and the resources available between the two groups of countries. This imbalance has important implications for designing and implementing efficient and equitable policies for development that can be sustained without damaging the environment.

The following conclusions emerge from Table 3.2 which summarizes the population growth rates of different countries.

- High population growth rate (2.2 percent and above) is being experienced by 45 percent of the world's countries, but this group accounts for only about 20 percent of the total world population. Most of these countries are in Africa and Western Asia. A few are located in other parts of Asia and also in Latin America. At a growth rate of 2.2 percent, this population will double in 32 years.
- About a third of the world's population, including India (1994 population: 913.6 million), Indonesia (189.9 million) and Brazil (159.1 million), are experiencing moderate population growth between 1.5 and 2.1 percent per year. As these countries continue to make progress in reducing their population growth rates, the possibility of reducing the growth rate of the population of the world will increase. At a growth rate of 1.5 percent, this population will double in 47 years.
- Nearly 45 percent of the world's population (with a per capita GNP much higher than the world average and generating 86.5 percent of the world income) is experiencing low rates of growth, i.e. less than 1.4 percent. If the present trend continues there might be a decline in the absolute number of people in some countries (Japan and parts of Europe), provided no inward flow or immigration occurs from other countries. At a growth rate of 1 percent, a doubling of the population will take 70 years.

The data in Table 3.2 have two important implications. First, as the countries with high population growth have higher per capita GNP than those with moderate

***Table 3.2** Population growth rate among different countries of the world, 1985–1994*

Population growth rate (per annum) (%)	Number of countries	GNP ($ billion)	Population, 1994 (million)	Per capita GNP, 1994 ($)
> 3	34 (16.3)	420 (1.6)	299 (5.3)	1410 (30.7)
2.2–3	56 (26.8)	1169 (4.5)	899 (16.0)	1300 (28.3)
1.5–2.1	34 (16.3)	1906 (7.4)	1884 (33.6)	1010 (22.0)
1.0–1.4	22 (10.5)	9223 (35.8)	1622 (28.9)	5680 (123.5)
< 1.0	56 (26.8)	13 069 (50.7)	892 (15.9)	14 660 (318.7)
No data	7 (3.3)	6	7 (0.1)	870 (18.9)
Total	209 (100)	25 793 (100)	5603 (100)	4600 (100)

Figures within parentheses are percentages.
Source: calculated from World Bank (1996a) The World Bank Atlas, *1996, p.10.*

population growth, it is important to examine the complex issue of why people want children and what are the appropriate incentives in this case to have fewer children. A high income may not necessarily lead to a low population growth. Second, population growth rates, income levels, and local environmental resources all vary between countries. People themselves may be regarded as an important resource in circumstances where the population growth rate is very low and the local resource base relatively abundant. When such conditions do not hold, a high population growth rate may adversely affect the capacity for sustaining development.

A widely used measure of fertility is the Total Fertility Rate (TFR). It is the average number of children a woman is expected to bear during her lifetime. A TFR of 2.1 maintains a stable population, assuming no net migration takes place. Many countries in Africa and Southwest Asia have very high TFRs, exceeding 5.0. In contrast, there were 37 countries in 1994 with a total population of 926 million and a per capita GNP of $15 380 which had TFRs below 2.0, indicating an absolute reduction in population. Countries such as Germany, Russia, Japan, Italy and Spain had especially low TFRs in 1994, between 1.2 and 1.5 (World Bank, 1996a). While the factors that influence TFR are complex, experience suggests that improving the status of women in society, level of education, economic growth and urbanization tend to reduce high fertility rates.

3.3 Future Population Trends

Future population trends are inherently uncertain. This uncertainty increases with the length of time beyond which projections are made. Short-term (3–5 year) projections of the growth rate of the world's population are, however, usually more reliable than other economic and social variables, e.g. the unemployment rate or the behaviour of the stock market.

Population projections are needed for several reasons. First, almost all planners, managers and administrators need to know about future trends for planning purposes, e.g. to determine the future financial viability of the existing health care system or the viability of future business expansion plans. Second, in many developing countries (and even for the world as a whole) it is essential to monitor food consumption requirements. So far, the world has escaped the gloomy conclusions of Malthus, who about 200 years ago argued that while food production grows arithmetically, population grows geometrically and sooner or later a food crisis is bound to occur. This escape has been possible due to increased use of technology and improvement in human knowledge and skills and slowing of population growth. There is, however, an obvious limit to population growth on a finite planet. Third, a large number of natural and social scientists working on the environment use indicators (e.g. energy consumption) on a per capita basis. By definition, these require population projections.

Usually population projections are first calculated on a country or a regional basis and then aggregated to derive the total for the world. Such projections need to assume fertility behaviour (the number of births), mortality behaviour (the number of deaths), and migration trends including population shifts among countries. Each of these three elements has a very complex background, and even a relatively small change may lead to large variations in the estimates. The main dilemma of the world population stabilization is that lower mortality (fewer deaths resulting in a rising life expectancy) is achievable in a fairly short period of time whereas declines in fertility (number of births) cannot be achieved rapidly. The reproductive behaviour of a large number of people does not change quickly. Rapidly declining mortality and early stabilization of population cannot be achieved simultaneously. This is the dilemma, as both are important objectives arising out of the population policies of the agencies of the United Nations such as the World Health Organization (WHO), the United Nations Fund for Population Activities (UNFPA), and many national population programmes.

Too rapid a decline in birth rate (as implied in the one-child policy of China) could lead a country into a sharp ageing of the population. It is widely anticipated that China will have the demographic profile of an industrial country in the near future in the sense that its population aged 60 years and over will form 21.9 percent of the total by 2030, up from 8.9 percent in 1990 (World Bank, 1994). This, however, does not necessarily indicate that a rapid decline in fertility is not desirable for China. It indicates that the benefits of such policies should be weighed against the costs imposed by a rapidly ageing population.

Figure 3.1 plots the estimates of the world population under different fertility and mortality assumptions. Population estimates for the year 2050 range from 7.2 billion to 13.3 billion. The most likely projected figure is a little below 9.9 billion. Such a projected figure implies an increase in the mean age of the world population from 28.8 years in 1995 to 36.9 years in 2050.

Furthermore, the percentage of population aged 60 and over would double, from 9.5 percent in 1995 to 19.6 by 2050. In contrast, the percentage of people below 14 years would drop from the 1995 figure of 31.4 percent to 22.2 per cent. The average life expectancy for the world as a whole is expected to rise to 71.4 years (Lutz, 1996). The world with an ageing population will be a different one for environmental managers. The population is also likely to become more and more urban.

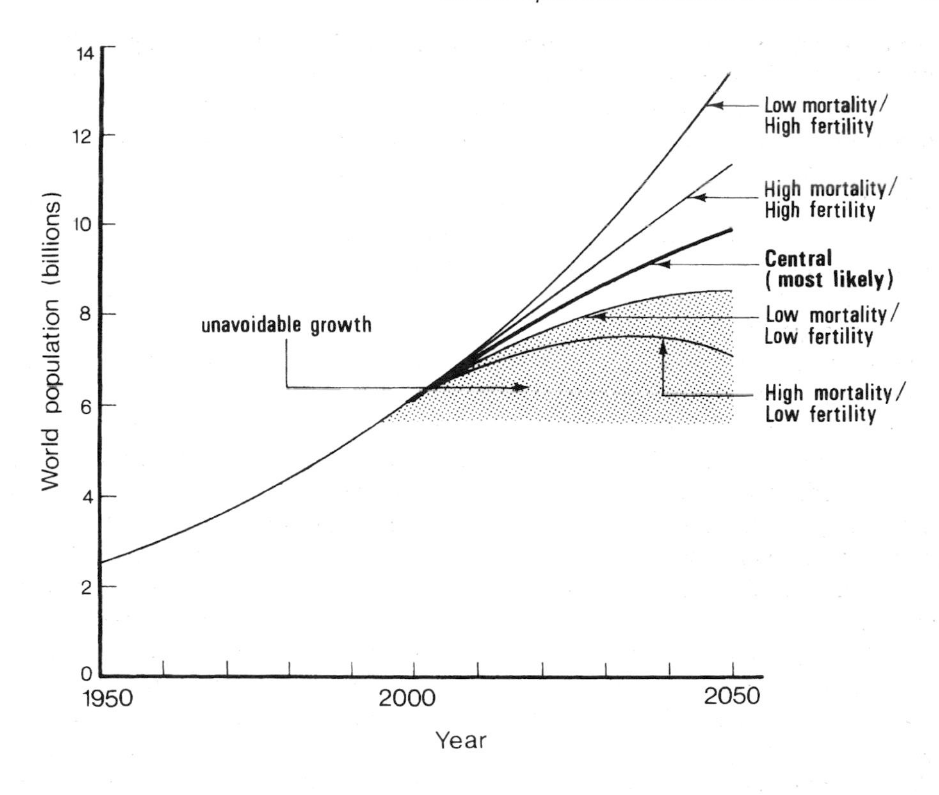

Figure 3.1 *Growth of the world population projected to the year 2050. From Lutz, W. (Ed.) (1996)* The Future Population of the World, *Kogan Page Limited, Earthscan Publications Limited, London*

3.4 Urbanization of Human Settlements

An urban settlement has two demographic properties: a large number of people and a high density. The majority of the residents have non-agricultural employment, and physically the settlement is distinguished by characteristic forms such as a dense network of streets, large buildings and (for old settlements) remnants of a boundary wall. In certain countries settlements with such characteristics are officially recognized as urban by government declarations.

A large urban settlement is called a city, the word implying a civilized place. Cities are centres with imposing buildings and streets; of political and administrative power; of education, culture and industry. Cities have evolved only during the last few thousand years of human history.

The first cities on record were founded about 5500 years ago in Mesopotamia, in the valleys of the Tigris and the Euphrates. Such cities illustrate the reason behind the evolution of rural settlements into cities. Cities evolved because of the presence of agricultural surplus from a fertile region; the use of the area as a crossroad which allowed repeated contacts among people of different culture, skills, and possibly with goods to trade; the concentration of religious and political powers with a small number of the residents of the city at the top of the social pyramid.

After Mesopotamia, the next urban community developed in the Nile Delta within a few centuries. A number of cities were well established in the Indus Valley and western India about 4500 years ago. The cities in the middle reaches of the Yellow River of China started about 3500 years ago, which is roughly the period when cities appeared in Mediterranean Asia and Europe as the ancestors of the Greek and Roman cities. It is believed that the concept of cities diffused over time, leading to their establishment in new areas. Such diffusion was not possible in the Americas, and the earliest cities there came later, probably not much earlier than 2000 years ago.

Cities, from the very beginning, exhibited at least some of the urban characteristics (both in form and function) described at the beginning of this section. Some showed astonishing levels of planning in terms of central edifices, street networks, and an outer wall for defence. Residents of Mohenjodaro and Harappa, the two best known cities of the Indus Valley, enjoyed excellent water supply systems and waste-water drains. The early cities are collectively known as *pre-industrial cities*. Their establishment and existence were based on regional agricultural surplus, location on trade routes, and functions as military, political and religious centres. Pre-industrial cities appeared and disappeared across the world, being replaced by other urban settlements in the neighbourhood. It is possible to trace a pre-industrial ancestry for a number of present cities.

The Industrial Revolution brought in the modern type of urban settlement based on industrial establishments and the associated trade and administration. This is the *industrial city*, which appeared first in Britain. The rapid growth of the number of cities and the number of urban residents in Western Europe and North America dates from this time. Not only did the cities grow in population, they also expanded in area, often converging with each other to form an agglomeration which extended for hundreds of kilometres. The best example comes from the north-eastern seaboard of the United States, where the present agglomeration stretches with a few interruptions from Boston to Washington, DC. A new term, *megalopolis*, has been coined to describe such an extended urban conurbation.

The cities of the developing countries are of mixed origin. A number of pre-industrial cities continue to exist although usually not with a very high population, unless they function also as the nation's capital or as a centre of some specialized function. A number of industrial and port cities have grown up, especially in recent years and largely after independence from the colonial powers. But many large cities of the developing countries (which by location are also tropical cities) started as trading or military posts of a foreign power only a few hundred years ago. The regional location of these cities can be explained but not the siting which was often fortuitous. For example, Kingston (Jamaica) started to develop only when a large earthquake destroyed the nearby harbour of Port Royal in 1692. Calcutta was sited on the bank of the Hooghly River at the location where the British trader Job Charnock's boatmen managed to moor the boat at the end of a rainy day in 1690. Nairobi grew up in the middle of a swampy plain where a tented depot of railway stores was established in 1899 to serve the rail line being laid across East Africa. Most of these cities were essentially entrepôt or trading cities with some administrative and military overtones.

A summary account of the urban growth of the developing world, region by region, is available in Gugler (1996). Once established, however, a number of these centres prospered and grew into large cities. Quite a few now have populations of millions and are still growing rapidly, unlike the industrial cities of the developed countries which have either stopped growing or show only marginal growth.

The term urbanization denotes the level of urban development of a country. It is measured as the percentage of the total population of a country or a region which lives in urban centres. Urbanization occurs mainly by migration of people from villages into cities. The number of city dwellers may be augmented by a high local birth rate but that does not necessarily imply urbanization. Urbanization compares urban population with the rural population. In the next section we summarize the present state of urbanization of the world.

3.5 The Size and Growth of the Urban Population

According to a recent United Nations estimate (United Nations, forthcoming), approximately 2.6 billion of the mid-1995 world population were urban-dwellers, i.e. 46 percent of the world population. This number is expected to rise to 50 percent by 2006, and above 60 per cent by 2030. The rise is overwhelmingly from the developing countries (Table 3.3). The developed countries (particularly in northern and western Europe, North America, Japan, Australia and New Zealand) display very high levels of urbanization, but their current rate of urbanization is extremely low or negative. For example, the annual urbanization rates for the United States in the four periods 1970–1975, 1975–1980, 1980–1985 and 1985–1990 were 0.01, 0.02, 0.2 and 0.2 percent respectively (United Nations, 1995). The figures vary across the developed countries. In certain countries the rate was negative; in certain countries the stagnation came only in the 1980s. In general, people have moved out of large cities into small towns and rural areas.

Table 3.3 *World urban population, 1970–2025*

Region	Urban population (millions)			Percentage of total population		
	1970	1994	2025 (est.)	1970	1994	2025 (est.)
Less developed countries	676	1653	4025	25.1	37.0	57.0
Africa	84	240	804	23.0	33.4	53.8
Asia (excluding Japan)	428	1062	2615	21.0	32.4	54.0
Latin America	163	349	601	57.4	73.7	84.7
Oceania (excluding Australia and New Zealand)	1	2	5	18.0	24.0	40.0
More developed countries	677	868	1040	67.5	74.7	84.0
World	1353	2521	5065	36.6	44.8	61.1

Source: United Nations (1995) World Urbanization Prospects: The 1995 Revision, Estimates and Projections of Urban and Rural Populations and of Urban Agglomerations, *United Nations Publications ST/ESA/SER.A/150, pp. 20, 23.*

The situation is different in the developing countries. Here the process of urbanization is progressing rapidly after a late start and is expected to continue for decades. The United Nations (1995) figures indicate an annual 3.5 percent growth rate for the urban population against a figure of less than 1 percent for the rural population. The urban rate is projected to decline to 1.9 percent annually between 2025 and 2030 (United Nations, forthcoming). For India, it has been estimated that the rural population will stabilize at around 800 million early in the next century and then start to decrease slowly, while the urban population will still continue to rise (World Resources Institute, 1994). This could be the pattern for a number of countries. The world rural population is expected to rise by small increments until 2020, when it is projected to decline slowly (United Nations, forthcoming). As Table 3.3 indicates, the urban population of the developing countries will exceed 50 percent of the total by early next century.

The level of urbanization varies across the developing countries. For example, it is highest in the Latin American countries where the figure is expected to rise to nearly 83 percent by 2050 (United Nations, forthcoming). The figures for Oceania are much lower: about a quarter at present and estimated to be 40 percent by 2025. The variations continue at the country level for all regions. For example, Argentina, Uruguay and Venezuela currently have urban populations of around 90 percent, whereas the figure for Guyana is less than 40 percent. The African countries range from below 10 percent for Burundi and Rwanda to over 80 percent for Libya. The urban population remains low for western, central and eastern Africa. Figures for Asia range from 6 percent for Bhutan to a number of countries in West Asia where the urban population climbs to more than two-thirds of the total population. This does not include the city-states of Asia. Certain countries such as Burkina Faso, Botswana or Nepal combine a very low level of urbanization with very high urban annual growth rates.

If we compare the urban populations of the developed and developing countries, a time-based change is obvious. Prior to 1970, the total number of urban-dwellers was higher in the developed countries. By 1970, both regions had almost equal populations. Since 1970, the developing countries lead, and the disparity is increasing. Estimates indicate an increase across the entire spectrum of the developing countries, but in sheer numbers, most of the urban-dwellers will be in Asia, and most probably in China and India. The current world urban growth rate is 2.5 percent per year, which is expected to continue until 2020 and then start to decrease. Meanwhile the world will have bigger and possibly more urban settlements, primarily in the developing countries.

3.6 Cities of the World

Certain trends can be identified regarding the cities of the world (United Nations, 1995).

- The majority of the urban population still lives in small cities (with a population of less than 500 000), although the proportion of the total urban population living in these cities is expected to drop by early next century.

- Cities of all size classes are increasing in number, especially cities which have a population of between 1 and 5 million.
- Over 30 megacities (defined by the United Nations as cities with at least 10 million people) are expected to exist by early next century, although the ranking of these cities by population is changing every decade.
- All cities, especially megacities, are growing much faster in the developing countries.
- The size of cities with a population of 10 million or more is rapidly increasing.
- Cities in the developed countries are hardly growing and there is a trend towards migration to smaller urban settlements and villages.

The changing nature of the megacities is best illustrated by ranking them according to population. Certain features emerge.

- Several decades ago, the largest cities of the world were located mostly in the United States and Europe; over time these have been replaced almost entirely by the cities of the developing countries.
- The sizes of the urban population for the biggest cities have increased tremendously over time.
- More and more of the large cities are in Asia.
- The large Latin American cities (Mexico City, São Paulo, Rio de Janeiro) are growing at a slower pace than the large Asian cities or Lagos.
- Ranking according to size changes within a few years (except for Tokyo), and therefore future estimates should be treated with caution.

The last point is illustrated by the dramatic change in the rankings of megacities according to population size among three United Nations surveys done about two

Table 3.4 *Ranked distribution of the 15 largest urban areas, 1950–2015 (city population in millions in brackets)*

Calculated			Estimated	
1950	1970	1995	2000	2015
1. New York (12.3)	1. Tokyo (16.5)	1. Tokyo (27.0)	1. Tokyo (28.0)	1. Tokyo (28.9)
2. London (8.7)	2. New York (16.2)	2. Mexico City (16.6)	2. Mexico City (18.1)	2. Bombay (26.2)
3. Tokyo (6.9)	3. Shanghai (11.2)	3. São Paulo (16.5)	3. Bombay (18.0)	3. Lagos (24.6)
4. Paris (5.4)	4. Osaka (9.4)	4. New York (16.3)	4. São Paulo (17.7)	4. São Paulo (20.3)
5. Moscow (5.4)	5. Mexico City (9.1)	5. Bombay (15.1)	5. New York (16.6)	5. Dhaka (19.5)
6. Shanghai (5.3)	6. London (8.6)	6. Shanghai (13.6)	6. Shanghai (14.2)	6. Karachi (19.4)
7. Essen (5.3)	7. Paris (8.5)	7. Los Angeles (12.4)	7. Lagos (13.5)	7. Mexico City (19.2)
8. Buenos Aires (5.0)	8. Buenos Aires (8.4)	8. Calcutta (11.9)	8. Los Angeles (13.1)	8. Shanghai (18.0)
9. Chicago (4.9)	9. Los Angeles (8.4)	9. Buenos Aires (11.8)	9. Calcutta (12.9)	9. New York (17.6)
10. Calcutta (4.4)	10. Beijing (8.1)	10. Seoul (11.6)	10. Buenos Aires (12.4)	10. Calcutta (17.3)
11. Osaka (4.1)	11. São Paulo (8.1)	11. Beijing (11.3)	11. Seoul (12.2)	11. Delhi (16.9)
12. Los Angeles (4.0)	12. Moscow (7.1)	12. Osaka (10.6)	12. Beijing (12.0)	12. Beijing (15.6)
13. Beijing (3.9)	13. Rio de Janeiro (7.0)	13. Lagos (10.3)	13. Karachi (11.8)	13. Metro Manila (14.7)
14. Milan (3.6)	14. Calcutta (6.9)	14. Rio de Janeiro (10.2)	14. Delhi (11.7)	14. Cairo (14.4)
15. Berlin (3.3)	15. Chicago (6.7)	15. Delhi (9.9)	15. Dhaka (11.0)	15. Los Angeles (14.2)

Source: United Nations, Population Division of the Department of Economic and Social Affairs, United Nations Secretariat (forthcoming) World Urbanization Prospects: The 1996 Revision, Estimates and Projections of Urban and Rural Populations and of Urban Agglomerations, *United Nations Publications ST/ESA/SER.A/170, pp. 20–21.*

years apart (United Nations, 1993; 1995; forthcoming). The estimated population for Mexico City or Shanghai (and their rankings) had to be drastically changed between the two studies because of their slow growth rate in comparison with the unexpected high growth rates of other cities. In contrast, the rankings of Dhaka and Cairo were significantly moved upwards.

The number of megacities is increasing rapidly and mostly in the developing countries. In mid-1994, 16 of the 22 megacities were in the developing countries. Twelve of these were in Asia. The estimated number for 2015 is 26, of which 22 will be in the developing countries and 18 in Asia, including two in Japan (United Nations, forthcoming).

The developing countries have to adjust to the phenomenon of rapid urbanization. Cities are striking anthropogenic modifications of the natural environment and their management and maintenance require special techniques, as discussed later in Chapter 10. The natural processes, as summarized in Chapter 2, operate even in cities, but their characteristics and relative intensity change. Urban management is rapidly turning out to be one of the major areas of environmental management.

4

Economics of Environment: Concepts and Tools

4.1 Introduction

Economics is the study of how a society utilizes its available resources to produce, consume and trade goods and services. Since the environment is regarded as one such resource, it is increasingly being analysed by economic reasoning. This has given rise to a new sub-discipline called *environmental economics* (Turner, Pearce and Bateman, 1994; Goodstein, 1995).

This chapter provides a basic understanding of the economic concepts and tools that are used to analyse the environment. In general, economists prefer market-type instruments over 'command and control' instruments such as government regulation. This is on the premise that a market is an impersonal and flexible instrument and is able to elicit information from participants with minimum possible costs for co-ordination of their individual decisions. However, incomplete information, uncertainty, distribution objectives, and environmental justice may require recourse to government and other non-market mechanisms.

Incorporating the environment into conventional GDP requires that we explicitly take into account the effects of economic activity on the depletion of natural resources and on environmental degradation. The environment can be affected by economic policies designed to influence the economic behaviour of individuals, firms, and of the government itself. Changes in economic behaviour in turn affect the demand for environmental services as well as controlling their supply (Helm and Pearce, 1990). Economic policies, however, vary in their impact on efficiency and equity. Society needs to strike a balance between the two when choosing between policy alternatives.

Economic reasoning is also essential in arriving at a proper valuation of the environmental services. This is because neither a zero nor an infinite price for these services is desirable. A zero price will provide no incentive to individuals, firms and governments to consider how their economic decisions could affect the environment. Assigning an infinite price would imply that the environment should not be disturbed at all. This would unduly hinder efforts to improve the average living standards.

The price for environmental services should therefore fall somewhere between the two extremes. Appropriate pricing for environmental services can be delineated by the application of economic reasoning. For this, an understanding of the term *economic efficiency* (discussed in Section 4.2) is essential.

Environmental effects may be regarded as externalities. An *externality* occurs when action by an economic agent, such as an individual or a firm, unintentionally affects the welfare of other economic agent(s) without being incorporated in the market price. Consequently, an externality is both an unintended and an uncompensated effect. As an example, unhygienic disposal of garbage by a household could harm not only the household itself, but could also render an unintended and uncompensated harm to other members of the community. A negative externality implies external cost while a positive externality implies external benefit. A production activity (e.g. a paper plant which lowers the quality of water in the river) or a consumption activity (e.g. a person smoking in a public place) which is not compensated for, represents a negative externality. A clean, well-kept neighbourhood providing benefits not only to its residents but also to passers-by is an example of a positive externality.

A widely used technique for evaluating physical projects as well as analysing a wide range of social programmes and regulations is called benefit–cost analysis (BCA). Integrating the environment into the BCA, and in particular how monetary values can be assigned to environmental impacts, is an essential component of the economics of the environment. While the BCA is essential for evaluation, public policy decisions, particularly those involving the environment, are unlikely to be based solely on the BCA. Limitations of the BCA technique should be given due weight in reaching such policy decisions. The relationship between national income and other macroeconomic variables and the environment has also been receiving increased attention in recent years (Gandhi, 1996).

4.2 Economic Efficiency

In 1776, Adam Smith enunciated the concept of the invisible hand. He argued that provided there are no impediments to the functioning of competitive markets, each individual pursuing his or her own self-interest would lead to the maximum welfare for society without any particular individual or a body being in charge. This represents a subtle but epochal insight. A competitive market is one in which there are many buyers and sellers, none of whom can influence the market price. Moreover, the cost of entering or leaving the competitive market is fairly low.

Two important functions of the market are the provision of information and the co-ordination of decisions made by numerous buyers and sellers. Prices and their behaviour provide the necessary information to economic agents to make appropriate decisions. Thus, a shortage of a commodity, such as coffee, drives up its price. The higher price could act as a signal to consumers to switch to other commodities such as tea, while providing information to the producers to increase resources devoted to producing coffee. In a competitive market, this information is conveyed impersonally and efficiently to everyone.

In the above example, a shortage of coffee represents an inconsistency arising from decisions made by numerous individuals. Market smoothes out this inconsistency by directing resources to where there is a shortage and withdrawing them from where there is a surplus, without any particular individual or body being in charge. This represents the co-ordinating role of the market. For maximum effectiveness in performing the co-ordination function, a competitive market is required.

Much of the work in economics has been devoted to refining and applying Adam Smith's concept of the invisible hand. It is this work which has led to the formalization of the concept of economic efficiency, referred to as Pareto efficiency after the nineteenth-century Italian economist-sociologist Vilfredo Pareto. Pareto efficiency represents a situation under which no alteration in production and consumption patterns will make someone better off without making someone else worse off.

Marlow (1995) provides a detailed discussion of the many conditions necessary to achieve economic or Pareto efficiency. It encompasses both *production* and *consumption efficiency*. A production efficiency requires that in a two-goods economy, re-allocation of available resources will not result in an increase in output of one good without reducing the output of the other good, assuming total available resources of the society and technology are fixed. This implies that given the quantity of one good, maximum possible output of the other good is being attained. A consumption efficiency is attained when, given their initial income and the market prices of goods, individuals have exhausted opportunities for mutually beneficial trading among themselves. Thus, no further improvement in the well-being is possible.

It follows that efficiency can still be attained even if the initial incomes are very unequally distributed. However, the outcome will not necessarily be regarded as fair or equitable.

Figure 4.1 provides a diagrammatic representation of the concept of economic (or Pareto) efficiency. A two-individual economy is assumed. If individual A's utility (defined as the satisfaction that an individual gets from consuming a good or undertaking an activity) is assumed to be OU_1, then the maximum utility individual B can attain is OU_2. Since this is true of all points on the Utilities Possibilities Frontier, all points on it are Pareto efficient. Point J represents an inefficient point, implying that through reallocation of resources, both individuals can potentially be made better off. Point K is unattainable by the society given its resources and technology.

Points E' and E" are also economically efficient. But in each case, one individual has an extremely high utility, while the other individual has a very low utility. The two points are therefore efficient but cannot be regarded as equitable. In this way economists separate considerations of efficiency from equity. Society, however, may prefer point J to point E' or point E" even though the latter are efficient, while point J is not.

Pareto efficiency leads to a situation where the marginal (additional) benefits of consuming an additional unit (which equals the price that an individual is willing to pay) are equal to the marginal (additional) social cost of producing an extra unit.

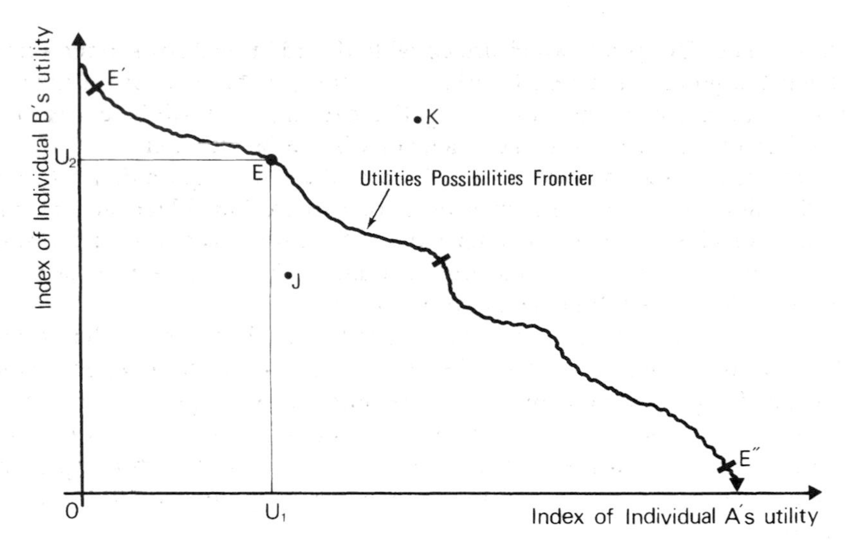

Figure 4.1 *The utilities possibilities frontier. All points on the frontier have the property that to make individual A better off, individual B must be made worse off and vice versa. So all points on the frontier are economically efficient. Any point on the frontier is achievable, provided appropriate initial income is ensured. Points not on the frontier are inefficient. All points beyond the frontier are unattainable given the present resources and technology. All points inside the frontier are inefficient, and potentially permit both individuals to be better off through reallocation of resources*

This implies that the price consumers pay should be equal to the additional resource cost to the society of producing an additional unit. This is known as the Pareto pricing rule.

This rule leads to efficiency because the decisions of the consumers and producers fully reflect the resource costs to the society. Any departure from this rule will lead to inefficiency in the sense that either too much or too little of the good will be consumed.

Suppose the price of a good such as a pen is $5 per unit but the costs of production to the society are $6 per unit. Then the price facing the consumers is lower than the resource costs of producing one unit of pen to the society. This imbalance will result in a level of consumption that is too high from the society's point of view. Similarly, if the price to the consumers is greater than the resource costs of producing one unit of pen to the society, the resulting level of consumption will be too low. Only when the Pareto pricing rule is satisfied will the consumption level of the pen be just right or efficient.

4.3 Environmental Effects as Externalities

Externalities are both unintended and uncompensated or unpriced in the market. As a result, when externalities arise (whether positive or negative), consumers and producers do not face prices which are equal to the additional costs to the society of producing the good. As a result, either too little (in the case of positive externalities) or too much (in the case of negative externalities) will be consumed and produced. In either case, economic efficiency will not be attained.

It should be stressed that economic efficiency does not require that externalities be totally eliminated. The efficient level is one where individuals and businesses have fully taken into account (i.e. internalized) the external effects. The problem is not that there is an externality, but that its quantity is 'wrong' for efficiency. This should be evident in the discussion on pollution below.

Negative environmental externalities arise in localized contexts such as air pollution from an industrial smokestack. They also arise in much larger international contexts involving several jurisdictions, large numbers of people, and different laws and regulations. In many cases, our understanding of the nature of externalities is incomplete, creating considerable uncertainty. Examples of such complex externalities include the effects of acid rain, global warming, loss of biodiversity, and deforestation.

A special class of externalities relevant to environment is referred to as common resource problems. Its central characteristic is a pool of scarce resources to which unrestricted access is permitted (Stiglitz, 1988). The high seas represent such a common resource. Unless its use is somehow regulated through an international treaty (such as the Law of the Sea treaty), there will be no incentive to use it efficiently, creating a negative externality for others. The key aspect is not a common resource *per se* but whether its use is managed in a sustainable manner. Thus, a village may have a community grazing land; but so long as its use is regulated in a sustainable manner it can be maintained in a productive condition.

To understand the basic concepts and instruments relating to environmental externalities, it may be useful to concentrate on pollution. There are four broad instruments which may be used to internalize the external costs of pollution:

- taxes (also called fines or charges)
- regulation
- subsidies for reducing pollution
- voluntary bargaining and exchange

Taxes

Figure 4.2 illustrates the use of emission taxes to tackle pollution externalities. Emission taxes may be defined as tax payments that are directly related to the measurement (or estimation) of the pollution caused, whether emitted into air or water, or deposited in the soil, or caused by generation of noise (OECD, 1996). These taxes generally deal with one type of emission at a time. DD' represents the demand curve for paper. Each point on the demand curve represents the maximum price consumers are willing to pay for a given quantity of a good. Thus, for quantity OQ_m, the maximum price consumers are willing to pay is OP_3. This price represents their assessment of the benefits from the last unit of the quantity OQ_m. Therefore, the demand curve may also be called the marginal (additional) benefit curve. This

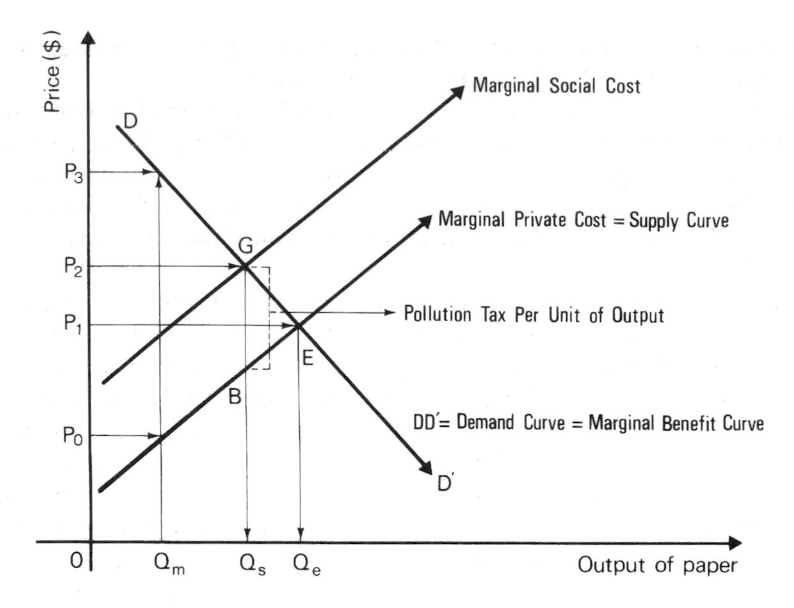

Figure 4.2 *Use of emission taxes to achieve an efficient output of paper*

curve is downward sloping because consumers would be willing to demand higher quantity only at a lower price. This is because as quantity consumed increases, each additional (marginal) unit of the good provides lower marginal benefit.

In Figure 4.2, private costs represent the costs to the producer of producing paper. The producer needs to pay for the raw materials, labour, electricity, water and the like. Private costs do not include costs of pollution generated in the process of producing paper. To obtain costs to the society as a whole, called the social costs, pollution costs should be added to the private costs.

Each point on the marginal private cost or supply curve represents the minimum price that must be paid to the producer to supply the given quantity. Thus, for output OQ_m, OP_o is the minimum price that must be paid. In a competitive economy, the willingness to supply is determined by the additional (marginal) cost of producing the good. This cost includes the minimum monetary reward that must be paid for factors of production such as land, labour and human-made capital to be available to produce this good rather than some other good. The supply curve is upward sloping because at higher prices, additional output becomes profitable.

In the absence of any government intervention, marginal benefit and marginal private cost will be equated at point E, yielding an equilibrium quantity and price of Q_e and P_1. But since paper manufacturing creates pollution, additional costs to the society generated by pollution should be taken into account. The pollution cost (GB) is assumed to remain constant regardless of the output. This implies a parallel upward shift of the marginal private cost curve by the pollution cost GB. The resulting cost curve is the marginal social cost. If a per unit tax equal to the pollution cost GB is levied, then the externality will be internalized as production and consumption decisions are now based on the social rather than the private cost.

The pollution tax of GB per unit of output is called a Pigouvian tax after the British economist A. C. Pigou who first formalized the analysis.

When the marginal social cost is equated with marginal benefit, the new equilibrium quantity and price are at Q_s and P_2 respectively. The quantity is now lower and the price higher. Since the output of paper is still positive, some pollution still exists, but this is at an efficient level since all costs, including pollution costs, have been taken into account in the market. Thus, the externality has been internalized.

For the emission tax to be effective in practice, the relationship of pollution cost to output must be known. If there are many firms producing paper, but their pollution behaviour is different, then levying one rate of tax will be inefficient. On the other hand, the cost of collecting information to levy differential taxes would be prohibitive. This is especially the case in many developing countries which are characterized by numerous small firms and by poor business registration records (Parikh, 1995). If the pollution spills over national boundaries, international co-operation may also be needed. Emission taxes are also not suitable for non-stationary sources of pollution because of high monitoring and administrative costs.

A slightly different problem in this context is as follows. Suppose the authorities would like a particular geographical area to have a certain air quality. Households (through their types of cars and level of car maintenance), industry (through production technologies) and government (through power plants owned by them) all contribute to the pollution. How can a given quality of air be achieved at the lowest possible cost to the society?

The general answer from economic reasoning is that a reduction in each unit of pollution should be undertaken so that the marginal cost is lowest. Thus, if costs are too high to alter the technology of power plants, subsidies (or credits) may be given to households who scrap their old cars, assuming pollution level increases with the age of the car. This will be an efficient solution. However, the impact of this solution on different income groups should also be considered before reaching a decision.

Similar reasoning can also be applied to the international arena. Thus, low-income countries may prefer to trade off higher levels of pollution for some income and more jobs through such devices as relocation of high pollution-generating technologies and activities from high-income countries. While this may be Pareto-efficient in some situations, it will also pose difficult ethical and political dilemmas. As a result, unqualified use of such reasoning remains quite controversial.

When the emission taxes shown in Figure 4.2 are not feasible because of such factors as high monitoring and administrative costs, other options are available. One option is to tax inputs (e.g. coal to produce electricity) which form a part of the final product and which bear some identifiable relationship to the pollution generated. Alternatively, where even the tax on inputs is not feasible, the use of certain inputs could be mandated.

Another option is to levy a product tax based on the units of harmful substance contained in a product (e.g. taxing different fuels on the basis of their carbon content). The carbon tax is considered appropriate as CO_2 is an important anthropogenic greenhouse gas. For maximum impact, this tax should be levied on a global basis. However, given the wide divergence in taxation of fuels around the world, finding a minimum rate of carbon tax which is efficient, equitable, and enforcable at the global level is not an easy task (Shome, 1995).

Some countries have levied an excise tax on ozone-depleting chemicals (Jenkins and Lamech, 1994). The tax is the product of the chemical's weight times the base rate times the ozone depleting factor of the chemical. This tax is also not without administrative difficulties. These include setting an appropriate tax rate and treatment of an ozone-depleting chemical that is imported.

The manner in which revenue generated from emission taxes is spent is also relevant. If the revenue is spent to fund a subsidy for finding ways to reduce pollution, there will be a stronger effect on the externality. Since the poor will suffer most from the price increases as a result of environmental taxes, another option is to spend resulting revenues disproportionately on expenditures benefitting the poor. In general, tax revenue should be spent where it generates the most benefits to society.

Regulation

In principle, what could be achieved through taxation could also be achieved through regulation. Thus, for example, in Figure 4.2, government could mandate the output of paper at Q_s, the socially efficient outcome. In some cases, as noted, government may simply need to mandate the use of certain inputs, particularly when relevant information is lacking.

Regulation is also appropriate concerning an event whose probability of occurrence is low but where the ensuing damage (a negative externality) is extremely high. Thus, regulation is quite appropriate for hazardous chemicals or for nuclear power plants, because of the severity of the damage caused by an accident and the subsequent prolonged and large-scale health care arrangements.

In some cases, economic agents do not respond sufficiently to feasible price signals, thereby rendering taxation ineffective. Regulation may then be more appropriate. As an example, any feasible tax on leaded petrol is unlikely to decrease quantity consumed sufficiently to obtain the desired improvement in air quality, or to give required incentives for developing alternative fuels. This explains why leaded petrol is simply banned in many high-income countries. Similar reasoning explains why CFCs are being phased out throughout the world.

Subsidies for reducing pollution

In Figure 4.2, the difference between marginal private cost and marginal social cost represents the pollution cost per unit of production. If encouragement is given to less polluting production technology through subsidization of research and development, pollution costs can be reduced.

Subsidies for pollution abatement by themselves, however, are unlikely to enable socially efficient output (Q_s in Figure 4.2) to be achieved. This is because firms do not take the cost of government subsidies for abatement into their calculations. These nevertheless are a part of the social cost of production (Stiglitz, 1988).

It is essential to distinguish between pollution abatement subsidies on the one hand and subsidies for commodities in the form of prices or user changes below costs. Commodities often subsidised in this manner include fossil fuels, electricity, water, pesticides, logging, land clearance and construction. Under such circumstances, reducing or eliminating subsidies could lead to reduced environmental damage and an improvement in the efficiency with which resources are allocated (Markhandya, 1997).

Choosing among taxes, regulations and subsidies

Choices between taxes, regulations and subsidies should be governed by consideration of the following factors (a more detailed discussion may be found in Stiglitz, 1988):

- effects on distribution of income and of consumption, particularly of essential goods;
- differing costs of administration and compliance;
- differing information requirements;
- differing performance under variability and uncertainty about costs and benefits;
- ease with which the instruments can be misused by different interest and pressure groups;
- consistency with social norms, ethical values and environmental justice.

A major fiscal difference between taxes and subsidies is that taxes generate revenue for the government while subsidies need to be financed. For governments in many developing countries with high budgetary deficits, this is an important consideration. Taxes are usually more visible, while subsidies can be more easily hidden and are more susceptible to pressure groups. Subsidies, once granted, are also more difficult to reverse.

In cases where regulation involves allocation of administratively determined quantity, such as the right to own a fixed number of cars during a given period, the use of auctions may prove to be attractive. There are many types of auction, each with differing rules governing the bidding process, how the winner is chosen, the price that the winner is expected to pay, and whether resale is permissible. Clarity in rules and integrity of the bidding process are essential for auctions to be effective instruments of public policy.

An illustration of how auctions work is provided by a scheme in Singapore under which fixed government-determined numbers of rights to own motor vehicles are allocated through an auction (Phang, Wong and Chia, 1996). Under a scheme introduced in May 1990, all purchasers of new motor vehicles are required to possess a Certificate of Entitlement (COE) which is valid for ten years of ownership. The COE must be obtained in a monthly public tender. Facilities for electronic bidding are available. The deposit, equal to half the bid, is deducted from the bidder's bank account. Successful bidders pay the lowest successful bid price in each category (determined by the engine size) of motor vehicles. The system delivers the bidding results and returns the deposits of unsuccessful bidders within two working days. Successful bidders then use their COEs to purchase and register their vehicles in the same month of bidding. The rules governing the bidding process have undergone many changes since 1990. The changes were primarily designed to address public concerns about unfair practices, inequities and speculation.

Auctions have two major advantages:

- Auctions provide information on how much people value the right in question, which in the Singapore example, is the right to own a motor vehicle for ten years.
- Auctions generate revenue for the government rather than for the private individuals who are able to secure the right. It is important, however to ensure that amounts available to different parties involved in auctions are not overly disproportionate. Otherwise, public acceptance of auctions as an instrument for allocating limited rights may be eroded. Auction rules also must not cause inequities or perceptions of unfairness.

In some cases, regulation may be used to establish standards and to control technologies, while taxes, user charges, marketable permits or abatement subsidies may be used to provide incentives for efficient behaviour. Thus, a combination of regulation and effluent charges are used in several European countries to control water quality (Goodstein, 1995).

Voluntary bargaining and exchange

Ronald Coase (1960), recipient of the 1991 Nobel Prize in Economics, has argued that if certain conditions are met whenever externalities arise, the affected parties can get together and make a set of private arrangements by which the externality is internalized. This is known as the Coase theorem. Market remedies to correct externalities are usually associated with it. Four conditions are necessary for voluntary bargaining and exchange to be effective in correcting externalities.

First, the rights of ownership or titles to resources such as land, called property rights, must be clearly defined, established, exercised and enforced. It is clear that two parties cannot engage in voluntary bargaining and exchange without property rights. Indeed, a market economy cannot function effectively without a strong system of property rights.

In many countries, property rights are not well defined or cannot be effectively exercised either because of arbitrary use of power, absence of an independent judiciary or the high cost in terms of time and money in seeking redress through the court system. Moreover, many property rights were established not through legislation but through common law or customary law. The common law system is based on judicial decisions over time, though occasionally parts of common law may be formally codified. A customary law system is based on the long-established practices widely used among the local population.

Second, the cause and effect of externalities must be known with sufficient certainty. But many environmental problems have multiple causes, and the contribution of each is not always easy to ascertain. Discontinuities (non-uniform environmental effects of an activity over time and space), non-linearities (environmental effects of an activity not necessarily increasing in a linear fashion with its level), and uncertainties further complicate the analysis.

Third, the number of parties involved and the costs of arriving at a voluntary agreement must be small. In technical terms, this condition requires low transaction costs. To the extent that externalities involve the provision of such goods as clean air or water, the number of parties involved is very large, information needed is costly, and individuals have incentives to let others bear the costs, i.e. to free-ride. The free-rider problem arises when individuals who do not pay for a good or activity cannot be excluded from its enjoyment. Thus, if the air pollution is to be reduced from motor vehicles, each individual has an incentive to let the others bear the costs as no individual can be excluded from enjoying the cleaner air if the pollution is reduced. Voluntary bargaining will not be feasible or will be too costly in this case. Indeed, the provision of services for people to make collective decisions is a function usually performed by the government rather than the market.

Fourth, society must be neutral concerning the distributional consequences of the solution achieved through voluntary bargaining and exchange. The Coase theorem suggests that voluntary bargaining can achieve efficient results regardless of the initial assignment of property rights. However, whoever has the initial property right is clearly better off financially because compensation accrues to it. If the two parties involved in tackling externalities have widely differing incomes and wealth, then who compensates whom is a matter of concern to the society.

This is illustrated by the following example. Suppose a corporation has set up a chemical plant near a river, and its operation creates water pollution. River water is used by surrounding communities for drinking, fishing and recreation. Suppose there are no laws against pollution. As a result, the corporation has the right to pollute the river unless it is compensated not to do so. Hence it has the initial property rights. Now suppose laws provide initial property rights to the communities in the form of the right not to have their river water polluted. It is now the corporation that would be required to compensate the communities for violation of its property rights.

If the corporation has initial property rights, then the owners of the corporation obtain monetary benefits from the communities. If instead, the initial property right is vested with the communities, then its residents will receive compensation. As owners of corporations and members of the communities are unlikely to have the same household income profile, assignment of initial property rights will affect the distribution of income among households.

Let us take an example at an international level. If high-income nations are assigned property rights to the global environment or if they appropriate these rights, then they can impose sanctions on developing countries to improve the environment. It is the developing countries who in this case bear the main burden of adjustment. Assuming the outcome is efficient, it can also be attained if property rights are assigned to the developing countries. Then the high-income nations will need to pay them to improve the global environment. Thus, distributional outcomes are very different, and they do matter, even if the outcome is efficient in each case. In the real world, political and economic power will have an important bearing on the assignment of property rights.

The above four conditions are not always satisfied. Therefore it is not true that a government can never do better than the market. Nevertheless, the Coase theorem represents an important contribution. It shifts the focus of public policies from the assignment of the property rights to the importance of having a well-functioning system of property rights. Allocation of clear property rights to resources usually provides incentives to use them efficiently. Thus, to address the common-property resource problems related to, for example, Antarctica or the deep seabed, some sort of property right allocation is required. This could be done through partition among individual nations, or through joint management governed by an international treaty.

The allocation of property rights could also help overcome the problem of limited information available to the policy-makers, and to provide incentives to efficient resource use. Suppose certain technical air quality standards are to be met as a matter of public policy. A property rights approach would assign rights (or permits) to qualifying existing businesses to pollute up to a certain level. Those businesses which are able to reduce their pollution below the specified level are allowed to sell the excess pollution rights to those who are unable to meet the specified level of pollution. This is often referred to as the *tradeable permits method*. The method will enable the country to meet overall pollution level standards, while giving each producer an incentive to reduce pollution in the least costly manner possible.

This method of tradeable permits, however, has not worked too well in practice due to many reasons (Goodstein, 1995). The market in such permits is likely to have only a few transactions and to suffer from imperfect information concerning trading possibilities. There is an inverse relationship between the pollution standards set and the possibilities of trade in pollution permits. Thus, if standards are stringent, existing firms will be preoccupied with finding ways to meet them, and therefore they will have very few permits available for sale. This in turn will result in too few transactions for effective working of the tradeable permits approach.

The necessity to acquire pollution permits may discourage competition as it would increase costs for new entrants. This would reduce competition, the main spur to efficiency. Even with permits, the problems of monitoring and compliance with the regulations remain. Finally, adherence to overall pollution levels does not preclude an unacceptable level of pollution in some parts of the region or a country.

The Coase theorem also brings out the importance of transactions costs. Thus, laws, rules and societal institutions designed to facilitate both the market and the government or public policy decisions should be subjected to scrutiny to ensure minimum possible transaction costs.

4.4 Benefit–Cost Analysis and the Environment

Benefit–cost analysis (BCA) was developed during the 1930s and applied to flood control projects. Since then, it has become a well-established technique with which to analyse the economic efficiency implications of a wide variety of government projects, policies and regulations. As a result, at least a working knowledge of BCA has become essential. For simplicity, the subsequent discussion refers only to projects. BCA requires considerable multi-disciplinary expertise in many diverse areas in the physical, biological, natural and social sciences. It is an expensive technique, but measured against the size of investment, usually well worth the cost if done competently and objectively.

BCA systematically accounts for all possible benefits and costs, including those relating to environment, which accrue to the society as a result of the proposed project. Since BCA concerns society as a whole, any benefits to one group at the expense of another (called pecuniary externalities) should not be considered.

In situations where benefits cannot be measured or when benefits are identical, but there are two alternative ways to accomplish a given objective, the project with the lowest cost should be chosen. Thus, if two programmes result in the same degree of improvement in soil quality, then the one with the lowest cost should be chosen. This technique is known as cost-effectiveness analysis.

If any benefit or cost cannot be monetarily valued, it should be listed separately. This ensures that decision-making does not focus only on those items which are amenable to monetary valuation. Since society consists of many individuals, benefits and costs to individuals need to be aggregated. Thus, individual-based evaluation underlies the BCA.

For some individuals, benefits may exceed costs, while for others, the reverse may be the case. However, what matters for the BCA is the sum of the benefits exceeding the sum of the costs to all individuals. The benefits and costs include those relating to the environment. The net benefit to the society can be written as:

$$NB = \sum_{i=1}^{n} (B_i - C_i) \tag{4.1}$$

where:

NB = net benefit to society,
B_i = benefit to individual i,
C_i = cost to individual i,
n = total number of individuals in a society.

If NB is positive then society as a whole gains from the project. If more than one alternative provides a positive NB, then the one with the highest net benefit should be chosen.

Positive net benefit to the society implies that those with positive net benefits can, in principle, compensate those whose net benefits are negative, and still be better off. The compensation, however, is *notional* rather than actual. This is known as the Kaldor-Hicks criterion after the two British economists (Gramlich, 1990). This criterion is widely used because the Pareto efficiency criterion, which does not permit any individual to have negative net benefits, is regarded as too stringent. The Kaldor-Hicks criterion, however, becomes less convincing when a project's costs are largely concentrated on particular groups. Much of the adverse environmental impact of such development activities as dams or mining projects is often on the lowest socio-economic strata while the primary beneficiaries are often the upper-income groups. Even if such development activities do pass the Kaldor-Hicks test, they could severely disrupt a significant number of communities (Sainath, 1996).

Equation (4.1) implicitly assumes that benefits and costs are known with certainty. However, many types of benefits and costs cannot be accurately predicted at the outset. For example, a major benefit of a project or a policy involving a reduction in air pollution will be a fall in the number of deaths and an improved health status for the population. But quantification of these effects can only be done with a certain margin of error. When probabilities of various outcomes can be estimated, the situation is said to involve *risk*. When such probabilities cannot be estimated, then the situation is termed *uncertain*. Risky outcomes can be insured against, but uncertain outcomes cannot.

Even in theoretical terms, there is no easy way to incorporate risk and uncertainty in the BCA, particularly when environmental effects are involved. These difficulties are compounded because when there is risk and uncertainty, the actual behaviour of people in practice is very complex (Turner, Pearce and Bateman, 1994). People's perceptions about the probability of an event occurring are governed by many factors, not all of them based firmly on the empirical evidence. As a result, when there is risk and uncertainty, the assessment of risk by experts and by the general population often differ greatly.

There is, however, widespread agreement in the literature that the use of a higher time discount rate, which relates the value of a dollar at one date to its value at a later date, is not appropriate in dealing with risk and uncertainty, when conducting the BCA (Stiglitz, 1988; Gramlich, 1990; Pearce, Barbier and Markandya, 1990).

Distributional considerations

In Equation (4.1), benefits and costs accruing to individuals at all income levels are given the same weight. But society may want to treat net benefits accruing to low-income groups as being more desirable than those accruing to other income groups. It is feasible for the BCA to take into account such distributional considerations (Gramlich, 1990). Two such approaches may be identified. The first approach is to attribute differing weights to benefits and costs accruing to different income groups. Suppose a society is divided into three income groups: low-income, middle-income and high-income. Assume that the society wants to attach higher weight to benefits and costs accruing to the low-income group than to the other two income groups. Then, a weight of 1.25 may be assigned to benefits and costs to the low-income group; 1.0 to the middle-income group; and 0.75 to the high-income group. The critical issue concerns the assignment of appropriate weights without being arbitrary, as arbitrary weights can undermine the objectivity and theoretical consistency of the analysis.

A complication of this approach is that it requires estimation of benefits and costs by income groups for each period. As a result, much more information about the impact of the project is needed, information that is difficult and costly to obtain.

A second approach entails undertaking the BCA in its traditional form, but then providing a separate analysis of the possible income distribution effects of the project, and incorporating it in the interpretation of the results. This could include an evaluation of how different income groups are affected by the proposed project (Stiglitz, 1988).

An important public policy question is whether a project-level analysis is the appropriate level for taking into account income distribution considerations. Economists generally prefer to address the distributional questions separately from efficiency issues which are the main concerns of the BCA. But political and social realities do not always permit such a course of action. For some programmes (e.g. social forestry), a redistributive impact in favour of the poor is usually an important objective, and therefore distributional considerations should be an integral part of the BCA related to such programmes.

BCA and discounting

Typically projects for which BCA is undertaken involve benefits and costs occurring over a period of time. This is true in the case of environment. Moreover, many environmental effects may be cumulative and non-linear. This would not matter if individuals had no preference concerning the timing of benefits and costs. However, individuals typically prefer benefits now rather than later, and vice versa for the costs. As a result, one dollar's worth of benefit (or cost) realized in the future, say two years from now, is worth less than one dollar in present value terms. The magnitude of the difference depends on the individual's time preference. It is this preference which governs the rate at which future values are discounted to arrive at the present value. It is important to keep in mind that the time-discount rate (henceforth referred to as the discount rate) relates the value of a dollar at one date to its value at a later date, and this is *not* related to differences in risk (or uncertainty) concerning magnitudes of costs and benefits (Stiglitz, 1988).

Equation (4.1) implicitly assumes that all benefits and costs are realized during the current period. However, as all benefits and costs are being discounted, it needs to be rewritten in terms of net present value (NPV). The formula for NPV is

$$\mathrm{NPV} = \Sigma_t\, (\mathrm{B}_t - \mathrm{C}_t)\, (1+r)^{-t} \qquad (4.2)$$

where

B, C = benefits and costs respectively,
t = time period,
r = the discount rate.

A positive NPV is required to ensure that society as a whole benefits. When more than one project performs the same function, the one with the highest NPV, keeping the total budget available in mind, should be chosen.

The NPV varies inversely with the discount rate (r). The higher the rate, the lower the NPV, and vice versa. Thus, the choice of a discount rate is critical. Two broad options are possible. First, funds for the project being subjected to the BCA may come either from foregone consumption or investments or a combination of the two. Depending on the source, foregone return from the source is chosen as the discount rate. If the funds come from a combination of the two, a weighted average of the foregone returns is chosen as the discount rate. Second, a social rate of discount may be used. This rate measures the valuation that society places on the present consumption to be sacrificed owing to the investment in the proposed project. Many variations of the two options are found in the literature (Gramlich, 1990).

While theoretical literature on the discount rate is extensive, no simple guidelines determining the appropriate discount rate have emerged. Empirical studies by such institutions as the World Bank use the same discount rate (10 percent) for all projects, thereby allowing comparison of many diverse projects involving different sectors and geographical regions. This rate, however, is on the high side.

The choice of discount rate also has implications for the environment. A high discount rate means that benefits and costs occurring far in the future count for little in the NPV. Thus, those projects which result in environmental costs far into the future may pass the BCA test too easily. This applies, for example, to nuclear power plants. The costs relating to the storage of waste and the hazards arising from the failure to safely handle the storage occur far in the future. Therefore, at any positive rate of discount, these costs have little impact on the NPV. Many environmental costs are of this type.

The problem cannot be addressed through the use of a zero or lower rate of discount for environment-related impacts, although this is often advocated by some environmentalists (Common, 1995). This is because intrinsic to the BCA methodology is the assumption that all inputs and outputs should be measured by the same standard, and therefore should be treated in the same manner. If environment, as an input, is accorded a special treatment in the form of a lower discount rate, then the internal logic of the BCA is violated. On the other hand, if environmental impacts are not regarded as commensurable, then the BCA is an inappropriate technique to use. What is needed is an appropriate valuation of environmental impacts before uniform discounting is applied.

Valuation of benefits and costs

Monetary valuation of benefits and costs for society as a whole is based on aggregation of individual preferences. Two key concepts in analysing individual preferences are willingness to pay (WTP) and willingness to accept (WTA). WTP reflects how much an individual is willing to pay to acquire various quantities of a good or a service, including environmental services. Figure 4.3 provides a graphical illustration of the concept of WTP. DD' is an individual's demand curve for a good. As explained when discussing Figure 4.2, each point on the DD' curve reflects the maximum price that individual is willing to pay for a given quantity of the good. This price represents the individual's assessment of the benefits from the last unit of the quantity demanded.

Suppose the quantity of OB is being consumed. Then, the WTP of the individual can be measured by the area ODCB, an area under the demand curve. If the individual is asked to pay OA, then the total payment is OACB. Its WTP (or benefit) therefore exceeds the expenditure by the triangle ADC. This triangle is called consumer surplus or net WTP.

Demand for a good is dependent on such factors as income, price of the good, prices of other goods, and preferences. In Figure 4.3, all the factors except the price of the good are assumed to be constant. Thus, the factors kept constant determine

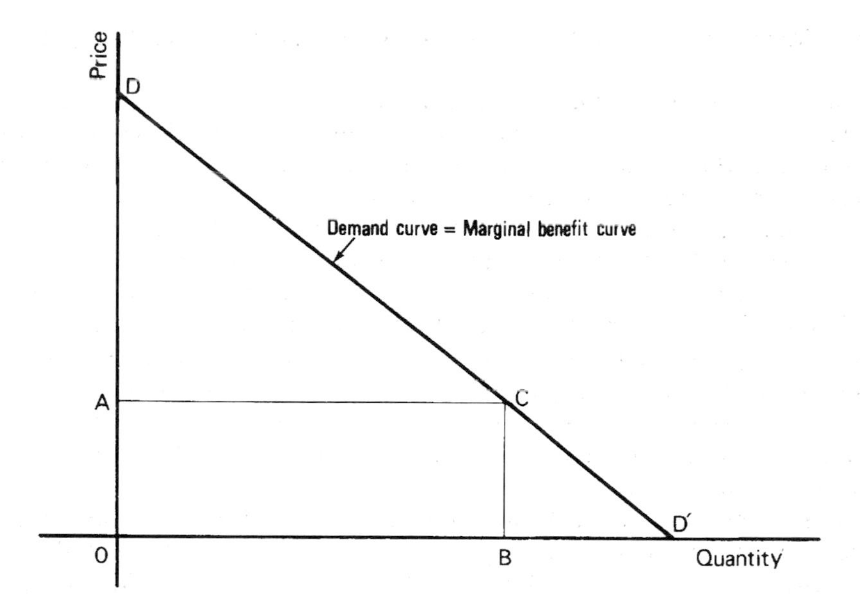

Figure 4.3 Willingness to pay as a demand curve. For quantity OB, the individual derives total benefit equal to the area ODCB. If the individual is asked to pay OA, then the total expenditure is OACB. As a result, total willingness to pay (WTP), or total benefit, exceeds the total expenditure by the triangle ADC. This triangle is the consumer surplus or net WTP

the position of the demand curve in Figure 4.3, while the effect of the price of the good on the quantity demanded is reflected in the slope of the demand curve.

In BCA, net benefits are often measured by valuing *changes* in consumer surplus (or in net willingness to pay) brought about by a project (Turner, Pearce and Bateman, 1994). A closely related concept is willingness to accept (WTA). It reflects the extent to which people are willing to pay to avoid the occurrence of something undesirable, such as the closure of a national park or the extinction of a rare bird. So WTA measures willingness to tolerate something people find unpleasant.

The nature of a particular problem, and the cost and availability of the required information govern the choice of whether to use the WTP or the WTA. There are many situations where the difference between the two is not large enough to make a material difference. Techniques for valuing environmental impacts are briefly discussed in Section 4.5. This includes situations where markets do not exist, as is the case for many environmental services.

4.5 Valuation Techniques and Environment

Valuation of benefits and costs in BCA should be based on competitive market valuations as far as possible. More complex techniques are required when such valuations are not feasible because markets are non-competitive or non-existent.

This is often true in the case of environmental services. Valuation through the use of such techniques is called shadow pricing. Without such pricing, most environmental effects cannot be incorporated in the BCA as there is often no market for them. As a result, some understanding of the techniques used for valuing environmental services is desirable.

Munasinghe and Lutz (1993) define the total economic value (TEV) of a resource that is not essential to sustain life on earth as consisting of use value (UV) and non-use value (NUV). The use value may be divided into direct use value (DUV), indirect use value (IUV) and option value (OV). DUV is defined as output that can be consumed directly, such as food or recreation services. The IUV of a resource confers functional benefits such as flood control or storm protection. The OV is based on the WTP of the individuals to pay today for the option of being able to enjoy environmental assets such as rain forests or endangered species. Thus OV refers to the potential use value.

Non-use value (NUV) may be divided into bequest value (BV) and existence value (EV). The BV refers to the value that individuals obtain from knowing that others may derive benefits from environmental assets in the future. The EV refers to the value that an environmental asset such as Mount Everest in Nepal or Lake Toba in Indonesia derives from its existence. It should be noted that particularly for EV and BV, differences between WTP and WTA in some cases could be fairly large.

The above discussion may be summarized as

$$\mathrm{TEV} = \mathrm{UV} + \mathrm{NUV}$$

or

$$\mathrm{TEV} = [\mathrm{DUV} + \mathrm{IUV} + \mathrm{OV}] + [\mathrm{EV} + \mathrm{BV}] \tag{4.3}$$

where all the terms are as defined in the text.

The above elements of the TEV need to be valued in terms of WTP or WTA. Where competitive markets exist, market valuation is relatively straightforward. Where this is not a feasible option, a variety of techniques may be used (Johansson, 1990; Munasinghe and Lutz, 1993).

Thus, DUV may be estimated by observing how production levels and efficiency of resource use change as a result of better water management. In the case of health, loss of earnings due to illness would provide an estimate of health costs of the environmental damage. For valuing death due to environmental accident or the saving of life due to improvements in environmental quality, the discounted value of earnings of the individual is often employed.

Another possibility is to observe differences in earnings in occupations with varying risks of accidental death, and thereby infer the value attached to life (Gramlich, 1990). Economists prefer to attach a monetary value to life explicitly rather than implicitly. An implicit valuation occurs when, for example, public health spending and access priorities are set. Valuing life, however, poses difficult ethical dilemmas.

Differences in property prices in areas with varying environment quality may be used to infer monetary values of environmental disamenities. The travel cost method (TCM), also known as the access cost-quantity demanded method, may be used to value non-marketed goods when consumers face different expenditures in time or money to get access to them. A frequently used technique designed to elicit information to estimate the WTP or WTA of environment-related impacts is called the contingent valuation method. It involves conducting a survey in which respondents are asked to value a hypothetical or actual policy or a programme. For example, such a survey may provide a scenario indicating the measures necessary to save the tiger population of the world from extinction. Based on the WTP or WTA derived from such a survey, aggregation of individual preferences can be ascertained and then compared to the costs of the policy or programme. An illustration is given below.

A 1995 study reported in Laxmi, Parikh and Parikh (1997) estimated willingness to pay (WTP) for setting up an autonomous organization to maintain Borivali National Park in Greater Mumbai (Bombay), India. This park, located in north-western Mumbai, is spread over 110 km^2 and attracts 2.5 million visitors annually.

The study used interviews to estimate the WTP not only in monetary terms (in Indian rupees) but also in labour-time. This was necessary as many residents of this city have low incomes, but value the services provided by the Borivali National Park in a crowded metropolis. The inclusion of labour-time in the WTP vastly expands the scope and relevance of environmental valuation studies in the developing countries. The study estimated a mean WTP of Rs 7.50 per family per month, with 28 percent willing to contribute in labour-time. The net present value (NPV) of the WTP of the city residents was Rs 1033 million, several times the NPV value (Rs 200 million) regarded as sufficient by the Borivali National Park authorities. A 10 percent discount rate was used.

The above study demonstrates both the practicality of the contingent valuation methods, and the fact that people with low incomes are willing to pay for environmental services, provided they have confidence that their contributions will be managed well, and serve the interests of the local residents. One disadvantage of this survey, and of the contingent valuation method in general, is the hypothetical nature of the situation. Consequently, individuals tend to visualize an ideal situation divorced from their actual willingness to assist in a concrete instance. A second difficulty arises in maintaining objectivity and minimizing the potential for bias (Turner, Pearce and Bateman, 1994).

BCA, public policy-making, and the environment

BCA focuses on the economic efficiency implications of a project. To be relevant, it needs to fit into the actual process of public policy-making which is multi-goal-oriented. While efficiency is not the sole criterion for public policy, it is nevertheless important.

Therefore, BCA, with its emphasis on economic efficiency and on systematic, internally consistent methodology is too useful a tool to be ignored. However, it should not be employed as the sole means of reaching a decision, particularly when non-quantifiable benefits and costs are relatively significant.

When environmental effects are significant, applicability of the following four critical assumptions of the BCA should be carefully examined. First, valuation in the BCA is based primarily on the willingness to pay of individuals. Such willingness need not necessarily coincide with sustainability. Moreover, as the BCA focuses on individuals, the impact on non-humans is ignored. Secondly, economists assume all resources to be substitutable. So exhaustion of one resource is not a matter of consequence. But as noted by Common (1995), it may be prudent to assume that for some environmental assets, there are no available substitutes. Third, BCA is appropriate when relatively small projects that do not affect market values or preferences in a fundamental way are involved. Large development projects such as on the Narmada River in India or the Three Gorges Dam in China require resettlement of a large number of people, with consequent disruption of their accustomed way of life. As a result, market prices and preferences may change significantly. This should be recognized in any BCA for such projects. Fourth, BCA has traditionally been carried out in the context of a nation state or at sub-national levels. But many environmental problems occur across national boundaries. Examples include rivers which flow across more than one country, and industrial air pollution drifting from one country to another. In such circumstances, to identify and then value benefits and costs poses additional challenges. A national or sub-national government has little incentive to take into account spillovers to other nations. The task of estimating WTP and WTA of individuals in nations at different levels of development is particularly complicated. Moreover, international political economy considerations become relevant when environmental problems occurring across national boundaries are addressed through negotiations and treaties. Transparency, public participation, and peer group review are required to ensure that those commissioning or undertaking the BCA do not bias the results in a particular direction.

4.6 Macroeconomics and the Environment

Macroeconomics involves economic aggregates such as the Gross Domestic Product (GDP), inflation, and total employment. The relationship between macroeconomics and environment is bi-directional (Gandhi, 1996).

Sound macroeconomic conditions and management are necessary, though not sufficient, for environmental preservation. This is because macroeconomic instability such as prolonged periods of slow growth, high rates of inflation or widespread unemployment could provide incentives to individuals and businesses to focus excessively on the short-term considerations. This could lead to excessive natural resource exploitation and neglect of maintenance of assets such as irrigation canals, thus damaging the environment.

In pursuing macroeconomic stability, it is essential to avoid reducing environmental, health care, and infrastructure maintenance budgets. This suggests that the quality and the manner of achieving macroeconomic stability are also relevant for avoiding environmental damage. Deliberate pursuit of slow growth in the developing countries is also not conducive to environmental protection. Instead, appropriate environmental policies and their competent implementations directed at particular sectors and activities are likely to be more effective.

Environment could also affect macroeconomic performance. Degradation of water, air or land reduces the productivity of people and natural resources, and thereby adversely impacts on economic growth. Such degradation could also have an adverse impact on food security, and lead to sharp rises in the prices of basic foodstuffs. As these carry considerable weightage in the commodity basket used in many developing countries to measure inflation, environmental degradation could also have significant inflationary impact.

4.7 Conclusions

Economic activities have environmental impacts. This implies that the environment can be affected by economic policies. Environment economics therefore has become an important sub-discipline in economics. Economics, however, provides a framework for reasoning rather than a body of settled conclusions.

Environmental impacts occur over long periods of time. They are also non-linear, interconnected, uncertain and sometimes intangible. While market-compatible instruments, such as taxes and pollution permits, may be used to address certain environmental impacts, regulation and other command-and-control techniques also have a role to play. In some cases, regulation and market-compatible instruments may be used in conjunction with each other. In choosing various instruments, efficiency, equity, biodiversity and environmental justice objectives need to be considered. These policies also need to be made consistent with a country's economic, social and cultural *milieu* and with its administrative and enforcement capacity (Parikh, 1995). The issue of who gains and who loses is important because emphasis on efficiency considerations alone could in some cases harm the very poor even in the long run. This in turn may elicit behaviour which may result in environmental degradation; and may reduce their capacity to deal with adverse environmental impacts. Moreover, efficiency should be seen in a system-wide context. For example, subsidizing kerosene as a cooking fuel used by the poor may be inefficient taken in isolation. If the subsidy is removed and the price of kerosene increases sharply, the poor may substitute wood as fuel. This may contribute to deforestation and could create dangerous levels of indoor pollution. This shows the complexity involved in designing appropriate policies.

In spite of its many limitations, benefit–cost analysis (BCA) remains an important tool for analysing projects and policies, including those with significant environmental impacts. However, it should not be regarded as a sole input in the decision-making process, but only as one of several inputs, albeit an indispensable one.

Environmental policies have the potential to benefit some regions and groups, and impose costs on other regions and groups. This explains why political considerations are an integral part of any policy affecting the environment. Sound knowledge of the political economy of a country is therefore essential in the formulation and implementation of environmental policies. Environmental management also requires a competent and accountable government with sufficient political will to pursue overall public interest.

Systematic application of economic reasoning has greatly advanced our understanding of the nature and causes of environmental impacts, and how to address them in an efficient and equitable manner. Thus, good economic policies and management do matter for environment. However, they must be based on an adequate database and an understanding of the subtleties of economic and environmental analysis in the context of a particular country or situation.

Exercise

1. Why is it inappropriate to assign either a zero or an infinite price for environmental services?
2. Explain the relationship between the concept of economic efficiency and efficient pricing of environmental services.
3. Explain how taxes and regulation represent alternative ways of achieving efficient output of a good with negative externalities. What considerations are relevant in choosing between the two? Can both be used simultaneously? Give an example.
4. Why cannot subsidies for abatement of pollution achieve an efficient level of output of a good? Under what circumstances are they useful?
5. Under what conditions are voluntary bargaining and exchange sufficient to tackle negative externalities?
6. How can environmental considerations be incorporated in a benefit–cost analysis (BCA)? Why is it inappropriate to suggest that a zero or lower discount rate be used when environment effects are significant?
7. What considerations should be kept in mind when evaluating the results of a BCA for a project with significant environmental impacts?
8. Enumerate the benefits and costs that would form a part of the BCA of a regulation requiring new cars (a) to have catalytic convertors and (b) to use unleaded petrol.
9. What is the role of Willingness to Pay (WTP) and Willingness to Accept (WTA) in the monetary valuation of environmental services? What are their limitations?
10. Briefly state the valuation techniques you would use to assign monetary values to the benefits identified in Question 8.

Glossary of Selected Terms

Benefit–cost analysis (BCA) A technique used to evaluate projects, policies and programmes by systematically accounting and valuing all possible benefits and costs to society, including those relating to the environment.

Coase theorem When parties affected by externalities can bargain among themselves to internalize externalities without any transaction costs, the resulting outcome will be Pareto efficient regardless of the initial distribution of the property rights.

Emission taxes Tax payments that are directly related to the measurement (or estimation) of the pollution caused.

Externality When action by an economic agent unintentionally affects the welfare of other economic agent(s) without having it reflected in the market price. An efficient level of externality is where the marginal social cost (MSC) of a good or an activity equals the marginal social benefit (MSB).

Pareto efficiency Pareto efficiency (also called economic efficiency) represents a situation under which no alteration in production and consumption patterns will make someone better off without making someone else worse off.

Pareto pricing rule When the price of a good equals the marginal social cost (MSC) of producing it. MSC includes both the costs incurred by private producers of the good for such items as labour, materials and utilities, and external costs such as pollution costs imposed on the rest of society by its production and consumption.

Time discount rate The time discount rate relates the value of a dollar at one date to its value at a later date.

Part II
Managing the Environment

5

Natural Vegetation as a Resource

5.1 The Distribution of Natural Vegetation

Prior to the spread of agriculture, the surface of the earth displayed a mosaic of vegetation types which were entirely natural. The mosaic was chiefly controlled by three factors: temperature, precipitation, and the history of vegetation change since the end of the Pleistocene. During the Pleistocene, the earth went through a number of cold (glacial) and warm (interglacial) periods. The vegetation adjusted to such climatic changes by moving to the warmer areas during the glacial period and returning to old positions during the interglacials (which resembled the present conditions). The present distribution was achieved some time after the end of the Pleistocene 10 000 years ago.

The twin control of temperature and precipitation determines the amount of moisture available to plants. Warm areas need higher precipitation to maintain a forest as a substantial part of the rainfall is lost by evaporation to the atmosphere. Where in the tropics the moisture is insufficient the tropical forest cannot be sustained, and in its place we find grassland and even desert. Away from the equator in the higher latitudes, the cold soils of the winter prevent the supply of ground moisture to plants. The local trees adjust by physiological adaptations that reduce water loss. The adaptations have given rise to deciduous forests where trees lose their leaves in winter, and coniferous forests where the leaves are reduced to needle-like forms. In even colder areas towards the North Pole, generally treeless vegetation (tundra) prevails.

Broadly speaking, the vegetation of the world is a combination of forests, grasslands, scrublands and deserts. However, when resolved at a finer level, the combination of temperature and precipitation gives rise to a much larger number of vegetation classes. Holdridge (1967), for example, classified the plant formations of the world into a large number of life zones. In the tropics, where most of the developing countries are, the vegetation types vary according to moisture availability and seasonality of climate, and include the rain forests, the tropical monsoon forest, tropical grasslands with trees, tropical grasslands, semi desert scrubs, and finally some of the largest deserts on earth. Altitudinal zoning of vegetation occurs on high mountains, the tropical rain forests being replaced by mountain rain forests, subalpine forests, alpine forests, and (on mountains that rise beyond 3500 m) a treeless vegetation community towards the top.

The destruction and replacement of the natural vegetation started significantly with the introduction of agriculture at the end of the Pleistocene period. Within a few thousand years, the demand for construction, shipbuilding and urbanization made further inroads into the forests. For example, the destruction of the famous cedars of Mount Lebanon started 5000 years ago. Deforestation accelerated early in the Mediterranean region and parts of Asia (particularly China). In Central and Western Europe deforestation for agriculture, industry and shipbuilding in the historical period accelerated the destruction of the European forests. The forests of North America were also partially destroyed for city building, mining, agriculture, and construction of the railways. The extent of deforestation ushered in scientific forest management (first in Europe and then in North America) for preserving and replanting the forests and for meeting the demand for timber by proper management of both natural and planted forests. By the mid-twentieth century, forests of the world had been reduced to between one-half and one-third of the original extent (Eckholm, 1976). A substantial part of this forest was in the tropics, mostly as rain forest. Although other types of tropical vegetation were also depleted to meet various demands of the local population, it is the destruction of the rain forests in the second half of this century (mostly to meet overseas demands) that has raised the spectre of an eco-disaster (Figure 5.1).

5.2 The Tropical Rain Forests

The rain forests of the humid tropics occur in three large regions: South and Southeast Asia, Equatorial Africa, Central and South America (especially the Amazon Basin). Rain forests also occur on Madagascar, Mauritius, various small tropical islands in the Pacific and Indian Oceans, and along the coast of Queensland (Australia). Such forests comprise tall trees with long, slightly tapering trunks, and a closed canopy overhead. Floristically, the humid tropics is very rich. Of the approximate 250 000 species of flowering plants in the world, about two-thirds are found in the tropics. The breakdown shows over 80 000 in Central America and the tropical part of South America, 40 000 in Asia, 35 000 in tropical Africa, and 8500 in Madagascar. Even excluding Indo-China, Southeast Asia has about 25 000 species (Whitmore, 1991). It is a luxuriant forest with the trees well festooned with climbers and epiphytes. The tropical soil under the rain forest is generally not fertile. The humus in its top layer is continuously circulated. New material is constantly added by rapid decomposition of the fallen leaves and branches, which replaces the old humus stored in the soil being used up by the species-rich forest. The extreme case occurs in the Amazon Basin where the fallen material decomposes on the roots themselves, which then directly absorb the nutrients. Very little material enters the soil.

The traditional inhabitants of the rain forest were hunters and gatherers like the Penan of Sarawak, and shifting cultivators like the Hanunoo of the Philippines or

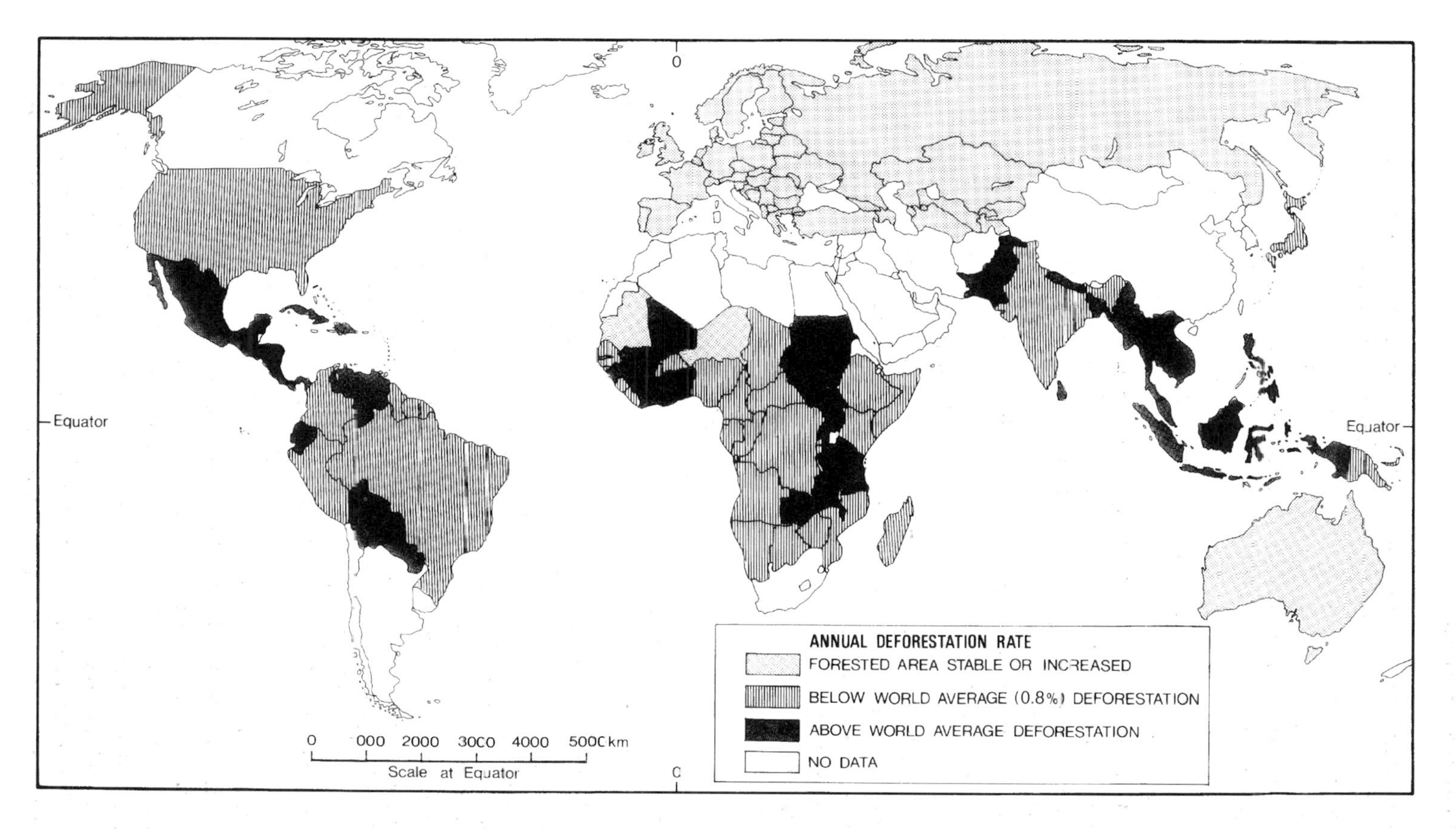

Figure 5.1 *Annual global deforestation rate. Source: World Resources Institute (1994)* World Resources Report 1994–95, *World Resources Institute, Washington, D.C.*

the Tsembaga Mareng of Papua New Guinea. The population density was low, the forest inhabitants did not remain long in one place, and their impact on the rain forests was limited (Box 5.1). Parts of the rain forest, especially on the plains and lower slopes, were cleared for plantation agriculture. For example, rubber was planted in Malaysia from the beginning of the twentieth century. The crisis, however, arrived about half a century ago.

Box 5.1
Shifting Cultivation

Traditionally, shifting cultivators clear small patches of the forest by cutting and burning, and then plant crops using a digging instrument. The plant nutrients are available to the crops as ash. The cultivators grow one or two quick crops of staples (rice in Asia; sweet potatoes in Papua New Guinea or Melanesia; maize, beans and squash altogether in Africa and the tropical America). Longer lived crops planted at the same time but harvested for several years include cassava (locally known as manioc or tapioca), bananas, chilli, and fruit trees. After two or three years, weeds proliferate, the soil is exhausted of the stored humus, and the yield drops. The shifting cultivators move to a different location but continue to collect fruits, and for the first few years, cassava from the old plot, which also yields firewood, medicines and building material. A wide range of crops are grown even within a small area. For example, about 120 species are planted by the Lua people of northern Thailand: 75 for food, 21 for medicine, 20 for decoration or ceremonies, and 7 for weaving or dyeing. Traditionally the farmers do not return to this spot for 10–15 years, giving sufficient time for the vegetation to be re-established and humus to build up in the soil. Traditional shifting agriculture displays sustainability. For example, the energy return for energy input is 20:1. In modern industrial agriculture the energy input is higher than the return when the whole package (sowing to marketing) is considered (Collins, Sayer and Whitmore, 1991). For a density of 10–20 people km^{-2}, shifting cultivation is a sustainable system of agriculture. A higher density creates a land shortage, resulting in the fallow period being shortened and/or the period of cultivation lengthened. The system then breaks down (Whitmore, 1991).

5.3 The Resources of the Forest

The rain forests are rich in resources and the rising demand for such resources has led to their drastic depletion in the last few decades. Some of the resources are used locally, but the extraction of timber to meet overseas demand is common. The perception of forests as potential foreign exchange earners and a debt-payment resource, has led to their rapid depletion in many developing countries. The various types of demand for the resources of the forest, including the land on which the forests are located, are summarized in Table 5.1. Different types of demands exist in different parts of the world. For example, the primary cause of deforestation in Southeast Asia is timber extraction, in Africa it is shifting cultivation, and in the Amazon Basin deforestation results from over-optimistic land clearance for farming.

Table 5.1 Causes of forest destruction

1. Extraction of forest products
   - (a) timber as industrial raw material
   - (b) fuelwood
   - (c) non-timber resources
2. Agricultural practices
   - (a) forest clearance for new lands, especially for transmigration
   - (b) pressure on shifting cultivation
   - (c) plantations
   - (d) cattle ranching
3. Miscellaneous causes
   - (a) mining
   - (b) highway construction
   - (c) dams and reservoirs
   - (d) development projects and fiscal incentives

Extraction of forest products

Tall trees belonging to the Dipterocarp family are common in the rain forests of Southeast Asia. When sawn through, the trunks of these trees provide large planks which maintain a homogenous high quality across the entire surface. Such tropical light hardwood trees are in great demand. Of all the wood used annually, 11 percent comes from the tropical rain forests. The timber is commonly exported to Japan, USA and the European countries, with Southeast Asia being the main source (Figure 5.2). About 80 percent of the world tropical hardwood exports come from Asia. Very high rates of timber extraction are found for Indonesia, Malaysia and the Philippines. It is not only the extracted material that depletes the forest; the felling of large trees destroys the vegetation near them, and the collection of timber also requires logging roads to be constructed into the forest. Until the 1980s, the usual practice was to export logs to Japan, and sawnwood and plywood to the European countries and USA. The conversion of logs into sawnwood and plywood was carried out in the timber mills of Hong Kong, South Korea, Taiwan and Singapore. The pattern has changed, and at present Indonesia has the biggest capacity for plywood production. Modelling of future production by Alan Grainger has shown a possible drop in timber export from Southeast Asia in the immediate future due to a combination of ongoing timber depletion, deforestation for agricultural expansion, and the rise in domestic consumption. Southeast Asian countries need to reassess their forestry and land use policies (World Resources Institute, 1988).

A limited amount of firewood is collected from the rain forest, but firewood collection is far more destructive in the drier forests. All possible forms of biomass (e.g. wood, twigs, crop residues, grass) are collected for both cooking and space heating. Fuelwood and charcoal are collected by a large number of poor people in China, Nepal, India, Egypt, Kenya, Brazil and many other countries, especially in their drier regions. Collection of fuelwood has particularly disastrous effects when carried out in marginal areas such as the desert fringes or the slopes of tropical mountains. In both cases the depletion of vegetative cover significantly increases

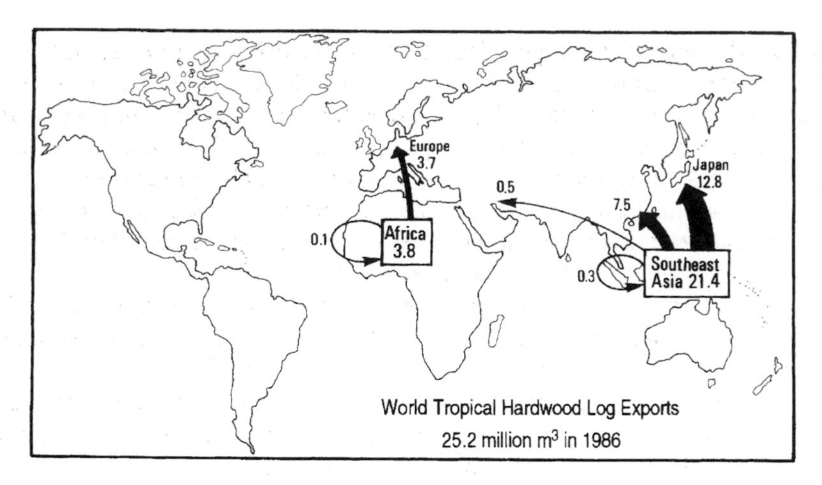

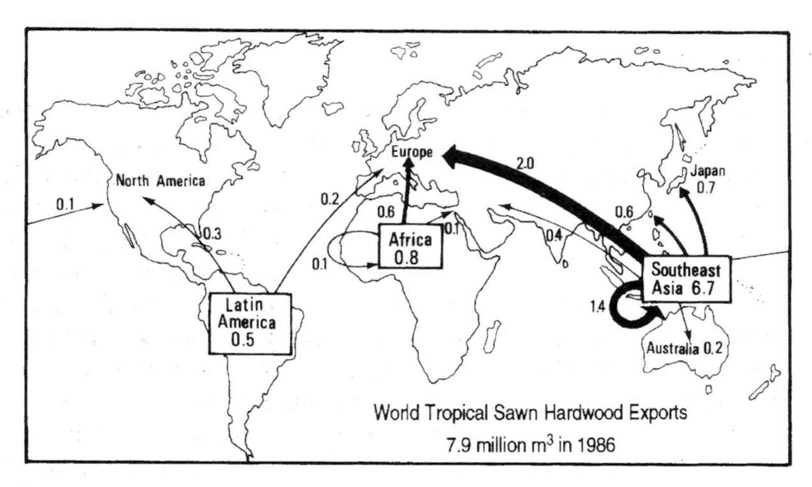

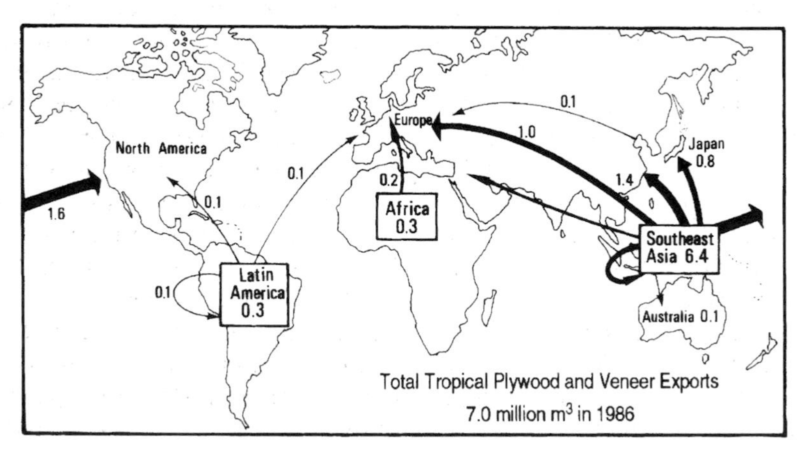

Figure 5.2 *Trade routes of tropical timber. From Collins, N.M., Sayer, J.A. and Whitmore, T.C. (1991)* The Conservation Atlas of Tropical Forests: Asia and the Pacific. *Reprinted by permission of Macmillan Press Ltd*

erosion. On the mountain slopes depletion of the original thin vegetative cover gives rise to slope failures, flooded streams and sediment-laden rivers. The depletion of forests on the mountain slopes forces local women to climb up from the villages in the valleys to the upper slopes in search of firewood. This, above a normal day's work for women, is not only fatiguing, but also reduces the time spent with the rest of the family. As Agarwal and Narain (1986) have pointed out, fuelwood is also brought in from long distances to meet the demand in the cities of India – a pattern that recurs in many developing countries.

The resources of the tropical rain forests have never been limited to timber. Traditionally, tropical forests have been the storehouse of dyes, rattan, rubber, oils, gums, fibres, resins, tannin, turpentine, varieties of fruits, herbal medicines, ornamental plants, etc. Earnings from these resources are, in many locations, a good argument for maintaining the forest. Collection does not normally deplete the forest and its maintenance can be economically sustainable.

Agricultural practices

The rain forests are destroyed not only for their resources but also for the land the forests occupy. Large-scale land clearance has taken place in the second half of this century to settle migrant farmers. This has occurred both under government planning and as a result of the spontaneous movement of poor and landless farmers. In Indonesia, government-sponsored migration resettled almost three million people from the congested islands of Java, Madura, Bali and Lombok to the low population areas of Sumatra, Kalimantan and Irian Jaya (the *transmigrasi* project). All forests in Indonesia have been designated for specific treatments: protection, production or conversion. Forests destroyed by *transmigrasi* generally belonged to the conversion category. However, the sponsored migrants are often followed by a much larger number who have heard the good news from friends and relatives who have already migrated. If the possibility of this influx is not recognized at the planning stage, then their presence leads to considerable forest destruction. The degradation of forests and slopes at Lampung Province, South Sumatra (across the water from Java) is an example of this type of environmental degradation. The FELDA project of Malaysia (where young rural people are trained and settled in virgin areas to become landholding farmers) has destroyed rain forests on Peninsular Malaysia (Collins, Sayer and Whitmore, 1991). For example, 70 000 people were settled in an area of about 600 km^2 of former lowland rain forest in Malaysia, known as the Jengka Triangle. In countries with a high rural density, poor landless peasantry move to the edge of the forest to clear vegetation and farm the land. The construction of a new road cutting through the forest also brings in such migration, as discussed below.

Sustainable shifting agriculture turns into a forest destruction activity when population pressure builds up. An increase in population density may follow an increase in the number of the people, but more commonly it starts to happen as the area under forest is reduced or as the shifting cultivators face more and more restrictions over the area they can shift for agriculture. This, as mentioned in Box 5.1, shortens the bush fallow period and lengthens the farming period. Given the fragile nature of the tropical rain forest soil, this leads to rapid land degradation.

Deforestation in Africa, to a large extent, is due to both the extensive nature of agriculture and its low productivity. A lot more land is required to feed its people than, for example, in Asia. African countries do not have to deal with the problems of high energy and fertilizer inputs but the pressure to clear more and more forested land grows (Sayer, 1992). An increase in the number of people practising shifting cultivation and pastoralism leads to its expansion onto marginal lands such as rain forests or semi-arid areas. This is happening in parts of the Sahel, the highlands of East Africa, and the drier parts of southern Africa. Countries which are losing natural vegetation significantly in this fashion include Ethiopia, Kenya, Rwanda, Burundi, Ghana, Nigeria and Togo (Sayer, Harcourt and Collins, 1992).

Plantations have been a significant part of tropical agriculture since the seventeenth century. For example, large areas in Southeast Asia, especially Malaysia, came under rubber in the twentieth century, usually replacing the lowland rain forest. The recent extension of oil palm continues to displace the forests. Plantations of bananas, cacao, coffee, tea, etc., continue to replace the forests, although these days plantations have a beneficial aspect, often being located on already degraded land and also used for watershed and soil protection. Government-sponsored conversions of forest lands to large agro-industrial estates usually modify large areas. Rubber and oil palm plantations in Peninsular Malaysia have replaced 12 percent of its forests (Repetto, 1988). According to an FAO estimate, 438 000 km^2 of forest plantations occur in the tropics, about 85 percent of which are in five countries: India, Indonesia, Brazil, Vietnam and Thailand. Forestry plantations for industrial purposes are mainly planted with eucalyptus, pine, teak and acacia trees (World Resources Institute, 1994).

Recently, the spread of pasture in Central America has grown at the expense of the rain forest. By 1981, a third of Costa Rican land was under pasture, for exporting beef to the United States (Jackson, 1983). The lean meat from cattle on the hoof meets the demand of the meat industry of the USA for hamburgers, hot dogs and canned meat at costs lower than those in the United States. The total area converted from forests to pasture in Latin America is more than 200 000 km^2. Pasture has also displaced forests in parts of Africa in order to meet European demands. Apart from the shortness of pasture life (in the Amazon, nutrient-demanding grasses are replaced rapidly by invading shrubs and non-forage grasses), pastures and degraded rain forests are fire hazards. Fire has destroyed 200 000 km^2 of the Amazon Basin, a large part of this area being exploited earlier for timber (World Resources Institute, 1990). Pasture by itself is unprofitable in Brazil. A Brazilian Institute of Economic Analysis (IPEA) study on Superintendency of Amazonia (SUDAM) ranches showed that the actual production and sale of livestock remained at 15 percent of the projected capacity even for large well-run operations. Unsustainable pastures are expensive to prepare and maintain, and the final earnings often do not repay the investment costs (Hecht, 1993).

Miscellaneous causes

A variety of demands and land development policies destroy huge areas of the rain forest. Large-scale mineral extraction such as the iron ores of Carajas (Brazil) or

Bailadila Ranges (India) destroys flora and fauna extensively, with not only the mining activities, but also the associated construction of roads and settlements, waste dumps, and air pollution.

The Brazilian Amazon has been sequentially destroyed as the coast-to-coast Trans-Amazon Highway was built, along with its feeder network. Peasantry from the east used the new roads to move into the Amazon forest. They were unfamiliar with the rain forest conditions, and especially the nature of the humus-poor soil. After a few years they had to abandon their attempts at farming and move further down the road to a new place originally deeper in the forest, a pattern repeated over and over again. A patchwork of forest destruction and soil erosion followed the migration of disappointed cultivators down the road (Fearnside, 1986). The Trans-Sumatra Highway cutting through the forests of Sumatra and bringing in settlers has also given rise to significant deforestation.

Large areas of the forests have been inundated by the construction of dams and reservoirs of multipurpose water projects, with loss of both flora and fauna. Examples include the 2160 km^2 reservoir of the Tocantis River (Tucuruí, Brazil), and the Batang Ai project of Sarawak which flooded 87 km^2 of land.

Rain forests, stretching across vast areas of sparsely populated land, are seriously threatened by development projects such as Brazil's Northwest Region Integrated Development Project (Polonoroeste). They are also under threat from misguided fiscal incentives. Brazil, for example, had a scheme for land ownership where land was distributed at a subsidized rate provided the claimant could prove that the land had been developed. The easiest way of furnishing the proof is to clear the land of vegetation. Such a scheme necessarily gave rise to large-scale clearance of the forest, even when the land was not used for settling of landless peasantry afterwards. The vast undeveloped region of northern Brazil is under threat from the Calha Nortes Project, with new roads being built into this strategic area. A substantial part of it falls within the Amazon Basin and its rain forests. The area is rich in mineral deposits and illegal mining for gold and tin is widespread. Road building into the Amazon Basin in the 1970s, which brought in the Trans-Amazon Highway and the road BR-364, led to rapid deforestation as described earlier.

5.4 The Amount of Deforestation

Causes of deforestation vary in importance from forest to forest. Rain forests are destroyed by logging and plantation agriculture in Southeast Asia; unsustainable shifting cultivation in Central Africa; settlement and ranching in South and Central America; and firewood collection in drier areas and on the slopes of the tropical mountains. Johnson and Cabarle (1993) have opined that tropical deforestation is essentially a matter of public policy, economic pressures and social conditions. In many instances the country government has taken control of the forest resources from the indigenous people and given private parties the property right to public forests (Johnson and Cabarle, 1993). This necessarily leads to the exploitation of the resources of the forest.

Considerable and continuing forest losses from the tropics were demonstrated when the Food and Agricultural Organization (FAO) of the United Nations estimated the area under forest for the world in 1980. Further mapping of the forests in the 1980s, often using remote sensing, suggested an even greater rate of forest loss than the original FAO 1980 estimates. Recently, FAO has come up with another assessment of the forests of the world. This assessment has been released in three parts. The first section, released in 1992, assessed the temperate forests in industrialized countries; the second, which came out in 1993, covered the tropical forests in developing countries; and the final release covered the forests of the nontropical developing countries such as Argentina and China (World Resources Institute, 1994).

Deforestation has been clearly defined for this mapping exercise. It is the permanent depletion of the crown cover of the forest to less than 10 percent. Tropical forests have been divided into three ecological types: rain, moist and dry. Within each of these categories, the countries that suffer from the highest deforestation have been identified. About 45 percent of the total global rain forest loss, according to this study, is happening in Brazil and Indonesia. Furthermore, degraded and fragmented forest covers a large area, indicating high potential for habitat loss, leading to a loss of biodiversity (World Resources Institute, 1994).

Certain reservations exist regarding the FAO findings. The reliability and existence of the information varied from country to country in the tropics, the data from Asia being the best. Sample surveys of tropical forest change between 1980 and 1990 were estimated for 117 sites from high-resolution satellite data. The present conditions of the forests of the world and the rates of deforestation are as follows (FAO, 1993 as in World Resources Institute, 1994):

- Tropical forests at the end of 1990 covered an estimated area of 17.56 million km^2. Of this, South America and the Caribbean carried 52 percent, Africa 30 percent, and Asia and the Pacific 18 percent. Four countries (Brazil, Zaire, Indonesia and Peru) had about half of the tropical forests of the world.
- In the ten years between 1980 and 1990, tropical forests have been lost at the annual rate of 154 000 km^2. Another study carried out by Trexler and Haugen (1994) estimated it to be 160 000 km^2. The figures are very close. The total world loss of tropical forests in these years amounts to more than 80 percent of Indonesia's land area. Almost all the tropical countries showed areal forest loss. Half of the total loss of tropical forests occurred in six countries: Brazil, Indonesia, Congo, Mexico, Bolivia and Venezuela. The ranked rate of deforestation is Asia and the Pacific (1.2 percent), South America (0.8 percent), Africa (0.7 percent). But as the total present forest area is much higher for South America, the total loss is the greatest there (74 000 km^2), followed by Africa (41 000 km^2), and Asia and the Pacific (39 000 km^2). The earlier (1980) FAO estimation of forest loss was much lower; the annual global loss was taken as 111 400 km^2. Other studies also have indicated an increase in the loss of tropical forests, although the figures do vary (Sayer, Harcourt and Collins, 1992; Colchester, 1993).

- The highest annual deforestation rates occur in two regions: (1) continental Southeast Asia, and (2) Central America and Mexico. The rate in these regions is roughly twice the global average, and has been primarily attributed to the clearing of small areas of contiguous forest (Figure 5.3). In these two regions, only Lao PDR and Costa Rica were not involved in this accelerated deforestation.
- Of all the different types of tropical forests, it is the easily accessed lowland forest which has been destroyed most between 1980 and 1990. The annual loss is estimated to include 61 000 km^2 of moist deciduous forest and 46 000 km^2 of rain forest. The fastest rate of loss (1.1 percent), however, involves the hill and montane forests. The depletion of lowland forest represents a potential high loss of biodiversity, while the removal of forest cover on the hillslopes (Figure 5.4) gives rise to accelerated runoff and erosion, especially in the humid tropics.
- For the same period, forests in the industrialized countries (roughly the temperate areas) increased slightly in area. In contrast to the tropics, forest degradation here has been ascribed to air pollution, pests, droughts and loss of nutrients.
- The average annual deforestation rate for the world for the 1980s was 0.8 percent of the total forested area. Figure 5.1 indicates countries with both above and below the average rate. This, however, is an indicator of the rate of loss, not the total area lost in this decade. Due to a lack of comparable data, it has not been possible to determine whether this rate is comparable, better or worse than what happened in the past.
- FAO has estimated that for each year the total loss of above-ground biomass from the tropical countries is 2.5×10^9 t. This converts to about 4.1×10^9 t of CO_2, which is 80 percent of the total annual CO_2 emission from energy use and cement production in the USA (World Resources Institute, 1994). Another estimate was a loss of 1.7×10^9 t for 1990 from tropical deforestation (Trexler and Haugen, 1994).
- Large areas of the surviving forests (including tropical and temperate varieties) are both fragmented and degraded (Figure 5.3), thereby creating a potential situation for biodiversity loss. Even in places where large areas are not being deforested, the edge of the forest remains vulnerable. In West Africa, forest fragmentation is reflected in the current pattern of small islands of forests in the middle of croplands.
- Areas under plantation agriculture rose from 180 000 km^2 in 1980 to over 400 000 km^2 in 1990.
- Sustainably managed forests did not show any areal increase between 1980 and 1990, although total log production rose steadily. Most of it came from the primary forests.

5.5 The Effects of the Deforestation of the Tropics

Evaluation of the effects of removal and degradation of the tropical forests has to cope with two problems. First, the extent and rate of deforestation and forest degradation are not always available with the desired accuracy. Small areas have

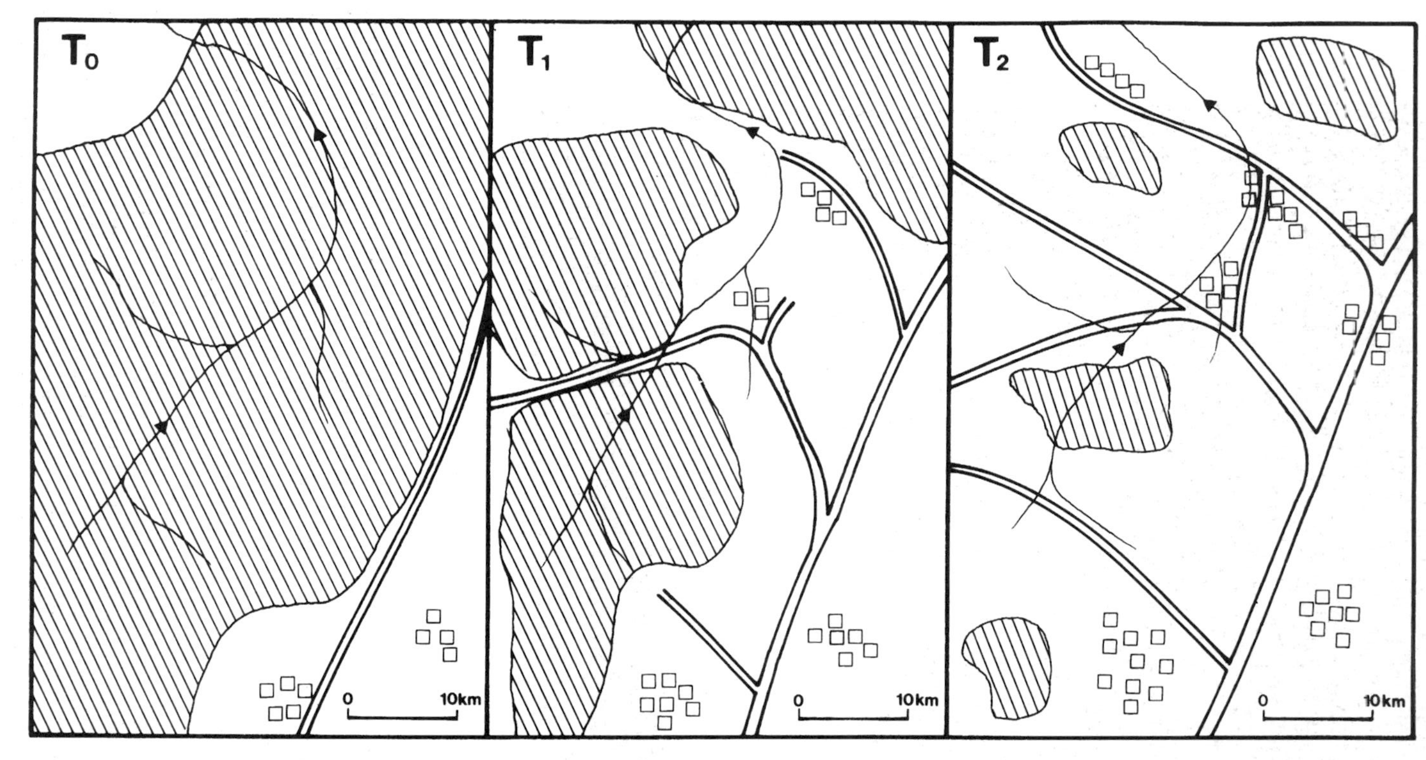

Figure 5.3 *Fragmentation of the rain forest over time: schematic diagram. Area under forest shaded. Note fragmentation due to logging, road building, settlements and expansion of agriculture. The increased edge to area ratio exposes the forest to faster destruction, leading to loss of biodiversity, increased erosion and sediment production, pressure on shifting cultivation and restricted space for forest animals leading to possible conflict between animals and farmers*

Figure 5.4 *Cleared watershed, southern Malaysia*

been well mapped showing the change in land use, but for larger areas we are dependent on approximations, such as the FAO estimates discussed earlier. With the rapid development of satellite imagery, especially radar imagery which can see through the near ubiquitous cloud cover over the equatorial mountains, it should be possible in the near future to (1) have more accurate measurements of forest depletion and (2) monitor forest loss over time. The general finding of drastic loss of tropical forests, however, is unquestionable.

Second, the effects of deforestation are both multiscale and interrelated. They range from local loss of soil fertility and small slope failures to alteration of the global climate. The effects can also be step-wise and cumulative. Forest clearance on slopes may give rise to a number of small slope failures and gully formation which in turn would contribute large slugs of sediment periodically to the river at the bottom of the slope. If deforestation continues over a large area, such sediment would accumulate in a number of rivers, thereby changing their flow pattern and physical character. Given enough river discharge and sediment supply, the eroded material from deforested slopes would, after a suitable timc interval, reach the coast as plumes of sediment (Figure 5.5), which in turn may affect the coastal beaches, mangroves or coral reefs. Table 5.2 lists the major effects of the deforestation of the tropical lands.

Local effects

The immediate local effect is the accelerated erosion of the ground surface by slopewash (a shallow layer of water running down the slope after rainstorms) and gullies (small natural channels). With the removal of forest cover and ground litter, the soil surface is exposed to the intense rain showers that occur in the tropics. Again, as the ground is compacted during logging, the infiltration rate drops considerably, and very little water enters the subsurface. Most of the rain water that reaches the ground surface is thereby converted into slopewash. The shallow soil on the slopes, the presence of considerable amounts of water (there is little evapotranspiration after the forest loss), and the removal of the binding force of vegetation on the ground, combine to cause a number of small slope failures. Although the data for the amount of eroded sediment from different types of land use are not easy to collect (the two common techniques for measuring this are (1) catching the sediment at the downslope end of a plot with raised boundaries, and (2) taking

Figure 5.5 *Coastal sediment plumes, Trengganu, Malaysia (Landsat image) from National University of Singapore Research Project RP3890040*

Table 5.2 *The consequences of tropical deforestation*

1. *Local effects*
   Soil erosion by gullies and slopewash
   Small slope failures
   Accelerated loss of sediment
   Less infiltration and increased runoff
   Loss of soil fertility
2. *Regional effects*
   Large slope failures via mass movements and debris flows
   Increased flooding in rivers
   Alteration of river characteristics
   Coastal sediment problems
   Loss of regional flora and fauna
   Alteration of the lifestyles of local inhabitants
   Climatic changes
3. *Global effects*
   Loss of biodiversity
   Contribution to global warming
   Possible climate changes

water samples from a river), enough information now exists for at least a qualified comparison between forested and cleared areas from the same region.

In Southeast Asia, the annual sediment yield data from forests range beween 0 to 460 t km^{-2}, with a mean of about 70 t km^{-2}. After logging, the figures rise to thousands of tonnes per square kilometre, a conservative estimate of 1500 t km^{-2} being acceptable (Gupta, 1996). Ruangpanit (1985) measured soil loss from 41 storms with a total rainfall of 1128 mm under a range of vegetation cover in Thailand. With a crown cover of 80–90 percent (forest), the total sediment yield in these storms was 28.5 kg km^{-2}. When only 20–30 percent of the crown cover was left, the figure rose to 65.3 kg km^{-2} (Ruangpanit, 1985). Similarly, a figure of 30 t km^{-2} $year^{-1}$ from the Javanese rain forest rose to 1590 t km^{-2} $year^{-1}$ following logging (Anderson and Spencer, 1991). Data provided by Rapp *et al.* (1972) from Côte d'Ivoire show an increase from about nil to 10 800–17 000 t km^{-2} $year^{-1}$ after vegetation clearance. The highest figure for Tanzania is 14 600 t km^{-2} $year^{-1}$. In the first couple of years during and after logging, most of this sediment will travel down skid roads in tropical downpours. With the abandonment of the roads and regrowth of secondary vegetation, sediment yield drops. Unless forest clearance continues, the rivers of the area will have to cope with a short period of very high water and sediment input. This may lead to local floods, bar formation in the channels, and in extreme cases, delta building and sediment plumes released in coastal waters. If the forest is cleared for agriculture or a plantation, a similar sharp spike of sediment will reach the rivers, the sediment supply then decreasing with the re-establishment of a vegetative cover.

In terms of nutrient loss, logging is a severe but short-term disturbance. With forest clearance for other types of land use, the nutrient loss continues for a longer period. Such losses, however, are extremely variable depending on the nature of the soil and the nature of the disturbance. If the forest is burnt for clearance, both nitrogen and sulphur are lost by volatilization but the rest of the nutrients are returned to the soil as ash. If the biomass is allowed to decompose naturally, the loss of nutrients from the original forest is smaller. Burning, which is often an early and necessary step associated with traditional agriculture (both shifting and more stationary types), not only controls nutrient mobilization but also deals with weeds and insects. Contiguous burning across a region, however, may cause regional air pollution (Chapter 9). The organic matter in the soil returns with the growth of secondary vegetation. According to Brown and Lugo (1990), this occurs rapidly, within a few years, under secondary forests.

Regional effects

The effects of forest destruction become regional when practised over a wider space and for a longer time, thereby not allowing the slopes, streams, soils or vegetation of the area any recovery time between disturbances. Landforms, climate and local inhabitants are all affected at the regional scale. The physical disturbances, discussed earlier, become regional in nature with widespread deforestation. The modifications become larger in scale and the problems more difficult to deal with. Their effects in the landscape also tend to be longer-lived.

For example, large-scale forest clearance (Figure 5.6) on steep slopes gives rise to extensive slope failures in the form of large landslides and debris flows, especially after a period of heavy rain. A huge amount of sediment reaches the rivers at the foot of the slopes, and rivers tend to flood more frequently because of the alteration of water balance in their catchments following widespread deforestation. The sediment is carried downstream to build features such as bars and floodplains. The general character of the rivers changes. Deforestation of tropical mountainous areas such as the Himalaya, the Ethiopian Highlands or the Andes has given rise to unstable slopes and local flooding. Where the sediment reaches the coast as plumes (Figure 5.5), its greater volume may modify beaches, mangroves and coral reefs. Disturbances in a number of neighbouring small basins may give rise to a lateral coalescence of plumes (Gupta and Krishnan, 1994) on the coast, causing considerable coastal alteration.

About 1000 mm of rain fell on southern Thailand in five days of November 1988. Thousands of logs, cut from the forests and left on steep hillsides, slid down the slopes along with numerous landslides, destroying a number of villages. The total loss of life came to 350. This resulted in a public outcry and in January 1989 the Thai government banned all commercial logging and furthermore revoked all logging concessions to prevent future recurrence of disasters of this kind. This, however, prompted the Thai logging companies to seek new sources of timber, which they found in Myanmar, Lao PDR and Cambodia, thereby bringing new areas under threat. Illegal logging has also been reported from the region (World Resources Institute, 1990). Floods and landslides occur in Southeast Asia when heavy rain falls on deforested slopes. In the same year (1988), deforestation in the Bengkulu Province of Sumatra (Indonesia) was followed by soil erosion, huge landslides, and flooded rivers. A large number of local people lost their possessions, and in the areas below the cleared slopes, flood control measures have turned out to be expensive and ineffective (Collins, Sayer and Whitmore, 1991).

Figure 5.6 *Bare soil on slopes following large-scale forest clearance, Jamaica*

Destruction of the rain forest on a regional scale may result in modification of the local climate, especially as a considerable amount of moisture reaches the atmosphere via evapotranspiration from the forests. The Amazon Basin provides an excellent if perhaps a unique example because of its size. It is the easterly trade wind that brings the rain from the Atlantic. However, it has been shown that only about half of the precipitation inland is directly from the Atlantic. The remaining half is collected by the easterly winds blowing over the forest canopy as recycled water released from the vegetation and the soil by forest evapotranspiration. A loss of the forest is therefore likely to alter the rainfall map of the interior of the basin (Salati, 1987). This tendency is also indicated by general circulation models (GCM) of the climate over the Amazon Basin assuming a much depleted forest (Lean and Warrilow, 1989; Shukla, Nobre and Sellers, 1990).

Forest destruction on a regional scale puts pressure on the original inhabitants of the rain forest or those who live on its periphery. Reducing the forested land may lead to environmental degradation (as discussed earlier) by forcing shifting cultivators to adopt unsustainable farming practices. For people who live to a considerable extent by hunting and gathering, a smaller forest is economically disastrous. For this reason, Penan, Kayan and other forest communities of Sarawak, Malaysia protested in 1987 and 1988 against logging of the forests over which they assert ancestral rights. Their protest took the form of blocking the logging roads. The blockades were dismantled by the army and the police, and a number of people were arrested (World Rainforest Movement, 1990). A Sarawak law recognizes interfering with logging as a crime (World Resources Institute, 1990).

Forests are storehouses of resources including food, medicine and various types of trade products such as rattan, rubber, fibres, dyes, resins, etc. (Box 5.2). The loss of these resources has a detrimental effect on the lifestyle of the collectors and their families. Fragmentation of forests, loss of migratory corridors for animals between forests, and replacement of woods by croplands all lead to confrontations between forest-dwelling animals and farmers. In both Africa and India, the expansion of croplands has led to periodic movement of elephant herds into the farming lands.

Global effects

The sharp increase in deforestation over the last few decades in the tropics has created worrisome environmental degradation on a global scale. The global effects are two: (1) loss of biodiversity and (2) possible change in climate including global warming. Both of these are alarming enough to bring in two binding agreements at the 1992 Earth Summit in Rio de Janeiro to halt global warming and losses in biodiversity. Details of these agreements are given in Chapter 15. It should be noted, however, that destruction of tropical forests to meet the needs of the local agricultural poor and the greed of the distant industrial rich has created two environmental problems that are global in nature and which affect everyone. Because of the tremendous floristic richness of the tropical rain forests (which is far from being satisfactorily investigated), their disappearance would mean a huge loss of flora and fauna. The potential effect of this alarming loss is discussed in Box 5.2. The global warming and climatic change are associated with an increase of CO_2 in the atmosphere due to the disappearance of one of the large global sinks of carbon, the tropical rain forests. Global warming and its relationship with deforestation are discussed in detail in Chapter 14.

Box 5.2

The Loss of Biodiversity

Biodiversity is a basket term indicating the range of variations or differences within the biological world. The term itself is a contraction of 'biological diversity', and it is commonly used to refer to the number, variety and changes of living organisms (Groombridge, 1992). Biodiversity is commonly measured in terms of genes, species and ecosystems. The popular use of the word biodiversity refers to species diversity, i.e. species richness (defined as the number of species in a given area or habitat). The tropical rain forests being extremely rich in species, their disappearance and the related loss of biodiversity is viewed with concern. This is particularly worrisome if we remember that no complete listing exists of all the species; we do not yet know them in detail, and we do not know what species have already disappeared, taking with them properties beneficial to humankind.

Holdgate (1996) has pointed out the enormous difference between the estimates of biological loss in the popular literature and the actual recorded loss, although both are much higher than the natural rate of about one species each of plants and animals per 100 to 1000 years. The impending extinction rate due to anthropogenic modifications is likely to surpass the natural rate displayed in the fossil records by a factor of at least 10 000.

The species richness of the tropical rain forests not only supplies us with various kinds of timber but also dyes, fibres, edible fruits, gums, oils, resins, tannin and turpentine. Plants possess medicinal properties. Tropical forests supply drugs for leukaemia and Hodgkin's disease and also strychnine, ipecacuanha, reserpine, curare, quinine and diosgenin. Plant species from the rain forests have been used for making contraceptive pills and also in hybridization to provide high-yield new cereals for the Green Revolution. This list only refers to the flora of the rain forests, but these forests are also home to a vast number of mammals, birds and insects. Forest destruction removes all these; even forest fragmentation or habitat destruction leads to faunal destruction. Various calculations indicate that in many parts of the tropical rain forest, sustainable yield of forest products can bring in an income per hectare that is higher than the once-off collection of tropical timber. Such estimates are of course difficult, and one does not know the value of a species in the future; for example, which gene from a wild species may prove useful in crop breeding.

The untapped resources of the tropical rain forests in 1991 led to an agreement being made between the National Biodiversity Institute (INBio) of Costa Rica, a private non-profit organization, and Merck & Co. Ltd., a US-based pharmaceutical firm. From the conserved wild lands of Costa Rica, INBio was committed to provide Merck with chemical extracts from wild plants, insects and micro-organisms. In return, Merck agreed to provide a two-year research and sampling budget of over $1 million and also royalties on any commercial product developed out of the samples. Furthermore, INBio agreed to contribute 10 percent of the budget and half of the royalties to the Government of Costa Rica for conservation of national parks, and Merck was amenable to provide Costa Rica with technical assistance and training in the field of drug research. This groundbreaking agreement shows that there is hope for the future of the richness and unexplored nature of biodiversity in the tropics (Reid *et al.*, 1993). This agreement also represents an acceptance of the principle that resources primarily belong to the region concerned, and are not for the taking (Durning, 1993).

5.6 Managing the Forests

Managing the tropical forests implies preserving the biodiversity-rich forests and at the same time meeting at least a substantial part of the demand for their resources. Deforestation has been accelerated in the tropics by the needs of the developing countries for foreign exchange and also to service international debt. About a decade ago, Myers (1986) connected the pressure of international debt payment to several types of environmental degradation: cattle ranching in the Amazon basin; reduction of logging restrictions in Ecuador, Côte d'Ivoire and Indonesia; expansion of cash crops at the expense of subsistence farmers in the dry Sahel countries. It has been suggested that control of deforestation requires a restructuring of property rights to forests, pricing of forest products, and political power over disposition of forests (Durning, 1993). Management of tropical forests, given the reasons for its depletion, clearly needs a combination of physical, economic, technical and political measures. A number of global instruments have failed to achieve this. The list includes the Tropical Forestry Action Plan (TFAP), the International Tropical Timber Agreement, and the United Nations' Statement of Forest Principles (Durning, 1993). The successes tend to be local rather than global.

Optimal management of the tropical forests requires a combination of multifaceted and innovative methods. The demand for tropical hardwood in the developed nations should not encourage the indiscriminate destruction of virgin forests in the developing countries. At least part of the demand could be met by forest plantations and by careful rotation of logged areas. Plantations of timber-bearing trees not only prevent destruction of the remaining natural forests, but also, by being located on already degraded and bare land, they protect the soil from erosion and improve the soil conditions. Most of the tropical timber is extracted to meet the needs of temperate countries (Figure 5.2), and Repetto (1988, 1990) has shown that timber extractions rise to unnecessarily high levels and costs due to faulty public policies. His illustrations include more than $2 billion worth of potential forest revenues missed by Indonesia between 1979 and 1982 to logging concessions and allied interests; $500 million worth of potential forest resource rents lost in the Philippines over the same period in inefficient timber mills; and the existence of more than 120 000 km^2 of uneconomic cattle ranches which replaced tropical forests in the Brazilian Amazon following generous tax and credit incentives. Not only do the tropical countries lose valuable timber, but they have in general failed to collect the full resource value in royalties, export taxes and other fees. The collected rent falls far short of the potential rent, and the amount collected is commonly only a fraction of the actual rent (Repetto, 1990). Deforestation is also encouraged by the existing policies which give out concessions for a period shorter than a single forest rotation. The logging concessions are held by groups which in some countries are connected to high-level military or ministerial personnel. There is little incentive for managing the forest. The forest departments in most developing countries are thinly and inadequately staffed, and are not in a position to protect the forest from illegal timber extraction. Logged forests are subsequently completely cleared by cultivators who arrive on the scene later. Barber, Johnson and Hafild (1994) describe how one single Indonesian company holds concession rights over an area of 55 000 km^2, enjoys tremendous backing from the national government, and has built the world's largest plywood operation. Enlightened forest policies and substitutions are needed to control deforestation and still meet the demand for tropical timber.

Both substitution and social forestry are used to meet the demand for fuelwood and other domestic requirements. The use of biogas in cooking and better designed stoves can drastically reduce the demand on firewood. Social forestry implies planting trees within reach of and for the needs of the local inhabitants. In India, social forestry programmes provide fuelwood, fodder, small timber and minor forest products. Indian social foresty has three faces:

- farm forestry, which supplies the farmers with free or subsidized seedlings to grow more trees on their land to meet the local demand;
- community woodlots, where village communities plant trees on common lands;
- forestry woodlots, where the forest department plants trees for community use on public lands such as the sides of the roads or on embankments.

The success of social forestry in India has been uneven, but certainly India is much greener now than it used to be, and innovative afforestation programmes such as entrusting village school children to look after newly planted trees have been unusually productive (Agarwal and Narain, 1986). The widespread planting of eucalyptus to provide fuelwood has been criticized in many countries as eucalyptus tends to lower the water table, depletes the soil, and hinders community participation (Shiva, Bandyopadhyay and Jayal, 1985; Agarwal and Narain, 1986; Kardell, Steen and Fabião, 1986). It is probably best to use suitable local species.

Forests, as described in Box 5.2, maintain a storehouse of resources. Replacing forests with agriculture does not necessarily improve economic conditions. A consciousness of forests as a useful resource and a better understanding of the physical geography of the tropics, including soil fertility, are both essential. Agroforestry is the combination of forest maintenance with farming in the same area. Shade-tolerant crops such as cocoa, tapioca, legumes, rubber, etc., are grown under the tall trees of the tropical forests and in small plots. Maize is sometimes grown under a scattering of trees (Figure 5.7). This mixed cultivation of trees with food crops which maintains yield is not new but has been practised for a long time in many areas such as Central America, Sumatra, Java and the Philippines. It sustains small numbers of local farmers, and is designed to lessen the need for forest clearance for agriculture. A special type of agroforesty is the planting of trees as hedges with the food crops being grown in the intervening space (Whitmore, 1991). Plantations of trees for timber or paper are spreading as the rain forests shrink. New plantations should be located on already cleared or degraded land, rather than replacing natural forests.

Figure 5.7 *An example of agroforestry: maize being grown under a scattering of trees, Kenya*

Plans for large-scale mining projects, highway building, or water reservoirs in the tropical forests should be carefully examined, and their impact on the environment carefully assessed. Often a careful analysis shows that the economic and social costs of these projects, which destroy huge areas of the forest, make the projects unprofitable. It is no longer possible to find an international donor for such schemes without proper environmental analysis (Chapter 16).

The extent of forest destruction prompted the formulation of the Tropical Forestry Action Plan (TFAP) in 1985 by the World Resources Institute, the World Conservation Union (IUCN), the World Bank, the United Nations Development Programme (UNDP), and the FAO. The administration of the plan is currently done by the forestry department of the FAO. This integrated plan for forest maintenance was accepted by the forest authorities in most tropical and industrial countries at the Ninth World Forest Congress in Mexico City, and was subsequently endorsed at a number of international meetings. The objective of the plan was to restore and manage forests for the continuing benefit of the rural people and the general economy of the country concerned. Countries of the tropics were expected to prepare a national plan for their forests. TFAP required a total expenditure of $8 billion over 5 years. The plan proposed accelerated improvement in the following areas:

- fuelwood and agroforestry
- land management on upland river basins
- forest management for industrial use
- conservation of tropical forest ecosystems
- institutions for forestry research, training and extension of services

Although TFAP was originally designed to involve local communities and encourage country governments to formulate national forestry plans, the impact of TFAP fell far short of the expected. The indigenous community of the rain forests has been marginalized, and in practice, the plan has concentrated on increasing investments in the traditional forestry sector. Studies by NGOs and the World Resources Institute, one of the architects of the plan, indicated that the cross-sectoral approach of the original plan had not been achieved (Colchester and Lohmann, 1990). At the national level, the effect of the TFAP has been variable. For example, the national TFAP for Tanzania stressed social forestry and conservation, unlike that of Cameroon where expanded industrial logging was stressed (Sayer, Harcourt and Collins, 1992). The plan has at least forced the national governments to consider changes in forest policy and greater conservation efforts.

The most recent of such global arrangements is the declaration on forest principles agreed at the Earth Summit in Rio de Janeiro in 1992. The agreed version is a diluted form of the original proposal, which shows the lack of agreement and diversity of interest at the international level. In contrast, the binding global treaty on biodiversity, also agreed on at the Rio summit, may have some significant impact (Chapter 15).

Local attempts to preserve forests, such as the *chipko* movement of India (Chapter 16), the control on the entry of outsiders into the forest by the Kuna Indians of Panama, and the recovery of degraded forest land by the farmers of Peru (whose practice of agroforestry is known as the Huerto Integral Familiar Comunal or the

HIFCO Project), have a better record. Barber, Johnson and Hafild (1994) quoted the example of the villagers of Dusunpulau in Bengkulu Province, Sumatra, who took a timber company that was destroying the local forests under government concession to court, claiming compensation for losses amounting to about $3.8 million. The losses are due to the destruction of local resin-bearing trees the villagers have traditionally tapped, and over which they claim they have legal rights. The Indian government has experimentally set up a formalized arrangement by which local people and forest dwellers are involved in forest protection, forest management, and sharing of benefits. This is done by setting up village forest protection committees (VFPCs). The forests are better protected and the relationship between the local people and the forest department staff has changed from one of confrontation to co-operation (Bahuguna, Luthra and Rathor, 1994).

One of the innovative attempts for forest preservation is the Biosphere Reserve plan of the Man and Biosphere Programme of UNESCO. Under this plan a core area of forest is preserved by building two buffer zones around it, which include degraded forest. Interference with the forest lessens as one goes from the outer buffer zone to the core. The buffer zones include existing settlements of forest dwellers, and research stations. They also support tourism which is kept away from the core, but this brand of tourism carries some educational properties and provides the plan with some income.

The tropical forests should be preserved to meet local needs for a long time and for the rich biodiversity. It has been estimated that total earnings from the combined sustainable harvest of forest products (including rattan, bamboo, oil, latex, tannin, dyes, pharmaceutical products, etc.) can be as high as $200 per hectare annually. This is higher than earnings from logging which can be done only once and the yield may be only as high as $150 per hectare. Wildlife that attracts tourism also raises more money than logging. It has been estimated that a lion in the Amboseli National Park, Kenya, brings in $27 000 per year in tourist money and the figure for a herd of elephants is $610 000 (Holdgate, 1996). Forests also need to be preserved as important sinks of carbon, which if released to the atmosphere will accelerate global warming (Chapter 14). In this sense, sustainability of forests, irrespective of their location, is beneficial for the planet. Their management and preservation, however, require financial adjustments, product substitutions, and forest management on global to local scales.

Exercise

1. Select a small country which still has some rain forest. Determine the extent and the nature of the forest from published sources. If the forest is under threat, rank the causes behind its destruction in order of importance, starting from the most serious cause.
2. If you are now given the job of protecting the forests of this country, what would be your strategy?
3. Would it be possible to save the rain forest without a global agreement?

6

Land Use and Environmental Impact

6.1 Introduction

Large-scale alteration of the landscape started about 10 000 years BP with the beginning of agriculture. Archaeological evidence collected so far indicates that domestication of various kinds of crops started at several different places. There was no single centre of innovation. The knowledge of farming subsequently diffused through the world, along with the modification of the local and regional environments. The three major modifications are as follows:

- dependence on, control of, and widespread diffusion of a limited number of cultivated species for food which gradually replaced a wide variety of native species;
- extensive modification of the natural landscape into agricultural fields, e.g. deltas into rice-producing areas and mid-continental grasslands into wheat fields;
- unintentional spreading of adventurous species (popularly known as weeds) on disturbed landscapes.

Farming techniques evolved over time, from simple shifting cultivation or digging stick agriculture to complicated widespread monoculture of grain crops. Large areas had been transformed throughout human history. For example, between 1855 and 1931 the area under wet rice in the Pegu Province of Burma increased from 2280 km^2 to 12 500 km^2. In Thailand, this happened later, with the wet rice areas expanding from 150 000 km^2 in 1907 to 690 000 km^2 in 1931 (Allen, 1993). The expansion occurred due to the migration of people into new areas and a rising regional market demand for cereals. In certain areas, the climate favoured mixed farming involving a variety of crops and livestock. Drier areas were cultivated with the aid of irrigation, or specialized dryland farming techniques, or they were utilized as pastures. Since the sixteenth century, plantation agriculture has replaced natural vegetation over large parts of the tropics. Recently increased market

nd improved transport and refrigeration systems have led to the substi-
of conventional food crops by high-price products such as fruit, early vegetables, and flowers on many agricultural fields. The present-day agricultural landscape is a complicated mosaic, incorporating examples of all these types.

Crops have been grown for hundreds of years in the fertile areas of the world without significant environmental degradation. Generally, environmental degradation occurs when the pressure on land becomes acute, due usually, but not always, to a rapidly rising population. The pressure on land may also be due either to a shortage of good land in the country or to socio-economic inequality, the usual pattern of which is a small number of large landowners and a large number of small landowners or landless peasantry. Under such circumstances, the traditional practice, which had sustained agriculture on the same fields for years, breaks down and both soil erosion and loss of soil fertility begin to occur. This was illustrated in the last chapter in the account of the shifting cultivators of the rain forests who were forced to rotate the fields quicker than had been the usual practice due to a reduction in the size of the forest. In areas where agricultural sustainability is marginal (e.g. steep slopes, subhumid areas, or dry lands), pressure on land leads to drastic deterioration of land quality at a rapid pace. As the population of the world rises, pressure on land rises also, and inequitable socio-economic conditions act as a catalyst speeding up land degradation. Some of the non-agricultural uses of land (e.g. mining) create land problems of special kinds. In this chapter, we discuss the different types of land degradation problems and also the choices of land management techniques that are available. The choices generally tend to be a combination of technical, economic and social management.

6.2 Agricultural Development, Population Pressure and Land Degradation

The total amount of land that has been seriously degraded since 1945, is about 12 million km^2 (Faeth, 1993). Although much of it involves land only marginally or not at all suitable for farming, intensive farming may also create problems even on good land (Box 6.1).

Population pressure on land is not a simple relationship depending on the absolute number of people and the area available for farming. Locally, it may be caused by poor peasantry being displaced from good fertile areas by rich landowners, or due to such areas being utilized for mining, urbanization or reservoirs related to river development projects. Such water reservoirs necessarily inundate useful valley-bottom lands. The displaced peasantry tend to move to other areas, thereby attributing a high density of the farming population to those places.

The current world population is about six billion. If we take the medium growth scenario of the United Nations, the world population by 2050 is expected to rise to 10 billion. Even if it stabilizes at this number, we are faced with two questions: (1) whether the present world produces enough food to feed itself, and (2) whether a future world would be able, given both the opportunity of technological advancement and the constraint of global climate change, to continue to do so. Kendall and

Pimentel (1994) estimated that current world food production is adequate to sustain seven billion people on a vegetarian diet when the food distribution is ideal and no grain is fed to livestock. However, they point out that possibly as many as two billion are now living in poverty, about half of whom are in extreme poverty with hunger. Feeding a large number of human beings with some kind of equity requires three achievements:

- the ability to produce an increasing volume of crops and livestock from the available land;
- improved food distribution;
- dietary shifts.

This implies an examination of the area available for food production, and an assessment of its quality.

Not even half of the total land area of the world (about 130 million km^2) is suitable for farming for crops or animals. Nearly all the good lands (flat, fertile, moist) are already in production and have been so for some time. Kendall and Pimentel (1994) estimated that it could be possible to extend the world's arable land at most by another 5 million km^2. An increase in crop production is only possible when

1. the yield from the better quality land is increased;
2. land marginally suitable for agriculture is put to use;
3. croplands are found at the expense of forests or grasslands.

The technological breakthrough of the late 1960s and the 1970s, known as the Green Revolution, falls in the first category. This was possible as a result of painstaking and outstanding research on crop genetics in several of the world's leading agricultural research institutions. Among these are Norman Borlaug's wheat programme in Mexico, and the contributions from the International Rice Research Institute in the Philippines. Research in these institutions succeeded in breeding high yield varieties of cereals, often via hybridization. The Green Revolution required the planting of these grains with large inputs of water, fertilizers and pesticides. This resulted in a phenomenal increase in grain production in a number of developing countries with large populations such as India, Indonesia, Mexico, Pakistan, the Philippines and Turkey. The high increase in grain output from the 1960s has enabled these countries to feed the increasing population but there were some shortcomings:

- The benefits of the Green Revolution went mostly to the rich farmers who could take greater advantage of the scientific breakthrough, and thus increased the economic disparity between them and the poorer peasantry.
- The production of fertilizers and a dependable supply of irrigation water demanded greater production of energy. This has the potential to degrade the environment. Furthermore, as most of the countries import oil and gas, the incidences of price rises of such commodities (c.g. oil in the 1970s and fertilizers in the mid-1980s) create problems for the developing countries.

- With high applications, fertilizers and pesticides tend to wash off the fields to pollute the surface waters and also to leach through the soil and lower the quality of groundwater.
- Biological complications may arise due to the reduction of genetic diversity in crops and the development of genetic resistance to standard pesticides.

In the tropics, cultivation leads to a rapid loss of soil fertility unless this fertility is renewed over and over again at brief intervals. Such renewal happens naturally by deposition of flood silts in the valley-bottoms or by weathering of freshly deposited volcanic material on hillslopes. Floodplains, deltas and volcanic slopes can support a high population density. Otherwise, tropical soils become low in fertility very quickly, useful constituents are leached out or used up, and soil becomes acidic. If the vegetation cover is removed, the intensity of tropical rainfall leads to erosion and gully formation. The soil therefore requires proper treatment, large doses of fertilizers, and protection measures. World-wide, about 4.3 million km^2 of arable land has been abandoned during the last 40 years, and the estimated annual soil loss is 0.7 percent of the total soil inventory (Kendall and Pimentel, 1994). In the tropics, this rate rises due to both intense cultivation to feed a dense population and the wrong choice of crops or agricultural practices. The classic case is the attempt during colonial times to impose a system of continuous cultivation on the soils of the cleared rain forest in central Africa.

Developing countries, located in the tropics and dependent on agriculture, use large volumes of pesticides. This is especially true for plantation agriculture and the Green Revolution crops. The area of India that is treated with pesticides, for example, increased from 60 000 km^2 in 1960 to more than 800 000 km^2 in about 25 years, and such rates of increase are quite common in the developing world. Although the developed countries still consume most of the pesticides produced, use has levelled off in recent years. In contrast, it is on a sharply rising trend in the tropical countries, although often from a very low base. Furthermore, pesticides such as DDT, chlordane or heptachlor, banned in many developed countries, are still a part of this rising trend. In fact, the export of chlordane from the United States increased tenfold between 1987 and 1990. Only a very small amount of the pesticides (as low as 0.1 percent in some cases) applied to an agricultural field reach the target organism. The rest contributes to environmental degradation (World Resources Institute, 1994). The problem of pesticides is discussed in detail in the next chapter.

Box 6.1

Agricultural Development and Land Degradation in the Uplands of Java

Located off a subduction trench, the island of Java in Indonesia has a central backbone of volcanic mountains skirted by coastal plains. Most of the rivers are small, less than 50 km in length, and a considerable part of the island consists of uplands with steep slopes. The annual rainfall except for the eastern end is high, i.e. between 1600 and 4000 mm. In 1995 it had a population of more than 110 million, with nearly 60 percent of the total population of Indonesia living on Java. As this is essentially an agricultural country, the rural density is extremely high. In 1960, there were about 700 people per km^2 (Geertz, 1963), which has risen to more than 1000 people per km^2 by 1980 (McCauley, 1991). Rice is grown in the lowlands with well-established irrigation systems, but in the uplands the crop pattern is more variable and includes rice on irrigated terraces and a mosaic of dryland farming systems (Figure 6.1). This mosaic consists of three basic patterns (food crop gardens, mixed gardens, and tree crop gardens), and covers about a quarter of the total land area of the island (McCauley, 1991). The rest of the uplands are under protected and degraded forests and large-scale plantations. During colonial times plantations of coffee, tea and cinchona replaced a large part of the upland forests, and also contributed to land degradation. Most of the plantations were later abandoned to other types of land use.

Figure 6.1 *Agriculture on steep slopes, Dieng Plateau, Java*

Pressure on land, steep slopes, and intense rainfall have led to active surface erosion by slopewash, gullies and mass movements. For example, about 4000 t km^{-2} $year^{-1}$ is the average amount of sediment eroded from the Cimanuk Basin of western Java (Aitken, 1981). Several thousands of tonnes of material are eroded out from each square kilometre every year, also from other river basins. The eroded material has partially filled up reservoirs behind dams and channels of lowland rivers, has complicated the maintenance of irrigation channels, and has led to the formation of coastal plumes of sediments on the north coast. The eroded surface material also implies the removal of the humus-rich upper soil, thereby reducing the effect of the fertility of weathered volcanic material. The soil conditions on non-volcanic uplands (e.g. limestone) are, as expected, much worse. This has led to attempts at managing the uplands of Java with programmes for reforestation, agroforestry, terracing of steep lands, preservation of ground cover, and social forestry. The success rate is partial (McCauley, 1991), and the extreme method of transmigrating people out of Java to less populated Indonesian islands such as Sumatra or Kalimantan under government assistance has been attempted. The uplands of Java illustrate the degeneration of land which should be fertile, by overuse and wrong use (plantations), over a number of years. The solutions to this are difficult to find.

Pressure on agricultural land leads to land degradation, usually by a loss of soil fertility or by erosion. The food shortage problem has been relieved to some extent by the Green Revolution, which in turn has created a different set of problems. However, given the wide range of slope and soil conditions and crop types, environmental degradation is not the same in every place. The Green Revolution in Bangladesh has increased crop yield substantially but the related developments illustrate the importance of understanding local variations in land management.

The increase in food production in Bangladesh has been based on intensified agricultural practices using high yield varieties of rice on almost half its rice fields, extensive use of double or triple cropping, and the application of chemical fertilizers (Pagiola, 1995). It has been suggested, however, that the progressively increasing use of fertilizers has become necessary to maintain yield in the face of declining yield, at least for the upland rice. Again, the high application of pesticides, not on rice but on vegetables, may have created health problems for the users and has certainly led to water contamination, as vegetables are usually grown near ponds or water channels. Land scarcity and high population growth provide Bangladesh and many other countries with no alternative to the continuous use of quick-maturing high yield crops with high inputs of water, fertilizers and pesticides, even though the practice of intensive agriculture is suspected to reduce natural soil fertility and increase contamination from pesticides. However, as the case of Bangladesh shows, the physical problems resulting from the Green Revolution do not seem to be serious at present, although they could be so in the near future. There are exceptions where local conditions, such as arsenic-rich groundwater pumped to the surface, create serious health hazards, as discussed in Chapter 7. Land management in the field requires the study of local conditions and inventive adaptations, as blanket management techniques will not work across the country. The two general rules that apply are as follows:

- the desirability of replacing the currently used pesticides by safer varieties and by techniques such as Integrated Pest Management (IPM);
- the need to integrate the high water use with a sustained water balance for the area.

6.3 Development of Marginal Lands and Land Degradation

The potential for land degradation increases significantly when farming is attempted in areas marginal to agriculture. This becomes possible either in times of good weather or by technical advancement such as the use of irrigation. Basically, farming is extended to three types of marginal land (Gupta, 1998):

- steep slopes;
- areas where there is a moisture deficit but agriculture is possible with irrigation;
- dry areas with rather limited potential for crop raising but where grazing is possible.

Agriculture on steep slopes

Farming steep slopes always has the potential hazard of accelerated erosion and sediment transfer to lower regions unless proper precautions are taken. The surface material on steep slopes is eroded in a number of ways: slope wash, gulleying, and mass movements. The sediment transferred downslope to the valley-bottom streams tends to be deposited in the channel and on the valleyflats. Furthermore, the removal of natural vegetation and disturbance of the ground surface may also decrease infiltration of water into the subsurface and increase the surface runoff. The valley-bottom streams, therefore, are not only partially filled with sediment; they also tend to have peaked hydrographs. This may happen on all steeplands but certain hillslopes – given the local geology, relief, and high and intense rainfall – are particularly vulnerable. The slopes of the large tropical mountains such as the Himalaya or the Andes, or steep plateaus such as the Ethiopian Highlands, are being eroded at a rapid pace. Ives and Messerli (1989) quote a study by Singh and Gupta (1982) which estimated an average sediment yield of 2800 t km^{-2} year^{-1} for the Himalayan slopes. Such figures could very well represent at least local conditions on tropical mountain slopes. Most of the sediment is probably stored for hundreds of years at the footslopes (Bruijnzeel with Bremer, 1989), and is gradually transported down the main river. The effects of land degradation (accelerated erosion, sediment storage, increased flooding) are most likely to be local. Sediment build-up behind dams could drastically shorten the life of reservoirs, as has happened in the Mangla and Tarbela reservoirs of Pakistan where sediment from the mountainous basin of the Upper Indus comes to rest.

It is of course the population pressure which causes these unstable slopes to be cultivated, as happens in the Bolivian, Peruvian and Venezuelan Andes or the Nepal and Indian Himalaya. Where rainfall is high, as on the south slopes of the Himalaya, or where agricultural fields on steep slopes are in the tropical cyclone belt, as in eastern Jamaica, the rates of episodic erosion and sediment transfer can be extremely high.

Agriculture with irrigation in moisture deficit areas

Agriculture has been possible with the help of irrigation in areas of moisture deficit, which could be year long or seasonal. Usually a barrier is put across a stream to pond water which is then transferred to the agricultural fields by a network of canals and progressively smaller distributary water channels. Water is also raised directly from rivers by a number of time-honoured and effective devices such as shadoof, Archimedian screws or Persian wheels. Large-scale irrigation has been carried out in an organized fashion for at least 6000 years. In areas of seasonal moisture deficit, or areas described as subhumid, irrigated agriculture is a sustainable practice and this technical innovation has made crop production feasible.

In areas that are not humid, the extension of canal irrigation and the practice of leaving standing water in the fields give rise to salinization of the land, which makes the area unsuitable for agriculture. In this process, water seeps down through the sides and base of the earth canals and from the fields to raise the level of the groundwater. When by cumulative addition the groundwater table is raised significantly close to the surface, subsurface water, via capillary action through the small soil pores, begins to reach the level of the fields; and given the prevalent high temperature, this water rapidly evaporates. The groundwater in dry areas often contains a high concentration of saline and alkaline matter, and as the water evaporates, such salts are deposited on the soil surface leading to salinization of the land. Over time the soil is crusted with a gleaming white coating of salts and agriculture becomes impossible.

This has happened ever since the beginning of extensive irrigation in semi-arid lands. About 6000 years ago, the land between the Tigris and the Euphrates rivers was cultivated by irrigation. This was part of the Sumerian civilization. However, over time, salinization of land forced a change in crop pattern from wheat to salt-tolerant barley, and then led to progressive abandonment of the once fertile fields which fed the prosperous empires of Sumer and Akkad. Eventually the kingdoms were depopulated and weakened; Babylon to the north, with less of a salinization problem, became important. Salinization following irrigation has affected a wide area of the semi-arid world (Table 6.1), ruining large tracts of agricultural fields in Australia, India, Pakistan, most countries of western Asia, Egypt, Mexico, Peru, Argentina, northeastern Brazil, and the southwestern United States. Any extension of canal irrigation in drier lands should be carefully monitored. Salinization is prevented at least partially by lining the canals and routinely draining out standing water from the crop fields.

Table 6.1 *Salinization of irrigated lands*

Country	Crop area (10^3 km^2)	Irrigated area (10^3 km^2)	Percentage of irrigated area salinized
Argentina	357.5	17.2	33.7
Egypt	26.9	26.9	33.0
Iran	148.3	57.4	33.0
Pakistan	207.6	160.8	26.2
United States	1899.1	180.1	23.0
CIS	2325.7	204.8	18.1
India	1689.9	421.0	16.6
China	969.7	448.3	15.0
Thailand	200.5	40.0	10.0

Source: Modified from Ghassemi, Jakeman and Nix (1995).

All five republics of Central Asia (Uzbekistan, Kazakhstan, Turkmenistan, Kyrgyzstan and Tadjikistan) suffer from salinization-related land degradation, which in 1985 affected between 35 and 80 percent of their irrigated areas (Ghassemi, Jakeman and Nix, 1995). This primarily lowland desert region has been cultivated since the 1950s with irrigation water from the two rivers Amu Darya and Syr Darya via a massive network of canals and also by withdrawal of saline groundwater to grow cotton, rice, vegetables, fruits, and fodder crops. The average annual contribution of the rivers to the Aral Sea used to be 56×10^9 m^3 between 1911 and 1960. By the mid-1970s, it had dropped to $7–11 \times 10^9$ m^3, and in the 1980s several years showed near-zero river discharges. The Aral Sea started to shrink along with other smaller lakes of the area; its level dropped by more than 14 m between 1960 and 1987, and its volume was reduced by 66 percent. The average salinity of the water increased threefold, from 10 000 to 30 000 mg l^{-1}. Wind blowing over nearly 30 000 km^2 of exposed sea bed deposited dust and salt over an area estimated to be between 150 000 and 200 000 km^2. Needless to say that little remained of a once-prosperous fishing industry. Furthermore, the concentration of chemical fertilizers, pesticides and defoliants that had been used in agriculture for years contributed to land degradation as well as having toxic effects on hundreds of thousands of local inhabitants. The two deltas of Amu Darya and Syr Darya have been seriously degraded, clean water has become scarce around the Aral Sea, and the reduced volume of this inland lake water is unable to ameliorate, as before, the harshness of the climate. The crisis of the Aral Sea was not recognized or understood for years and it was only in the late 1980s that it became a priority and publicized issue (Ghassemi, Jakeman and Nix, 1995). The shrinkage of the Aral Sea is another illustration of the effects of the application of technology on a massive scale without working out the consequences or keeping the public well-informed.

Alkalization of soil may occur in association with salinization. In alkalization, the clay part of the soil becomes saturated with sodium, which then disperses the clay particles and destroys good soil structure. This results in a less permeable condition and therefore waterlogging of fields. Clay-rich irrigated soils require careful monitoring as reclaiming alkalized soils is difficult and expensive (Ghassemi, Jakeman and Nix, 1995). The environmental costs of salinization are difficult to translate into monetary terms but several estimates do exist. In the extensively irrigated Colorado River Basin of the United States, the estimated annual loss is $750 million. In the Upper Indus Plain of Pakistan, such figures are $300 million, and in the Murray-Darling Basin of Australia, $208 million. Preventive measures which drain the water out of the fields in Egypt cost about $30 million annually (Ghassemi, Jakeman and Nix, 1995). Such figures indicate the extent of the problem.

Use of dry lands

Deserts such as the Sahara, Takla Makan or Atacama, with extremely limited and localized farming potential, cover about a third of the total land area of the world. Nomadic herdsmen may survive around the fringes of the desert in years of good rainfall. In other years, even survival becomes difficult. The Sahel of Africa, which includes parts of Mauritania, Senegal, Niger, Mali, Burkina Faso and Chad, is the best known example of this type of land where dry years are associated with starvation and famine. The dry years are often associated with an ENSO event (Chapter 2). Such droughts occurred during 1968–1973 and 1982–1984. The Sahel famine of the early 1970s is still remembered as it claimed the lives of thousands of people, although the estimate of more than 100 000 prevalent at that time has been questioned (Olsson, 1993). The drought certainly destroyed a very high percentage of the livestock.

Rainfall on the humid southern side of the Sahel is higher and more dependable, although most of the area is under stress from above-optimal numbers of people and livestock. In Africa, land of this type includes parts of the countries listed for the Sahel plus Sudan, Somalia, Ethiopia, Kenya and Tanzania from eastern Africa and parts of southern Africa, especially Botswana. In South America, parts of Argentina, northern Chile and northeastern Brazil have a similar semi-arid environment. In Asia it is found in a belt stretching from Israel to the Thar Desert of western India. Areas around the Gobi Desert, parts of central Asia, and Australia are also at risk. This type of area has supported people for centuries, but the current increase in population pressure, overgrazing by livestock, and experimentation with cash crops (cotton, groundnuts and soyabeans) since colonial times have degraded the land drastically. The traditional food crops could survive harsher years but the cash crops that are replacing them have brought in a feast or famine pattern. When the cash crops fail in drier years, especially when a string of dry years occurs, the farmers do not have any money to purchase food that they no longer grow themselves. The situation worsens as the purchasing value of their cash crops or livestock depletes due to the simultaneous drop in quality in a bad year and increased amount available at the market as people generally try to sell off their products in order to purchase food, a situation which has been described as the loss of entitlement (Sen, 1981).

Olsson (1993) discusses the drought and famine of 1984 in Sudan when the provinces of Kordofan and Darfur and the Red Sea Province were particularly affected. The famine affected about half of the people in these areas, where the total population was 20–25 million. Locally, a death rate of 3 percent per month has been reported. Over 2 million people migrated to urban or better areas (Ibrahim, 1988). The establishment of irrigation schemes since the 1920s has led to the expansion of cultivated areas and interest in large schemes, ignoring the traditional production of food (sorghum and millet) and cash crops. However, the rainfall in semi-arid areas is variable, and the balance between production and need is always fragile. The 1984 drought, land degradation and famine occurred due to a combination of reasons:

- several years of below-average rainfall culminating in the especially low figures of 1984 (Figure 6.2);
- an accumulated grain deficit from 1982;
- maldistribution of food leading to a very high price rise;
- inflated grain prices in combination with depressed livestock prices and the opportunity for traders to profit (Olsson, 1993).

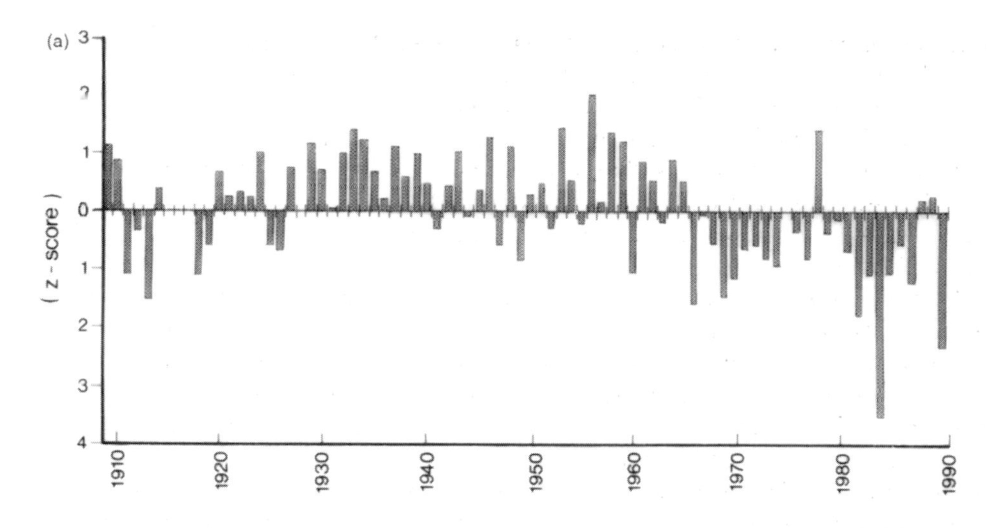

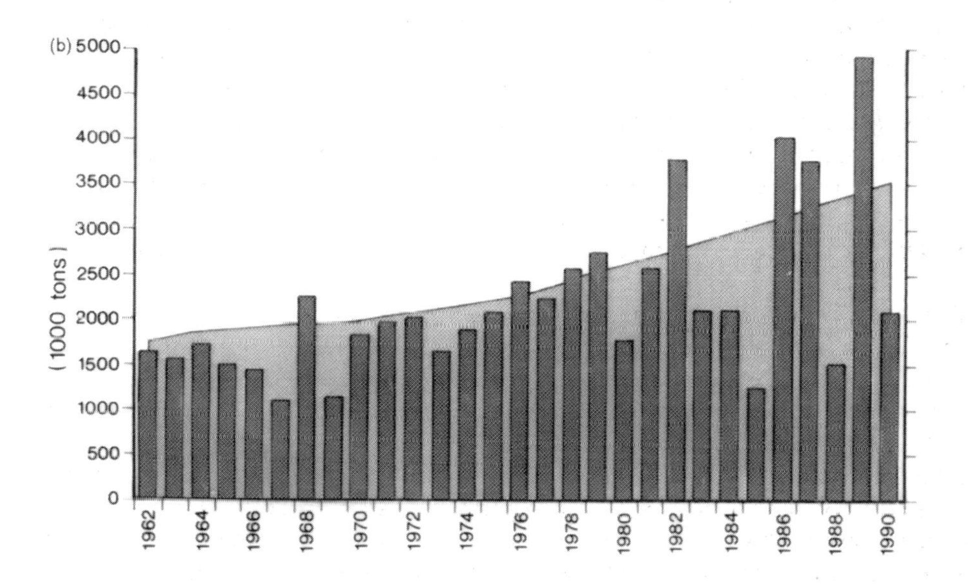

Figure 6.2 *(a) Rainfall deviation from normal (z-score) in western and central Sudan, showing the mean for four stations. (b) Production of sorghum and millet in Sudan (bars) against the estimated total grain needs (145 kg per capita per year). From Olsson, L. (1993) On the causes of famine – drought, desertification and market failure in the Sudan,* Ambio, *22, 395–403*

Olsson points out that famines may occur in Sudan as a combination of drought, market failures, and unjust loan systems, where a small farmer is committed to deliver a fixed amount of crops against the loan, irrespective of price fluctuations. In bad years, it can be disastrous. Famines may therefore occur in marginal lands even without significant land degradation.

The overall effect over years is that of disappearing natural vegetation, erosion of the surficial material, windblown sand, and salinization. This phenomenon is known as *desertification*. Although the term has become popular (especially since the United Nations organized the 1977 UN Conference on Desertification), and has been used for just about any kind of land degradation including destruction of the rain forests, it essentially refers to the extension of desert-like conditions in semi-arid areas. The boundaries of deserts do fluctuate, although controversies now exist as to whether desertification is really an expanding phenomenon at present and Hellden (1991) has attributed desert extensions to climatic variations rather than anthropogenic land degradation. Desertification is, however, a popularly accepted notion, attention to which is frequently drawn by the United Nations Environmental Programme (UNEP), and it even formed part of the Rio Declaration of 1992 (Chapter 15). The story of the accelerated southward spread of the Sahara perhaps needs to be toned down, but the threat of desertification does justify proper and careful monitoring of agricultural land in dry areas.

Nelson (1990) argues that desertification is a poorly characterized and improperly understood phenomenon whose control is often seen as damage retention rather than improving land management and dealing directly with the causes. It is essentially dryland degradation, a very serious problem but not necessarily the popular spectre of rapidly moving sand dunes. Good data, suitable for time-based comparison, are difficult to come by, although studies by the Central Arid Zone Research Institute (CAZRI) in India, UNESCO in Chile, and individual researchers in Australia have shown clearly evidence of serious dryland degradation and the need for careful monitoring and possible recovery. We need to understand the causes of desertification before attempting proper land management and recovery. The causes so far seem to be a combination of factors which include rainfall variability; high population growth; shortsightedness of residents, governments and donor agencies; and inappropriate social and economic structures (Nelson, 1990). Dryland management therefore needs to take all of these into account.

6.4 Land Management and Degradation Mitigation

Good lands in the tropics are currently being used to sustain several harvests of high-yield varieties of grain crops every year. This is possible only through high inputs of water, fertilizers and pesticides. As described earlier, such practices also create socio-economic inequalities, but physical problems also arise because of such high inputs. The environmental problems can be listed as follows:

- demand for irrigation water which has to be supplied either from rivers or groundwater;
- demand for power as pumps are used to lift water from the canal or the subsurface to the fields;
- draining of excess fertilizers and pesticides to surface water and groundwater.

Apart from the problem of providing pumps and fuel to the farmers, the supply of enough water and also the quality of both local surface and subsurface water needs to be ensured. Issues of water management are discussed in the next chapter but it has to be stressed that fields with high-yield crops need to be carefully monitored as sources of pollution.

The marginal lands require intensive management, and standard techniques exist for a number of problems which are frequently encountered. On steep slopes, soil erosion, accelerated runoff, and loss of surface material downslope could be controlled by a variety of embankments, terraces (Figure 6.3) and ground cover. The size and the slope direction of the terraces vary depending on crops grown and the nature of the hillsides. There are many varieties of terraces which are constructed differently for different objectives:

Figure 6.3 *Rice terraces in the valley of the Cimanuk River, Java*

- to create flat surfaces on steep slopes (bench terraces)
- to slow down surface runoff (contour terraces)
- to prevent slope erosion and pond back water

Terraces can be reversed-sloped, flat or outward-sloping. Long slopes are often broken up into small segments by a series of reverse slopes which act as ditches draining away excess water after the rains. Besides preventing erosion, terraces have a secondary role. They increase infiltration and both surface and subsurface storage. Where this increases the chance of slope failures, the terraces are outward-sloping without an embankment to allow rainwater to flow down the hillslopes in steps. Otherwise, the inward-sloping variety with an embankment to pond water provides a higher yield.

Similarly, embankments vary from a fence of brushwood grown at the downslope end of the plot to labour-intensive earth or stone embankments which not only act as a barrier to downslope sediment transfer, but also pond water on the terrace, ensure adequate supply for the crop, and prevent slopewash. A combination of terracing and embankments allows slopes up to 30° to be farmed. Large-scale afforestation of the slopes or agroforestry may also be tried as remedial measures against accelerated erosion.

Erosion by gullies on sloping lands becomes a large-scale problem (Figure 6.4), especially in the drier tropics when the vegetation is destroyed or the land is left fallow. This is a difficult problem and can be arrested only by labour-intensive methods such as putting barriers at several places across the gullies using wire

Figure 6.4 *Gully erosion, Baringo, Kenya*

bolsters, netting dams or brushwood dams, and letting the gulley fill up with sediment. Control of grazing by animals helps, especially when in conjunction with revegetation of barren fields.

In drier areas cultivated with irrigation the standard management techniques for preventing salinization and alkalization include lining the watercourses and providing adequate drainage to the fields to prevent accumulation of standing water. Both techniques prevent seepage to the groundwater and hence salt encrustation of the farming lands. Drainage for these fields is about as important as irrigation. Fields may be sloped to direct the surface water to a waterway. Such drainage ditches are locally deepened so that groundwater flows into them. Subsurface drains currently have started to replace surface drainage. These are designed according to the hydraulic conductivity of the local soils and the required drainage, and are installed as a network at an appropriate depth. The pipes used to be made of tiles or concrete but corrugated plastic pipes (50–200 mm in diameter) have started to replace them. Installation of pipes, however, is very expensive and making the tiles or concrete pipes is labour intensive too (Ghassemi, Jakeman and Nix, 1995). Salinity control has also been attempted by revegetation such as replacing short grass with deep-rooted alfalfa in the United States. Salinized land is being rehabilitated in the state of Uttar Pradesh, India, by planting the saline-tolerant tree *Terminalia arjuna*, which lowers the water table by evapotranspiration. It may take about 25 years for the water table to be significantly lowered. The economic incentive of growing silkworms fed with the leaves of this plant helps during this period. Alkaline soils have been rehabilitated in India by draining water through vertical tubewells and constructing drainage canals and holding ponds. This is followed by the application of a local green manure and the planting of alkali-tolerant vegetation. Over time, the soil improves enough to support vegetation successions which are less tolerant of alkaline soil conditions. The project is a demonstration of land rehabilitation at low cost using much hand labour and local materials (World Resources Institute, 1988).

Dryland degradation, being climate driven to a large extent, is episodic and difficult to ameliorate. Standard techniques include the following:

- planting of green belts as a barrier to wind transport of fine sand and silt;
- stabilizing sand dunes;
- dependence on a variety of crops and also hardy traditional food grains that can survive drought to some extent;
- controlled grazing;
- government action to prevent pastoral herdsmen and dryland farmers from losing their entitlement during bad years.

Proper farming of the steep lands can be carried out to a large extent by individual cultivators who collectively may sustain agriculture on slopes. Prevention of salinization and dryland degradation both, however, need large-scale, planned and continuous management which requires state participation. It is of course best not to farm badlands, but given the socio-economic conditions, that choice is usually difficult to execute in a developing country. From the viewpoint of sustained agricultural use of the land, it is best, however, to plan its use properly and according to its carrying capacity.

Box 6.2
Land Management in East Africa

The highlands of East Africa provide excellent examples of both anthropogenic degradation of land and its recovery (Ståhl, 1993). These highlands, stretching across Tanzania, Uganda, Kenya and Ethiopia, have a high agricultural potential, and have supported large numbers of people throughout historical times. At present, rising population, a history of land mismanagement, and land scarcity have led to extensive land degradation, resulting in both erosion and loss of soil fertility (Figure 6.4). On the other hand, recognition of the seriousness of the problem has led to concerted attempts to arrest land degradation, often with low-cost technologies and at times with considerable success.

Even during the late 1800s, traditional social and agricultural practices were organized on an ecologically sustainable basis, and were capable of producing crops surplus to subsistence. In the past, these areas have supported kingdoms with stratified social structures. Currently, traditional farming practices with long fallow periods have become impossible due to the rising population (East Africa has an annual growth rate of 3 percent) and expansion of impoverished small landowners. Furthermore, landless people have moved into dry areas which suffer from crop failures due to drought. In turn, the nomadic herders have been pushed further into the drier barren lands, thereby putting their production system under stress and making themselves vulnerable to climatic vagaries.

Local groups, national governments and international donor agencies have initiated a number of attempts to arrest land degradation. Often the initiatives are based on low-cost technologies, technical and agroforestry trials, and training and extension programmes. Such initiatives include national conservation strategies, environmental action plans, community forestry, and soil conservation campaigns. For example, in Kenya alone, more than 18 000 agricultural officers have been trained, and conservation measures are found in more than a million small farms (Ståhl, 1993). In Ethiopia, the total length of terraces constructed between the late 1970s and the 1980s, was 200 000 km and the number of trees planted during this period exceeded 45 million (Pretty, Thompson and Kiara, 1995). These are impressive figures, but land degradation is still continuing and Ståhl suggests that conservation measures have only reached a third of all small farms. Conservation measures also include farming the slopes in hand-dug terraces with an embankment and a ditch at its downslope end. Eroded soil material is washed across the terrace surface to build a level bench terrace. In dry areas, these terraces trap rainwater whereas in more humid areas the excessive surface runoff following intense rainfall is drained off the fields by diversion ditches. Soil fertility is increased by vegetating the land, for example, by agroforestry (Figure 5.7). Regional institutions, such as the International Centre for Research on Agroforestry (ICRAF) in Kenya, study and advocate appropriate measures. Soil conservation strategies also include livestock management, e.g. controlling the free grazing of cattle, goats and sheep on agricultural land after harvest.

It has been suggested that an overhaul of the landownership system is necessary in Kenya to motivate small farmers in soil conservation. The situation is particularly bad for farm families headed by a woman. As Ståhl points out, farms owned by poor women may degrade faster because of a shortage of labour, cash income and technical services. There are several success stories, especially where local communities have become involved. One of the success stories comes from the Machakos District of southeast Kenya where land degradation which started in the 1930s has been arrested due to a long-term political, technical, social and economic commitment to water and soil conservation (Tiffen, Mortimore and Gichuki, 1994). Similarly, the long-continuing and extensive soil erosion problem of Dodoma area, Tanzania, has been resolved by the HADO Soil Conservation Project. Previously, this area was losing 1 cm of soil annually by erosion which translates into 13 500 t km^{-2} annually, and regional water reservoirs had a life expectancy of not more than 45 years. The project recognized the interdependence of agriculture, forestry and livestock; the importance of popular participation in soil and water conservation; and the importance of the introduction of conservation practices and training of farmers. Perhaps the most dramatic event of the project was the government's effort at closing the intensely degraded areas to cattle in 1979 and relocation of about 85 000 heads of livestock on the neighbouring plains (Christiansson, 1988; Blackwell, Goodwillie and Webb, 1991). Such long term commitments from both government and community are required across the tropical world. The present conditions could be appalling at places. It is estimated that from the island of Madagascar 25 000 t km^{-2} of sediment reaches the oceans every year (Randrianarijaona, 1983).

6.5 Conclusions

A substantive part of the tropical land is still in reasonable condition and producing a large amount of foodcrops, but the extent of soil degradation is rising. The International Soil Reference and Information Centre (ISRIC) in 1991 produced a revised version of their Global Assessment of Soil Degradation (GLASOD) in collaboration with UNEP, out of a co-operative effort of about 250 soil scientists from various parts of the world (Oldeman, Hakkeling and Sombroek, 1991). Soils of the world were mapped on a scale of 1:10 000 000 to indicate anthropogenic soil degradation of recent times. The three major causes of soil degradation appear to be overgrazing, deforestation and agricultural mismanagement. The total area of degraded soil is 19.64 million km^2, i.e. about 15 percent of the total land surface. More than half of this is due to water erosion. Other estimates indicate an annual loss of 60 000–70 000 km^2 of agricultural land by erosion with another 15 000 million km^2 affected by waterlogging, salinization and alkalization (World Commission on Environment and Development, 1988). Figures like this indicate that we should take much better care of our agricultural fields. These are global figures, but the local data can show even greater losses. As stated earlier, the mean value of 42 measurements carried out in the volcano-bounded basin of the Cimanuk River of Java gives a sediment yield figure of about 4000 t km^{-2} year^{-1} (Aitken, 1981). That is a lot of sediment.

Techniques of land management that will reduce erosion, salinization and desertification have been discussed earlier. Appropriate techniques are often difficult to apply in countries with pressure on land created by physical or socio-economic conditions or both. Use of land according to its capability is the best way to manage land and avoid soil degradation. The best known classification of this type is an old one proposed by the Soil Conservation Service, United States Department of

Agriculture (Table 6.2). Similar classification schemes, modified for local conditions, can be adapted for individual countries. It should be stressed, however, that along with land-use zoning and appropriate agricultural practices, a degree of equity in socio-economic conditions of the country might also be needed.

Table 6.2 *Land Capability Classification, Soil Conservation Service, USDA*

Land class	Characteristics	Primary uses	Secondary uses	Conservation measures
Land suitable for cultivation				
I	Excellent, flat, well-drained	Agriculture	Recreation, wildlife, pasture	Not needed
II	Good land with minor limitations: slight slope, sandy soil, poor drainage	Agriculture, pasture	Recreation, wildlife	Strip cropping, contour farming
III	Moderately good land with limitations of watershed: soil slope or drainage	Agriculture, pasture	Recreation, wildlife, urban, industry	Contour farming, strip cropping, waterways, terraces
IV	Fair land, severe limitations of soil, slope or drainage	Pasture, orchards, limited agriculture, urban, industry	Pasture, wildlife	Limited farming, contour farming, strip cropping, waterways, terraces
Land unsuitable for cultivation				
V	Used for grazing or forestry, slightly limited by rockiness, shallow soils, wetness or slope	Grazing, forestry, watershed	Recreation, wildlife	No special precautions if properly grazed or logged; must not be plowed
VI	Moderate limitations for grazing and forestry	Grazing, forestry, watershed, urban, industry	Recreation, wildlife	Grazing or logging, limited at times
VII	Severe limitations for grazing and forestry	Grazing, forestry, watershed, wildlife, recreation, urban, industry		Careful management required when logged or grazed
VIII	Unsuitable for grazing or forestry: steep slope, shallow soil, lack of or excessive water	Recreation, wildlife, watershed, urban, industry		Not to be used for grazing or logging. Steep slope, lack of soil

Note: The Soil Conservation Service is currently known as the National Resources Conservation Service

Exercise

1. The agricultural yield of an area in the humid tropics with rolling topography and year-round rainfall has increased remarkably with inputs of better quality seeds, irrigation, fertilizers, pesticides, etc. A very large fraction of the pesticides applied is escaping into the local streams and thereby creating an environmental problem. How would you deal with this problem? Locate several areas which have suffered from problems of this type.
2. You have the following land use data for the imaginary Dhansiri River Basin.

Type	Area (km^2)	Sediment yield ($t\ km^{-2}\ year^{-1}$)
Forest	410	30
Cropland	820	1500
Cleared land	30	2500
Urban	150	25
Total	1410	

What is the total amount of sediment being eroded from this basin in one year? If the Dhansiri River is 40 km long, and the distance from the basin boundary to the river is around 15–20 km, do you expect all the sediment to be transported out of the basin at the same rate as the erosion? If not, where is the sediment being stored?

The rainfall is seasonal, about 1400 mm a year. What effect would the sediment have on the environment?

7

Development of Water Resources

7.1 Introduction

On our planet life exists because of water. Our bodies are mostly water, and we require a considerable amount of it for our survival. The demand for water is worldwide, although it increases with the affluent and industrial lifestyle. There is, as we have seen in Chapter 2, a finite limitation on the volume of water on earth. Of this, only a very small fraction is available to us from the atmosphere, the land surface, and the shallow subsurface. It is technically feasible to increase this volume by collecting water from the ice sheets of Antarctica or Greenland, or from the deeper subsurface, or by desalinizing sea water, but the cost becomes prohibitive.

Technical innovations to increase the water supply and possibly optimize its use have been in vogue for thousands of years, possibly since the beginning of agriculture in the Early Holocene. In Chapter 6, we discussed the complicated canal irrigation system that supported the agriculture of the Sumerian civilization. We have also reviewed the associated environmental degradation such as salinization from such projects in the same chapter. The ruins of the ancient cities of the northwestern Indian subcontinent (usually referred to as the Harappan civilization) display 5000 year old engineering feats in keeping prosperous cities supplied with water. Techniques of water utilization have evolved across the centuries. We see this in the carefully irrigated rice terraces of Java, the tank irrigation of South India, the large dam across the Nile at Aswan, and the pipelines and international agreements to keep the three million people of the affluent small island city of Singapore adequately supplied with water. This chapter is built on three themes associated with water utilization:

- the demand for water
- the techniques for water utilization
- the related environmental problems and their management.

7.2 Types of Water Demand

The major types of demand for water are listed in Table 7.1. It is not enough to supply a required quantity of water; the water that is available should also be of sufficient quality. The constraint of quality of course changes depending on the type of use. Drinking water requires water of very high quality, the standards for which have been stringently determined by various international and national agencies. In contrast, the crucial factor for hydroelectricity generation or navigation is the minimum volume of water that could be expected during the dry season; quality is less important. Again, the volume of demand varies from country to country directly depending on the size of the population and its affluence. The post-utilization condition of the water is also crucial. Dirty effluent from cities, for example, should not be simply discharged into a stream or lake. This, however, does happen. Similarly, industries should treat their water for chemical or biological content and ambient temperature before release. The control of water running off agricultural fields, as it drains to a river or seeps into the subsurface to join the groundwater, is more difficult than the control of urban or industrial wastewater which is released only at specific points. Like other natural resources, the availability of water is non-uniform across the geographic space. In general, it is easier to meet the water demand in humid areas, unless the demand exceeds availability as happens for large cities such as Jakarta, Bangkok or Dakar. It is then necessary to find multiple sources of water and to transfer it across river basin boundaries in pipes. This is not a new technique – the Romans were famous for water transfer along their aqueducts – but the practice tends to be expensive and certainly requires administrative arrangements and, possibly, international agreements. Such practices may also lead to environmental problems as shown later in this chapter. Like management of land, management of water is a demanding occupation.

7.3 Irrigation and Water Projects

About 5000 years ago, King Menes of Egypt used clay bunds to hold back the floodwaters of the Nile for basin irrigation when the annual flood receded. The long history of irrigation using the seasonal floodwater of the Nile to inundate the agricultural fields of Egypt (a technique known as basin irrigation) continued successfully for thousands of years. Among the early technical innovations were devices for lifting water from rivers or wells for irrigation. About 3000 years ago, the shadoof, which is essentially a bucket and a counter weight at two ends of a pole balanced on a vertical stand, was well-established for lifting water from rivers to irrigation channels at higher levels. Other time-tested simple implements for lifting water and pouring it down a channel to reach the fields by gravity flow include the mot, the Archimedian screw, and the Persian wheel. Different instruments for lifting water came into practice in different parts of the world, and some of these are still in operation (Figure 7.1). A hand or diesel pump for lifting water from

Table 7.1 *Types of water demand*

Type of use	Desirable quality	Quality after use
Potable water used for consumption	Highest of all use, predetermined standards must be met	Needs treatment
Other kinds of municipal use: fire fighting, street cleaning, etc.	Not of high quality, but in many cities the potable water is used to avoid the complications of two water networks	Needs treatment
Generation of electricity	Low in sediment, usually reservoir water behind dams	Reusable
Industrial	Low in sediment, mineral and biological constituents	Could be seriously polluted or of high temperature after use; mostly resusable after treatment
Agriculture	Not high in saline or alkaline concentration	Polluted by pesticides, herbicides, fertilizers, and salts; immediate reuse needs treatment
Navigation	Not applicable except for aesthetic reasons	All reusable
Recreation	Same as for navigation	Reusable, low level of pollution possible
Wildlife support	Not applicable	Reusable, low level of pollution possible

wells, rivers and canals is the modern equivalent. In contrast, the common practice over large parts of southern India was to put in place a large number of small tanks, taking advantage of the natural relief of the land, and use the water from the tanks for irrigation. Modern day small-scale irrigation uses canals and furrows of various sizes and type, and water sprinklers. The rising demand for water, especially for the high-yield crops of the Green Revolution, requires a large-scale intake of water from rivers or lakes and deep tubewells which pump water out from aquifers that occur at depth in the subsurface.

The second half of this century has seen a proliferation of large-scale river valley projects in the developing countries. These are usually multipurpose projects designed to meet a variety of needs such as flood control, irrigation, and production of hydroelectricity. The modern technology was developed in the western countries. The 221 m Hoover Dam over the Colorado River, and Lake Mead with a storage capacity of 38 billion m^3 upstream of it, was engineered in the 1930s, as was the Tennessee Valley Project. This 90 000 km^2 basin of the Tennessee River, which flows into the Ohio, had earlier suffered from a series of floods and extensive erosion of arable land following uncontrolled logging and strip mining practices. The Tennessee Valley Authority (TVA) was established in 1933 as a scheme for

Figure 7.1 *Lifting of water from a river in central Sumatra*

generation of hydroelectricity in a relatively poor agricultural part of the United States but later developed into an integrated network of flood control, power, navigation and irrigation. A number of dams and reservoirs were constructed in the Tennessee River Basin as part of a multipurpose project which treated the drainage basin as a functional unit, and recognized the necessity for state interference in order to achieve social welfare. Management of land use in the basin, especially soil conservation, was considered an essential part of this development project. The concept of a state implementing a large, integrated and multifaceted engineering project in order to save the people from floods and soil erosion and to provide them with water and power, attracted the governments of a number of newly independent countries in the second half of the twentieth century. Even back in 1938, the need for such projects was stressed in the presidential address given to the Indian National Institute of Sciences by the physicist Meghnad Saha. In 1954, when seven years after independence India's first prime minister Jawaharlal Nehru inaugurated the Nangal canal as part of the massive Bhakra multipurpose project, he described such places as modern day temples (D'Monte, 1985). Large projects like this were completed across the world, the developing countries not excepted. These were not only engineering marvels but also essential steps towards national economic development and flood hazard control. The number of new dams inaugurated in a year increased fivefold between 1940 and 1970. Their sizes also got larger. Ninety-three of the 100 largest dams of the world were completed after the Second World War and most of these are in the developing countries (World Resources Institute, 1988). The three well-known projects of the 1970s from Africa illustrate this: Lake

Kariba, Zambia (160 million m^3); Lake Nasser, Egypt (known as Lake Nubia in Sudan; 157 million m^3); and Lake Akosombo, Ghana (148 million m^3). According to the International Commission on Large Dams (ICOLD), more than 35 000 large dams (defined as over 15 m in height) exist world-wide, excluding China. Between 1951 and 1982, an average of 344 dams were constructed each year, although this number has shown a decrease since 1975 (Dixon, Talbot and Le Moigne, 1989). On certain continents, about 20 percent of the total runoff is currently controlled by reservoirs. It is therefore logical to wonder about the impacts of so many dams and reservoirs, especially such large reservoirs, on the environment. Table 7.2 lists such impacts, which could be either beneficial or detrimental, local or large-scale. The size of the new dams in the developed countries, however, has generally declined; a number of small dams is now considered preferable to one large dam. In developing countries the picture is mixed, with China and India demonstrating the expertise in small dam technology (World Resources Institute, 1988). Large dams, however, continue to be planned and constructed. In 1986, the Guri Dam was completed in Venezuela with a 10×10^9 watts capacity for production of hydroelectricity, the largest in the world. The Itaipu Dam at the Brazil and Paraguay border area is designed to produce 12.6×10^9 watts. The Three Gorges Dam on the Yangtze River in China is planned to generate 13×10^9 watts when constructed (Dixon, Talbot and Le Moigne, 1989).

A multipurpose project produces power and water for irrigation. Dams have also been used for flood control by holding up the high discharge in the reservoir and releasing it slowly downstream. Proper flood control, however, requires keeping the drainage basin under vegetative cover so that the surface runoff does not increase drastically. Reservoirs also serve as a focal point for recreation and wildlife propagation. All of these are good reasons for constructing dams and reservoirs, but also associated with dams and reservoirs are various kinds of environmental degradation.

The environmental problems can occur at various scales. Certain problems are local: sedimentation in the reservoirs; increased slope failures around reservoirs; displacement of people who previously resided in the river valley. Problems can be regional in nature, such as the triggering of earthquakes, salinization around reservoirs and irrigation canals, and transmission and expansion of parasitic organisms which cause health hazards such as schistosomiasis or malaria. The schistosomiasis parasite is carried by a snail vector whose habitat increases with the establishment of perennial irrigation systems in the dry tropics. For these reasons, it is crucial that controlling measures be taken (usually at the regional scale) simultaneous with the dam construction. This is not always successfully carried out, and the standard example of this type of failure is uncontrolled upland erosion and sediment transfer to the reservoirs. The filling up of the reservoirs shortens their life expectancy, and reduces the number of years these dams (constructed at huge costs) can function effectively. Such shortening of the life-time of reservoirs has been described from Anchicaya (Colombia) or the Mangla Reservoir, Pakistan. The nature and speed of environmental degradation vary according to the ambient geology, terrain characteristics, climate and land use. Repeated seismic activities are on record after the closing of a number of large dams such as the Vaiont (Italy), Hoover (USA),

Table 7.2 *Environmental impact of dams and reservoirs*

Beneficial effects towards hazard mitigation, economic production, and recreation

- flood control
- power generation
- ensuring availability of water for irrigation
- recreation
- navigation
- supporting fisheries and wildlife
- increased tourism

Potential environmental degradation

- reservoir sedimentation; this could be controlled by using the dead storage capacity of reservoirs as a sediment trap
- deepening and narrowing of the downstream channel
- land subsidence
- triggering of earthquakes by the weight of the reservoir water
- changes in groundwater level, which could lead to slope instability around reservoirs
- salinization around reservoirs and irrigation canals
- waterlogging
- reservoir-related biological problems: fish kills, algal production, aquatic weeds, loss of wildlife
- inundation of forests
- inundation of archaeological and historical sites and monuments
- transmission and expansion of parasitic organisms which cause schistosomiasis, malaria or river blindness
- thermal changes in the reservoirs
- environmental problems during construction
- displacement and resettlement of local residents – at best, a difficult proposition
- disruption of the local economy and transport links

Kariba (Zambia) and Koyna (India). In spite of all this, large dams are still designed for seismically active areas such as the Tehri Dam or the dams on the Narmada River of India. The proposed Three Gorges Dam over the Changjiang (the Yangtze) is another large-scale and highly controversial project. A number of these problems can be envisaged at the project design stage, especially as the problems are often standard ones.

In 1971, the Aswan High Dam with Lake Nasser upstream was constructed over the Nile in Egypt with Soviet assistance. Apart from producing billions of kilowatts of hydroelectricity, it increased Egypt's potential arable land from 28 000 km^2 to 36 000 km^2 by storing floodwater and regulating its release for irrigation. This dam has enabled Egypt to feed its population during years of drought. It is also a flood-control device. Regulation of the flow has improved river navigation on the lower Nile by reducing fluctuations of water level. The Aswan Dam has been studied in depth regarding its effect on the local and regional environmental conditions. Table 7.3 summarizes the findings. This list may also be seen as one which includes the possible environmental problems that may arise when a large dam is built in the drier tropics. Some of the problems are general and are shared with dams in more humid areas, although each location may have its characteristic hazards. Raising the

Table 7.3 *Negative environmental impacts of the Aswan Dam*

- Mismanaged population displacement from the reservoir area
- High evaporation and seepage loss of stored water (estimate: 10 km^3 year^{-1})
- High flood silt sedimentation in the reservoir, leading to a reduction in the water storage capacity
- Loss of Nile silt downstream, forcing the farmers to be dependent on fertilizers; traditional brickmaking affected
- Channel erosion downstream of the dam to a depth of between 25 and 70 cm
- Erosion of the Nile Delta due to a reduction in river sediment downstream of the dam
- Saltwater intrusion in the delta area due to a reduction in Nile discharge
- Loss of offshore sardine fishing due to a lack of nutrients reaching the Mediterranean; partially offset by increased fishing in the reservoir
- Increased pest invasions, including vectors of human onchocarciasis (river blindness) and schistosomiasis
- Proliferation of aquatic weeds in the clear water downstream of the dam
- Increased salinization and alkalization in the agricultural fields following irrigation
- Rising groundwater levels requiring improvement in local drainage
- Approach of desert sands eastwards
- Possible cause of the 1981 earthquake

Sources: Hammerton (1972), Kishk (1986), Shalaby (1988), White (1988).

Note: The Aswan Dam did provide Egypt with a number of beneficial effects as discussed in the text. This is true for most dams and reservoirs. A careful determination of the benefit–cost analysis and also an evaluation of the non-quantifiable and intangible factors needed to be done before the construction of the dam could begin. It is also possible to ameliorate the expected negative impacts by redesigning the project before its initiation.

water level in a limestone gorge, for example, gives rise to slope failures as limestone can be dissolved. All seismic areas are suspect, especially as the weight of the reservoir water behind dams in the past has given rise to a series of earthquakes even in previously quiet areas. Table 7.3 illustrates how far-reaching the effects of building a dam on a large river may be.

Of the regional problems, the displacement of people, the disruption of the local economy, the increased hazardousness of the environment, and especially the loss of sediment downstream need to be reviewed at the project pre-feasibility stage. Large dams like the Volta in Ghana or the Kariba in Zambia displaced tens of thousands of people from the valley-bottoms to be inundated by the large reservoirs. Such numbers reached 120 000 for the Aswan Dam, and are expected to rise to well over a million for the proposed Three Gorges Dam of China (Goldsmith and Hildyard, 1984). Similar figures have been estimated by the critics of the Narmada Dam projects of India (Dixon, Talbot and Le Moigne, 1989). The reservoirs also displace wildlife, inundate valuable forests, and (sometimes) flood archaeological sites and monuments. The accumulation of sediment in the reservoirs means that the water released downstream of the dam will carry very little sediment, thereby being prone to spending the excess energy on eroding the river bed and building up the sediment load. This results in bed degradation, channel lowering, and locally a cover of gravel being deposited across the bed of the river. Evidence for this (Figure 7.2) has been well-documented (Williams and Wolman, 1984). The lowering of the channel bed may be enough to disrupt the water intake of irrigation canals.

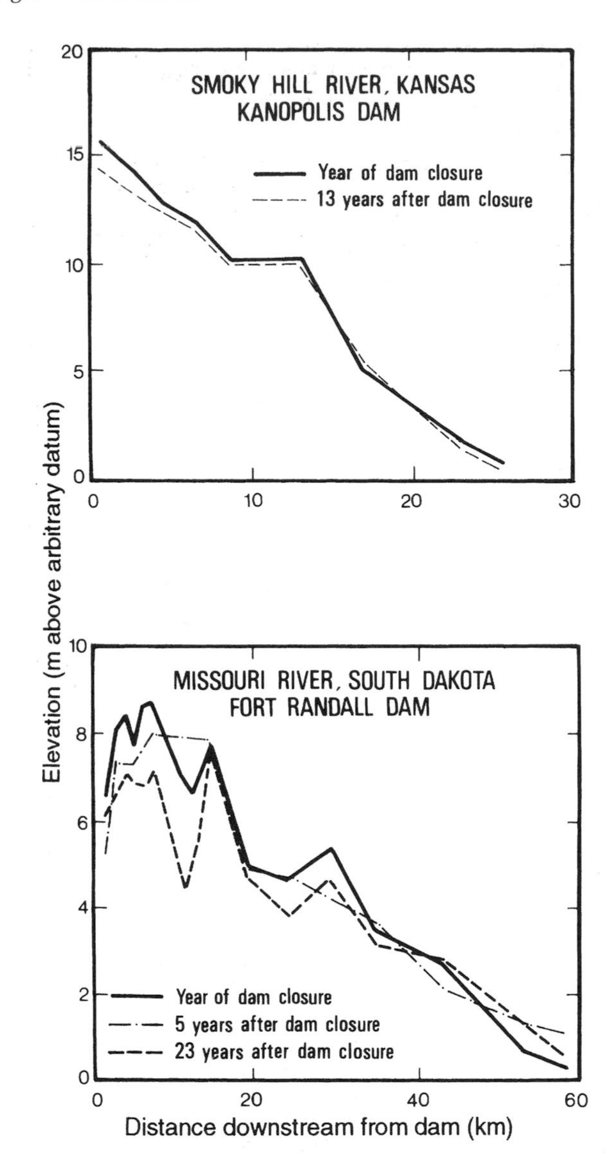

Figure 7.2 *Changes in the longitudinal profiles of rivers downstream from four dams in the United States. From Williams, G.P. and Wolman, M.G. (1984) Downstream effects of dams on alluvial rivers,* US Geological Survey Professional Paper 1286

Given the number of dams and reservoirs constructed during the last 50 years, it is perhaps not surprising that a very large volume of sediment are estimated to be in storage behind the dams world-wide. There are several obvious follow-up questions. Does this lead to the deltas and beaches being starved of sediment and subsequently prone to erosion? This has happened to the Nile Delta after the closure of the Aswan Dam. Given the possibility of global warming forcing sea level

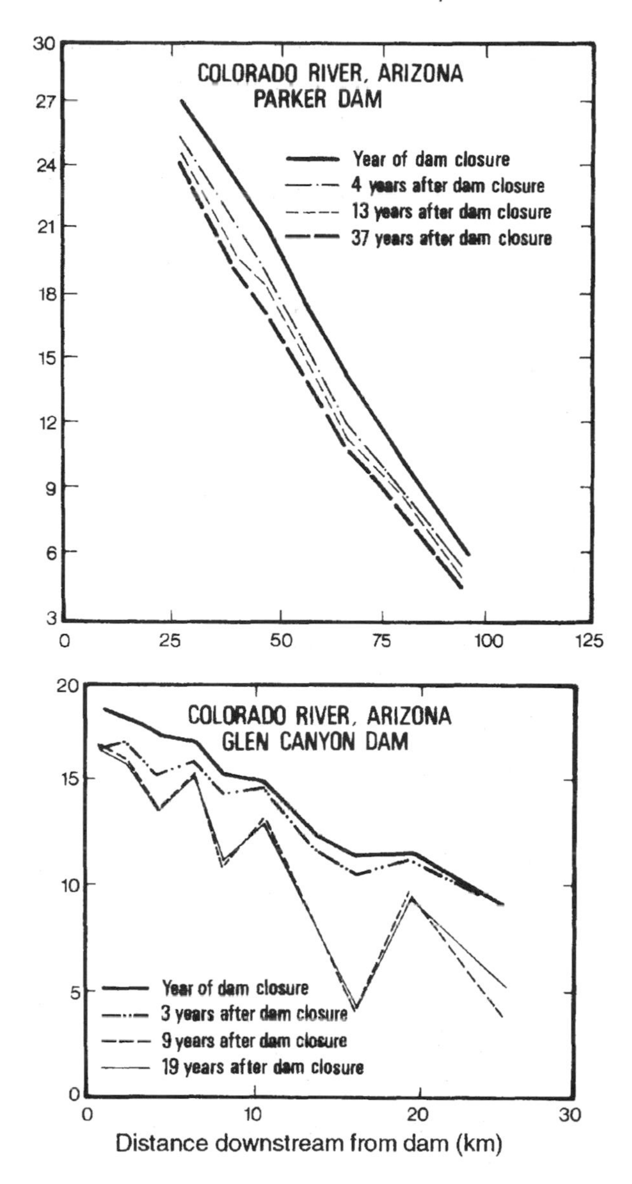

Figure 7.2 *(continued)*

rises (Chapter 14), are such areas going to be especially vulnerable? Given that the dams and reservoirs reduce both water and sediment downstream, what effect does such a phenomenon have on river pollution?

Engineers are extremely concerned about the safety of dams, and standard procedures exist for guaranteeing safety. The evaluation of the environmental impact of a proposed large-scale water project should be carried out with the same concern, and at the pre-feasibility stage of the project design. The majority of the possible environmental impacts are standard expectations to which a number of local variations (depending on the terrain, climate and land use) need to be added. It is not a difficult task, based on the experience gathered over the last few decades when a very large number of dams were constructed. Public pressure has stopped the construction of at least two proposed dam projects (Nam Choan in Thailand and Silent Valley in India) because it was possible to show that when all the factors were taken into account, the net impact on the society would be unfavourable. Unlike in the past, international donor agencies such as the World Bank are now increasingly reluctant to support projects of this type (Chapter 16). An exercise at the end of this chapter invites you to practise working out the possible environmental outcomes of a multipurpose water project in a humid tropical setting. This is the best way to understand the issues involved.

Box 7.1
The Three Gorges Project, China

This project has been on the drawing board since the 1930s. If completed, the proposed Three Gorges Dam on the Changjiang (Yangtze River) will be the biggest in the world. The 180 m dam is designed to be located at Sanduping, 40 km upstream from the Yichang (Hubei Province), with a narrow, 500 km long reservoir along the gorges of the river upstream. The surface area of the reservoir is expected to cover 1150 km^2, with a volume of about 450 km^3. The storage volume is not expected to be very large given the scale of the project, and hydroelectricity is planned to be generated using the flowing water of this large river (Dixon, Talbot and Le Moigne, 1989). The project is expected to produce at least 13×10^9 watts of hydroelectricity, control the floods on this long river, and improve navigation almost up to the city of Chongying (Sichuan Province). Irrigation is not a major objective. The dam has attracted world-wide attention on various accounts: its size; the expectation of hydroelectricity replacing the use of very large amounts of fossil fuel; and the high potential for environmental degradation. At 1990 prices, this dam is estimated to cost over $10 billion.

The catchment of the reservoir consists of areas with considerable relief: rocky outcrops, steep gorges and gullied mountain slopes. Flatlands are very limited. With both population and the demand for food increasing in this area, a considerable area of forest was converted to agricultural fields from the late 1950s. At present, only 10 percent of the catchment is under forest. Soil loss from Zigui County towards the lower end of the area has been reported to have increased from 865 000 ton $year^{-1}$ in 1956 to 1 259 000 ton $year^{-1}$ in 1980. The thin soil cover over most of the basin, especially over the steeper parts, is particularly vulnerable to erosion. This is not good agricultural land, and the holdings are very small even for China, although the population density of 300 people km^2 is about three times the national average (Chau, 1995).

The project has the potential to provide China with a large amount of clean energy which is much in demand – a demand that is expected to rise considerably in the future. This may replace burning of huge quantities of fossil fuels and biomass. However, the project is also perceived as potentially damaging to the environment.

- The biggest concern is possibly the huge displacement of people who need to be moved in stages from the river valley to other areas. Estimates of their numbers vary but a figure well in excess of a million by the year 2008 has been discussed. It is planned that this population will be resettled mostly on the upper slopes to minimize social and economic upheavals, but the uplands are ecologically rather fragile and may degrade rapidly under such population pressure. A striking part of this project is the inundation of towns and cities (covering more than half of the total displaced population) along with the rural areas. Such a large-scale city inundation is extremely unusual. The resettlement of the people is estimated to cost nearly a third of the total cost of the project. Chau (1995) has discussed in detail the resettlement problems.

- Concerns exist regarding the reservoir functioning as a sediment trap, especially in the early years after its closure. This is due to the physical geography of the area: steep slopes, shallow soils, destroyed forest cover, and increasing rainfall with altitude which reaches 2000 mm a year towards the top. The resettling of farmers on the upper slopes would accelerate the problem. Sediment trapped behind the dam would alter the behaviour of the river downstream and deprive its estuary of nutrients (World Resources Institute, 1994).
- The loss of a number of endangered species which live in the Changjiang and its basin is feared. The list includes the Chinese sturgeon, the Yangtze dolphin, and the Siberian crane.
- The long reservoir will inundate the famous gorges of the river, known for their natural beauty and which attract many visitors, some of whom come from outside China.
- Unless proper precautions are taken, the construction of a reservoir may increase malaria which is endemic in this region.

The Three Gorges Project presents the usual conflict between meeting the needs for both development and environmental protection. Even in the general area of environmental management, it is a trade-off between (a) saving fossil fuels and biomass, and clearing the skies of air pollution, and (b) degrading the basin environment as described above. The project will take 17 years to complete and electricity generation is expected to start 12 years after project initiation. This indicates that at least for a decade, the negative aspects of the project will be very much to the fore, while China waits for the benefits to appear. This complicates the perception of the project (Dixon, Talbot and Le Moigne, 1989; Chau, 1995).

7.4 Meeting the Demand for Water in Settlements

Great disparities exist regarding the availability of safe drinking water between the developed and developing nations and even among the developing countries. In some of the urban areas of the developing countries, the availability of potable water may approach 100 percent and indeed be comparable to the expected situation in the developed world. Not all the citizens of the developing world, especially the poorer ones, are this fortunate, but the really striking disparity between the two worlds occurs in the rural sector. In a WHO survey, 75 percent of the urban communities of the developing world had access to potable water, a number which dropped to 20 percent for rural settlements. These are average figures which mask the far wider range of disparity. Biswas (1978) showed that when individual units are considered, the figure ranges from 100 percent for some cities to less than 1 percent for some rural areas. The numbers may have improved since this survey, but the general pattern has remained similar. The concern over the lack of adequate water supply prompted the UN General Assembly to declare 1981–90 as the International Drinking Water Supply and Sanitation Decade. The goal was to urge national governments to provide every resident with clean water and proper sanitation by 1990. This was a difficult task, as in 1980 the number of people without a proper water supply and sanitation was about 2 billion. At the end of the decade, it was estimated that the number of people without an adequate drinking water supply had dropped to about 1.2 billion, while the number of people without adequate sanitation arrangements did not change significantly and is currently estimated to be as high as 1.7 billion. In Latin American countries as a whole, apparently as little as 2 percent of sewage receives any treatment (World Bank, 1992). The original goals were not reached, and it became apparent that the improvements that did occur varied significantly across regions. It is the poorer sections of the population in the developing countries that are at the receiving end of water-related diseases, by collecting water from and washing clothes and dishes in open channels and pools which also carry sewage.

The situation is not helped by increasing population and the limited local availability of water for consumption. It is possible to increase the amount of water that is available by bringing it from a long distance, pumping it from deeper in the subsurface, or by desalinization, but these methods are costly and unfeasible for supplying water to the villagers of the developing world. Often, such techniques give rise to new environmental problems such as land subsidence and pollution. Figure 7.3 correlates total water availability with population growth, illustrating the growing scarcity of water over the years. Figure 7.3 also illustrates the climatic forcing of the problem, the scarcity of water not being entirely dependent on the absolute number of people (Falkenmark, 1986). Water scarcity is most acute for a number of countries in West Asia, North Africa and Sub-Saharan Africa due to both dryness and vagaries of climate and fast-rising populations. The problem exists in lesser forms for certain areas of northern China, western and southern India, and Mexico (World Bank, 1992).

Only a fraction of the available water could be used for consumption, and it would be necessary to underwrite the quality of water. One of the great benefits of science and technology is the production of potable water, which reduces health hazards, and especially reduces child mortality from gastro-enteretic diseases. Polluted water in the developing world is primarily responsible for the outbreaks of cholera, typhoid, amoebic and bacillary dysentery, hepatitis, etc. In the dry tropics, villagers usually walk to a water course to collect drinking water, which results in the spread of vector-related diseases such as schistosomiasis, onchocerciasis (river blindness), guinea worm (*Dracunculus medinensis*), and trypanosomiasis (sleeping sickness), as the vectors of these diseases (such as the snail which spreads schistosomiasis) accumulate near the water courses. This also happens with the establishment of perennial water channels such as irrigation ditches, which may also cause higher incidences of mosquito-borne diseases (malaria, filaria, yellow fever, and arborivorous encephalitides). About 3 million people, most of whom are children, die each year from diarrhoea-related diseases. The annual number of people being affected by such diseases world-wide is 900 million. More than 900 million people suffer from roundworm infection each year. For schistosomiasis, the annual

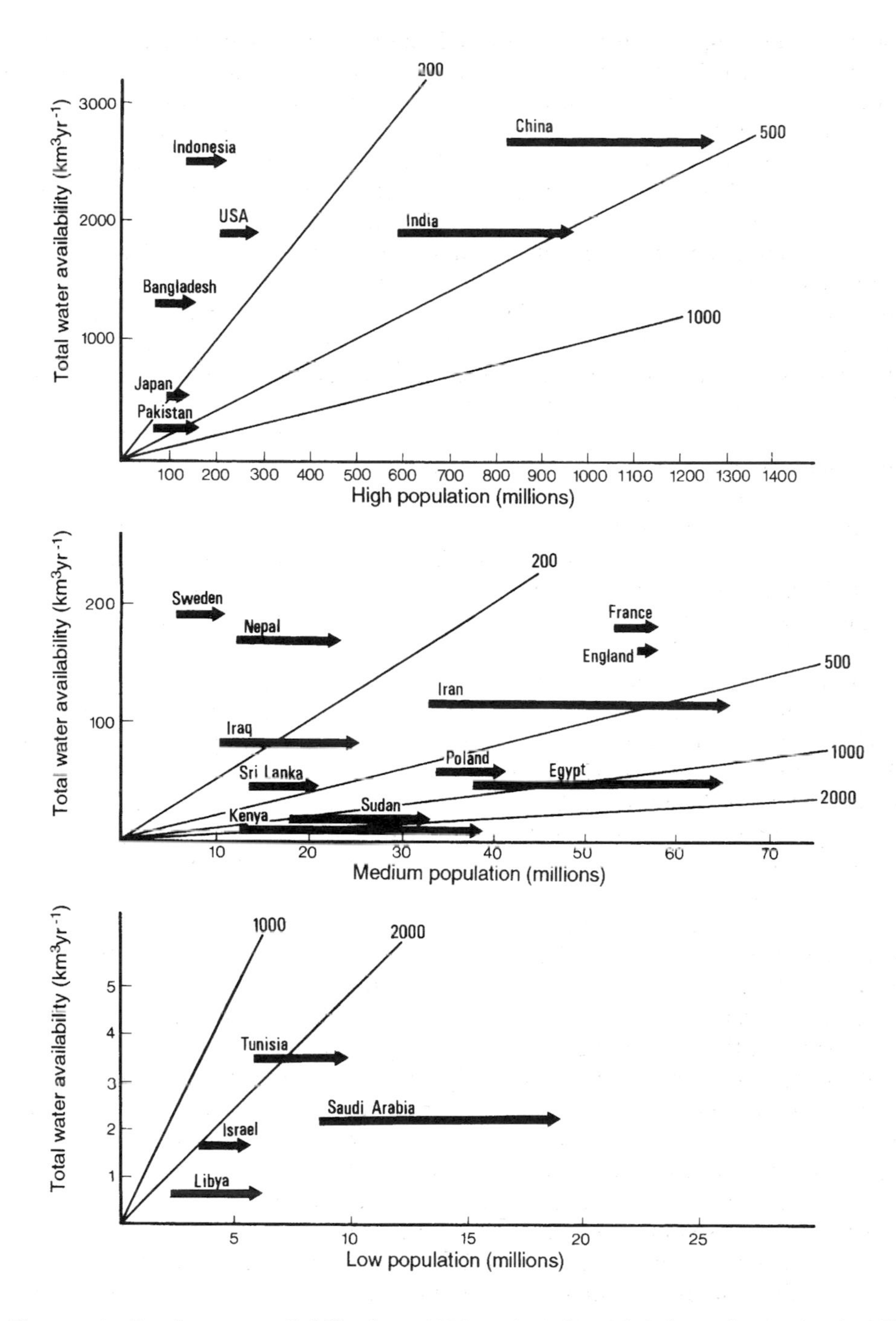

Figure 7.3 *Total water availability from 1975 projected to 2000 for various countries. The diagonal lines represent the number of people to be supplied by 1 × 10⁶ m³ of water per year. From Falkenmark, M. (1986) Fresh water – time for a modified approach,* Ambio, **15**, *192–200*

figure is 200 million (World Bank, 1992). Clean drinking water and adequate sanitation can cut this figure down significantly. Undegraded rain forests are often seen as diagnostic signs of a clean and well-managed environment; so should be settlements supplied with clean drinking water and proper sanitation facilities.

In certain countries of the developing world (e.g. Malaysia, Mauritius and Jamaica), it has been possible to serve the rural population with clean piped water. An alternative is to lift the local groundwater by handpumps (tubewells). This is less expensive, and is reliable, unless the groundwater is contaminated by leaching of fertilizers, pesticides or human wastes. The problem in most cases is ensuring the quality of potable water. The fringe benefit of laying on a dependable water supply is that it frees the village women from walking long distances to fetch water. Like the collection of firewood (Chapter 5), the drawing of water increases physical stress for usually the least-nourished member of the family. Most of the food is often given to the men and the children.

The supply of potable water to urban areas is more organized. Ideally, a city should have enough water for the following:

- drinking water of acceptable quality
- water under enough pressure for fire fighting and street flushing
- water for industrial purposes

This requires (i) a source, (ii) a treatment system, (iii) a distribution system, and (iv) arrangements for treating wastewater and its disposal. With the expanding population of the cities, water may have to travel for hundreds of kilometres and from a variety of sources to meet urban demands. Bangkok and Calcutta, for example, augment the local supply from rivers by groundwater. Singapore, which is not located on aquifers, adds to the supply from local rainfall by bringing in considerable amounts of water across the border from Malaysia. In the future, Singapore may transfer water from nearby Indonesian islands as well.

The practice of treating drinking water started in 1802 at Paisley, Scotland, with the filtering of potable water. The process has developed and improved over the years, and at present in many cities (given the local geology) it is not enough to supply purified water only; fluorine is added to it for dental health. The volume of the water demand is determined by the population of the city, its economic prosperity, the pricing structure for water, and the industrial demand. In fact, in the United States, the demand from industry far outpaces the demand for domestic consumption. As the developing world becomes more industrialized, a similar pattern should be expected to emerge.

The quality of the water is maintained according to national guidelines. The WHO also provides a water quality standard. Details of these are listed in Table 7.4 which shows the water quality standards required and also the long list of contaminants that need to be monitored. The basic steps for water purification are similar world-wide, although variations are introduced when dealing with sources of water of low quality.

The basic steps are as follows:

Table 7.4 *A selection of WHO standards for treated potable water*

Colour	15 true colour units
Turbidity	5 nephelometric units
pH value	<8
Taste and odour	Should be acceptable
Chlorine	0.6–1.0 mg l^{-1}
Chloride	250 mg l^{-1}
Total dissolved solids	1000 mg l^{-1}
Arsenic	0.05 mg l^{-1}
Copper	1 mg l^{-1}
Iron	0.3 mg l^{-1}
Lead	0.01 mg l^{-1}
Manganese	0.5 mg l^{-1}
Sulphate	250 mg l^{-1}
Zinc	3 mg l^{-1}
Total coliform bacteria	0 in any 100 ml sample

Source: Summarized from WHO standard of 1993. World Health Organization (1993) Revision of the WHO Guidelines for Drinking Water Quality, *World Health Organization, Geneva.*

1. *Spray aeration and pre-chlorination.* Water piped in from a river or a lake first passes through a screen to keep out debris, and is then sprayed into the air as jets in order to remove CO_2 and H_2S gases. Activated carbon may be added to the water for subsequent filtering out of some of the organic material. If the original water is not of a high quality it could be chlorinated at this stage to destroy algae and bacteria. Lime may be added to adjust the pH of the water for the next step which is coagulation.
2. *Coagulation and sedimentation.* Alum [$Al_2(SO_4)_3.18H_2O$] is usually added as a coagulating agent and the water is then piped into a sedimentation basin. Sometimes $FeCl_3$, $FeSO_4$ or other coagulants such as polyelectrolytes are used. Water carries colloids (very small particles) in suspension due to the negative charges that surround them. These charges repel these fine particles from each other and they stay small and suspended in water. Alum (or in some cases iron ions) added to water, being positively charged, destabilizes this pattern and the particles come together to form larger units called flocs. The process is known as flocculation. Flocculated material grows bigger in size and settles to the bottom of the sedimentation basin.

   After flocculation has taken most of the sediment to the bottom, the partially clarified water leaves the sedimentation basin through an arrangement known as the rapid sand filter. This comprises concrete basins which contain several layers of graded sand with the finest sand at the bottom and the coarsest at the top. A layer of gravel overlies the arrangement. Water is passed through the rapid sand filter, and colloidal materials and bacteria are removed. The filter is kept clean by backwashing the filter with clean water, which is then drained off.
3. *Chlorination.* Chlorine is added to the water as the next stage and prior to distribution. Chlorine destroys most of the micro-organisms, and the amount to be added again depends on the organic material present in the water. Lime is

added to hard water which does not produce proper soap lather because of a high content of Ca or Mg salts. Fluoridation is carried out in many cities before distribution, which safeguards children's teeth by adding sodium silicofluoride (Na_2SiF_6) to the treated water. Ozone is used at times when a more elaborate treatment process is necessary due to the poor quality of the source water.

However, not all the residents of a large number of cities in the developing world have access to treated potable water that meets the national standards. In the better cases, the supply is augmented by pumping water from underground aquifers which may be safe but which gives rise to land subsidence and associated problems (as discussed later in Chapter 10). Disastrous situations are created when shallow groundwater, often polluted, is used, especially in those cities where industries operate within residential locations due to a lack of zoning regulations and enforcement. Such industries may discharge various synthetic chemicals as industrial effluents to shallow wells. The poorer slum-dwellers and squatters are often reduced to buying water from vendors or collecting it from any nearby waterbody with no expectation of quality. The urban poor pay a very high price for such water of extremely dubious quality. The poor of Port of Prince (Haiti) and Onitsha (Nigeria), for example, spend 20 and 18 percent of their income on water respectively (Munasinghe, 1992, quoting a 1987 World Bank report by Whitington and others). People connected to the municipal water systems pay much less. Degradation of the urban environment is discussed in detail in Chapter 10.

7.5 Pollution Problems: Degradation of Water Quality

Wastewater from irrigated fields, industries and urban settlements is released into rivers, lakes, coastal waters and groundwater. This pollutes the waterways, at times with serious consequences. For example, our current awareness of environmental degradation developed partly from the disastrous conditions of Lake Erie and Lake Ontario in North America following unrestricted dumping of industrial and urban wastes and discharge of agricultural drainage earlier in this century. As late as the 1980s, many beaches there were highly polluted and posted unfit for swimming. Similar problems occurred in the Malacca Straits, off the coast of West Malaysia, due to unrestricted urbanization without proper sewerage (Sharifa Mastura, 1987). Shallow aquifers are often polluted by seepage of wastewater from latrines and rubbish dumps, and the discharge of untreated industrial effluents by unconcerned industrial establishments.

A river collects polluted water along its course. Luckily rivers and lakes have the ability to purify themselves unless the amount of pollution exceeds the limits of this natural cleansing process, as happens for many rivers flowing through mining or industrial landscapes. The Rhine is a well-studied example. The 2525 km Ganga in India, draining a 1 million km^2 basin with 500 000 km^2 of agricultural land and more than 600 towns and cities, is repeatedly polluted along its course by many sources of pollution: agricultural drainage carrying pesticides, herbicides and fertilizers;

industrial wastes from tanneries, petrochemicals and fertilizer complexes, pesticide factories, mills producing rubber, jute, textiles and paper, and distilleries; and untreated municipal wastes. The water of the river is in heavy demand at the same time for irrigation, domestic purposes, industrial use, navigation, and religious purposes. The problem of maintaining the quality of the river water is enormous and requires considerable planning, investment, effort and political will. The problem is accentuated at times by the seasonal nature of the river and at reaches below the points of withdrawal of water for irrigation. The Ganga clean-up project started in 1986 but its implementation involved enforcing the industrial establishments to process their effluent properly, and also treating before discharge the wastewater of hundreds of cities, some of which have a population of hundreds of thousands (DasGupta, 1984; Agarwal and Narain, 1986). This is a mammoth task and progress has been slow.

The common pollutants of water are described in Table 7.5. Given enough time and flow, rivers can cleanse themselves of some of the pollutants. The oxygen-demanding wastes are the best examples. These organic wastes are utilized as food by bacteria and protozoa present in the water. This is done using the oxygen dissolved in the water, which, depending on ambient temperature and salinity, ranges between 8 and 15 mg l^{-1}. The amount of oxygen necessarily drops below points of entry of organic wastes, such as the outfall pipes of municipal sewage, as decomposition of the organic wastes uses up the dissolved oxygen. The difference between the amount of oxygen necessary for this decomposition and the amount actually present in the river is known as the *biochemical oxygen demand* or BOD. It is used as a surrogate measure of water pollution. Another measure of the oxygen demand is the *chemical oxygen demand* (COD), which is defined as the amount of oxygen necessary to oxidize the waste material chemically. The third standard indicator of water quality is the faecal coliform bacteria count.

The level of the oxygen drops downstream of the source of the waste, but oxygen is also taken up by the water from the air (reaeration). If plotted against distance downstream of a point source of waste, the level of oxygen therefore shows a concave-upwards curve, described as the oxygen sag curve (Figure 7.4). Most fish require a dissolved oxygen (DO) content of around 5–8 mg l^{-1}, below which only certain types of fish (such as carp) and aquatic fauna may survive. As the amount of DO drops, not only do the aquatic organisms change, but the water turns foul and smelly. With the downstream rise in DO, the quality of water is restored and the previous aquatic fauna, including various kinds of fish, returns.

It is therefore essential that (1) the organic wastes are treated before their discharge, and (2) repeated addition of such wastes should not take place while the oxygen content of the water is returning to an acceptable level. The urban wastewater should be treated properly before its release. The standard technique for this is described below. The standard practice for measuring the BOD involves testing the oxygen demand over five days in the laboratory. This is really only an indicator, as a complete recovery test for the oxygen in water would take several weeks. In five days, however, nearly 90 percent of the total oxygen required is consumed. The five-day test (BOD_5) determines the amount of oxygen used up by the micro-organisms in water in the first five days of recovery. Basically it involves the following:

Table 7.5 *Major pollutants of water: summary observations*

Type	Source	Effect
Oxygen-demanding wastes	Biodegradable organic material from sewage and industrial wastes (food processing, paper mills, oil refining, tanning), decaying vegetation, agricultural runoff	Drop in dissolved oxygen, stress and even death for fish and other aquatic organisms, foul odour, taste and colour of water
Pathogens	Untreated organic wastes and domestic sewage	Spread of waterborne and water-contact diseases of gastro-enteritic tract, infectious hepatitis, polio, schistosomiasis
Nutrients (mainly C, N and P)	Wastewater discharges, industrial wastes, chemical fertilizers, phosphorus from detergents, mining	Algal blooms, aquatic weeds, loss of oxygen, degradation of colour, odour and taste of water
Synthetic organic compounds	Pesticides, herbicides, detergents, fuels, plastics, fibres, etc.	Toxic effects, suspected carcinogens and mutagens, growth of algae and aquatic weeds, foul odour
Petroleum-related material	Discharges from machines and vehicles, tanker spills, pipeline breaks	Ecosystem disrupted, aesthetic degradation of environment
Inorganic chemicals	Mineral acids, metals and metal compounds, especially lead and arsenic, urban runoff	Increased acidity, salinity and toxicity of water
Radioactive material	Uranium mining, nuclear power plants, weapon testing	Carcinogens and mutagens
Thermal pollution	Released effluent used for cooling industrial and power plants	Usually harmful to aquatic life, decreased O_2 solubility, increased chemical reactions
Sediment	Deforestation, poor agricultural practices, mining, urbanization	Deposition in channels and reservoirs, increased turbidity, harmful for certain aquatic life

1. taking a water sample from the polluted part of the river
2. putting it in a 300 ml bottle which is then filled with water
3. keeping the temperature constant at 20 °C
4. measuring the difference in DO between the start and the end of the test five days later

There are three standard steps for controlling pollution in municipal wastewater (effluent) before discharging it into a body of natural water, and, not surprisingly, they are known as the primary treatment (removes mainly solids), the secondary treatment (removes mainly biological wastes), and the tertiary treatment (special processes, used on *ad hoc* basis).

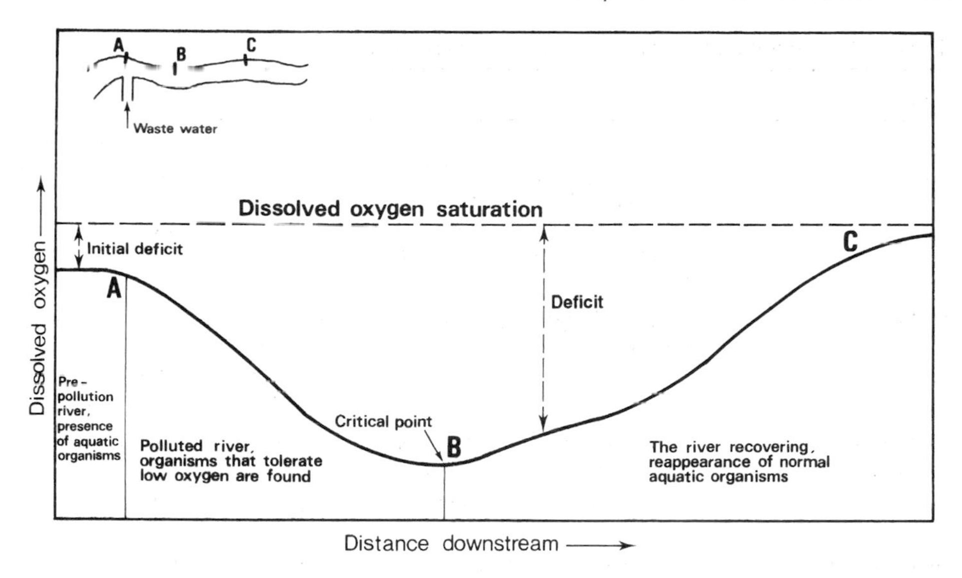

Figure 7.4 *The oxygen sag curve*

- *Primary treatment.* The wastewater is first passed through a screen to remove solid material that could be burnt or used as landfill. The water then enters a circular or rectangular settling basin where the suspended solids are precipitated as sludge to the bottom and removed. At this stage, most of the solid material but only a fraction of the biological wastes have been removed.
- *Secondary treatment.* The removal of the biological wastes can be carried out by either of the two following techniques. In the *activated sludge process* the water from the settling basin is passed to an aerated tank where it comes into contact with a suspension of micro-organisms, protozoa and metazoa. The dissolved organics are thus removed as bottom sludge; a process which takes between four and eight hours. The water then enters a sedimentation tank where the thick suspension at the bottom is removed and the clearer water from the top is then passed on for the tertiary treatment. Most of the thick bottom suspension is returned to the activated sludge basin for recycling. In the *trickling filter process* the water after primary treatment is passed through a round concrete basin filled with crushed stones with large aerated voids in between. The stones are coated with a biological slime which removes the organic wastes from water as it trickles through the stone filter. The flow is then taken to a sedimentary tank for settling out of solids and slime masses. The thicker bottom material is sent back to the primary treatment tank.

At this stage the water is free of nearly all solids and organic wastes but other polluted matter such as nitrogen, phosphorus, pesticide remnants, toxic metals, synthetic organic compounds, etc., remains. This requires the next stage.

- *Tertiary treatment.* Tertiary treatment is expensive and not universally used. It is a set of specialized techniques that deal with water which carries a particular type of pollution, such as nitrates or phosphates.

Water is also disinfected from disease-carrying bacteria by chlorination, and chlorine is added to the effluent before it is discharged into a river, lake or sea. A properly treated effluent will, of course, have a low BOD. The described method for treating water is the basic technique and variations of it exist, depending on the nature of the pollution and the amount of finance available. Details are available in standard environmental engineering books (Fair, Geyer and Okun, 1968; ReVelle and ReVelle, 1974; Masters, 1991). At the other end of the scale, it is possible to use low-cost waste treatment techniques, either as used in Calcutta (Chapter 10) or on an individual basis such as septic tanks. Successful water management requires both supplying settlements with clean drinking water and properly treating wastewater from settlements and industries before releasing it into a natural body of water.

Wastewater from settlements and industries reaches a river at specific outfalls or points along its course which are described as point sources of pollution. Agricultural runoffs carrying pesticides, herbicides and fertilizers tend to reach streams over an area, and as such are known as non-point sources of pollution. Given the prevalence of their use, it is not amiss to examine what type of polluted material is reaching the rivers or lakes from agricultural fields. We use pesticides as an example to illustrate this type of hazard.

Many types of pesticides exist. The three major groups are

- chlorinated hydrocarbons
- organophosphorus compounds
- carbamates

Inorganic pesticides containing copper, lead, arsenic, mercury, etc., are also in use. The use of pesticides has given rise to three types of problems:

- cumulative residues in water, soil, air and organisms
- increased concentrations higher up the food chain
- increased and essential use of pesticides in modern agriculture

Chlorinated hydrocarbons (composed mainly of carbon, hydrogen, oxygen and chlorine atoms) are the earliest synthetic chemical pesticide developed. DDT is the most common and still widely used member of this group, which also includes aldrin, dieldrin, chlordane and heptachlor. These are very effective pesticides but the problem lies in the very factors which make them effective: their persistence and fat solubility. DDT, for example, has a half-life of 20 years or more, and it collects in soil, in air, and in the fatty tissues of organisms of all sizes and types. Organophosphorous compounds are toxic because they combine with the enzyme cholinesterase and affect the nervous system. Unlike chlorinated hydrocarbons they do not accumulate in the soil or the human body, but because of their high toxicity, even a small dose such as 0.068 g could be a poisonous dose. Use of these compounds therefore requires extreme safety precautions. Carbamates are chemical compounds from carbamic acid and they also combine with cholinesterase, but the combination rapidly breaks down and therefore a toxic dose needs to be a large one.

These days, contamination of water by pesticides occurs by direct spraying, from sewerage effluents, from rainfall, and finally as runoff from agricultural fields. The amount could be very high, as indicated by measurements from the rivers of the developed world. In the developing world, the use of pesticides is spreading, including, alarmingly, those pesticides banned in the developed countries. Other compounds that pollute water include herbicides, plastics, polychlorinated biphenyls (PCBs), detergents, and chlorine compounds. Certain herbicides, e.g. Agent Orange (2,4,5-T), could be carcinogenic, teratogenic or mutagenic. PCBs are found in lubricants, hydraulic fluids, asphalts, plastics, adhesives, etc.

Industrial wastewater is a common polluter in the developing world not only by raising the temperature of the rivers and lakes, but also because of the unregulated discharge of effluents containing toxic metals and metal compounds. At times, it is released into shallow groundwater which is often used as a source of drinking water in urban areas. Even in rural areas, groundwater should be checked from time to time, as the example from the basin of the Hooghly River (a distributary of the Ganga) near Calcutta indicates.

This area saw an increase in demand for water following the farming of quick-yielding crops of the Green Revolution and a rise in population. Part of the demand has been met by putting into operation pumps, locally known as deep tubewells, which bring water up from sand aquifers at depth within the Quaternary clayey alluvium of the Ganga River. Unfortunately, due to their geological history, such sediments contain arsenic in soluble form, the source of which is probably old volcanic rocks upstream. The water that is brought up and used therefore contains toxic doses of arsenic, which has led to serious arsenic poisoning of a large section of the community, as shown by Dipankar Chakraborti and his colleagues (School of Environmental Studies, Jadavpur University, 1994), both in the state of West Bengal, India and in Bangladesh. As revealed at an international conference in Dhaka in early 1998, the number of people using arsenic contaminated water in Bangladesh is in tens of millions.

Such cases have been reported from other parts of the world but when it happens in a densely populated rural area, a solution is almost impossible to find, as it is neither possible to relocate the inhabitants nor to supply them with safe water in sufficient amount.

7.6 Conclusions

Managing water resources is an extremely important sector of environmental management. The demand for water of sufficient amount and required quality increases progressively with increasing population and affluence of the people. Large sectors of the world are not supplied with clean drinking water, nor is the wastewater treated adequately before its release. This is reflected in the health of the population, especially of the children. Large-scale engineering projects may solve some of the various types of water-related demands, but such projects need to be extremely carefully evaluated and monitored because of their high potential for environmental degradation. The dilemma could be solved locally by small-scale projects which involve the participation of the local community, as in the *subac* organization of Bali where water for agriculture is supplied via small-scale irrigation channels and by community participation. Each farmer who owns at least one ricefield has to be a member and receives water in proportion to his holdings. *Subac*, which is really a rural agricultural society, collectively determines the distribution of irrigated water among its members. The society also assures the farmers of a water supply, guards against pilferage of irrigated water, and is responsible for repairing damages to the irrigation dikes and channels. This being Bali, *subacs* also arrange community banquets at the completion of a harvest. The paddy-fields of Bali on steep volcanic slopes have been supplied with water in this fashion for centuries without environmental degradation. Unfortunately, for the rest of the world the problem of environmental degradation by overuse and the use of low-quality water will continue into the future.

Exercise

1. The Sungai Lumpur Basin is located in a developing country. A large earth-filled dam will be constructed on the river to provide hydroelectric power for a number of planned future industries and the capital city located about 100 km away. The background information is given below. The basic characteristics of the basin are shown in Figure 7.5.

   Sungai Lumpur Basin

Area	196 km^2.
River length	68 km.
Relief	Land over 1000 m is generally mountainous. The central part of the basin has a rolling topography.
Geology	The higher areas are underlain by a suite of clastic sedimentary rocks: conglomerate, sandstone and shale. The lower central area has a thin alluvial fill over sandstone. The alluvial fill is wedge-shaped, and increases in thickness in the downstream direction.
Climate	Seasonal humid tropics. Annual rainfall about 1500 mm. Most of the rain (about 80 percent) comes from the southwest monsoon (June–September). Infrequent tropical storms bring in spells of very heavy and intense rainfall, usually in August and early September.
Population	About 280 per km^2 in the central area. Very low density in the mountains.
Land use	The mountains are still under forest although signs of past disturbances are common. Recently, timber concessions have been

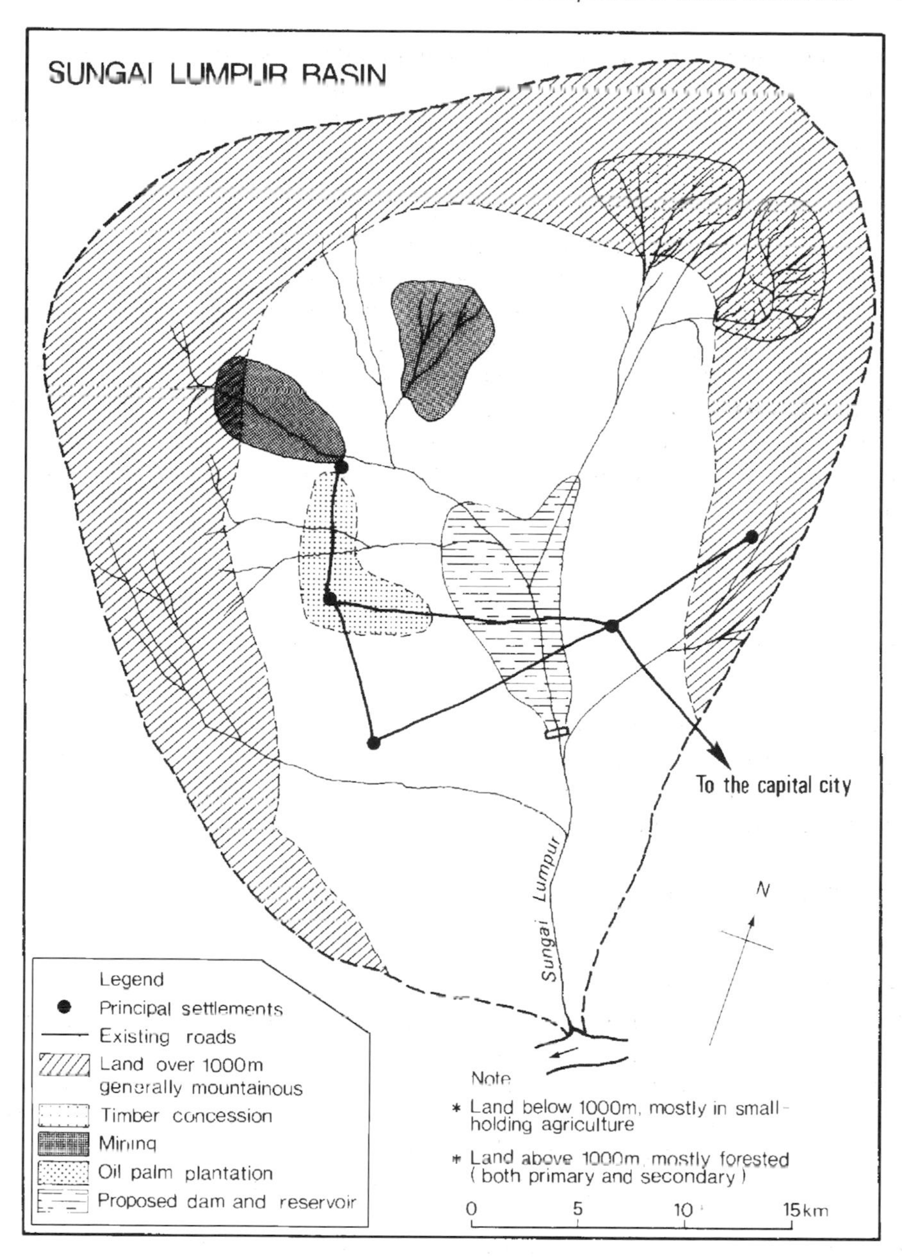

Figure 7.5 The Sungai Lumpur Basin

granted to a Japanese and a Malaysian timber company to earn foreign exchange. Two opencast coalfields occur in the basin. Oil palm plantations are fairly new and localized. The rest of the central area is under small-holding agriculture. River water is used for irrigation in the lower third of the basin.

Project	The project consists of a 40 m high earth-filled dam and a 60 km^2 reservoir behind it. About 500 MW of power will be generated. The project has been planned in haste and now the donor agencies are asking for a feasibility study. The local NGOs working with the farmers who will be displaced are demanding an Environmental Impact Assessment and compensation.

Evaluate the environmental problems that the project will create. Use the data supplied and the sketch map. If you were involved with the feasibility study, what would have been your recommendations?

2. Select a city you are familiar with. Find out the nature of its water supply including sources, quantity and quality. Do you think the supply is adequate to meet the various demands of the city?
3. How does the city of your choice dispense with wastewater?
4. Select a river flowing through a settled region. What are the different types and amounts of pollution that are discharged into this stream? How does it affect the quality of the water? Draw a map or a flow diagram of the river and on it plot in a summary form the sources of pollution and polluted reaches.

8

Energy and Development

8.1 Population, Prosperity and Energy

All human activities require energy. The nature of the energy used varies from simple physical efforts to the use of the following:

- biomass
- solid fuels (coal, lignite and peat)
- petroleum products and natural gas
- electricity generated from coal, oil, gas and running water
- nuclear power

Other sources of energy, such as geothermal heat, wind, tides and solar heat, are less common, but could be significant in a suitable environment. The demand for energy rises with population and also with prosperity. Industrial and prosperous nations such as the Organisation for Economic Co-operation and Development countries (OECD countries: Australia, Austria, Belgium, Canada, Denmark, Finland, France, Germany, Greece, Iceland, Ireland, Italy, Japan, Luxembourg, the Netherlands, New Zealand, Norway, Portugal, Spain, Sweden, Switzerland, Turkey, UK and USA) tend to use a very large amount of energy per capita. Energy use in the transportation sector of the OECD countries nearly tripled in the 30 years between 1960 and 1990, the fastest rise being in the area of air transport and road freight. Gasoline vehicles (mostly passenger cars) use most of the energy used in road transport. This has been associated with a sharp rise in the use of oil. In 1990, 43 percent of the OECD total primary energy supply came from oil, the share for the transport sector being 60 percent of the final consumption of oil products. Put another way, oil provided 92 percent and more than 99 percent of transport energy use in 1960 and 1990 respectively. This happened in spite of efforts by various OECD governments to substitute other fuels for oil (International Energy Agency, 1993).

In contrast, the energy consumption of the developing countries is low, even for those with very large populations, such as China and India. These countries are not yet highly industrialized, plus a majority of their residents cannot afford energy-consuming equipment such as refrigerators, central heating, or automobiles. In these countries the demand for energy concentrates on generating electricity for lighting, acquiring fuel for running public vehicles including trains, powering pumps for lifting water for irrigation, and running large refrigeration units for food preservation. The per capita energy use in the developing world is less than a sixth of that of the developed countries (World Resources Institute, 1994).

The rise in commercial energy consumption in this century has been phenomenal, a tenfold rise leading to a consumption of slightly above 320×10^{18} J against an annual global production of slightly less than 340×10^{18} J. The 1991 global per capita commercial energy consumption was 59×10^{9} J, which hides the striking differences in per capita energy consumption across the globe (World Resources Institute, 1994):

- 12×10^{9} J in Africa
- 136×10^{9} J in Europe
- 320×10^{9} J in the United States

Four factors were primarily responsible for this rise in commercial energy consumption:

- increases in population
- increased energy-intensive consumption
- rapid industrialization
- the growth of mechanical transport

This rise has serious consequences as both production of energy by conventional means and its utilization are directly associated with environmental degradation. The degradation may arrive in many ways:

- deforestation for fuelwood (Chapter 5)
- pollution of air (Chapter 9), land and water by both accession and burning of solid fuels and petroleum derivatives
- environmental problems associated with large hydroelectric projects (Chapter 7)
- release of radioactive matter

The last of these has the potential to be extremely dangerous, even when from a single plant.

The accumulated pollution from solid fuel utilization has given rise to the deposition of acidic material in soil, water and vegetation. The total accumulation of CO_2 since the time of the Industrial Revolution is responsible for warming the global climate, with associated consequences (Chapter 14). In the developed nations, the rising cost of energy (especially since 1973) and progressive environmental degradation in combination led to an arrested rise in energy demand, achieved by widespread use of energy-conserving technology and energy substitution. Such technologies could be costly and are not common in the developing countries. But as these countries become populated, and as some of them become relatively prosperous and partly industrialized, they also suffer from the consequences of energy-related pollution. As wind and water driven pollution from a country knows no national boundaries, the effects could accumulate and diffuse to be degrading on a global scale. It is therefore necessary to examine the types, production and utilization of different types of energy in order to understand (1) problems of environmental degradation and (2) strategies for environmental management.

8.2 Energy Sources and Their Distribution

The different types of energy can be broadly classified into two groups: non-renewable and renewable. Non-renewable energy is defined as energy from sources which are not being regenerated. In contrast, the renewable type of energy comes from sources which are either constantly renewed or not used up or transformed in the production of energy. Examples of the two types are given in Table 8.1.

Table 8.1 *Different sources of energy*

Non-renewable types	Renewable types
Petroleum and its derivatives	Hydroelectricity
Natural gas	Biomass
Coal, lignite, peat	Geothermal heat
Oil shale, oil sand, etc.	Wind
Nuclear fission	Tidal
	Solar

Note: Technical advances may produce other sources of energy in the future. For example, recent experimental studies have shown that sugar could possibly be an energy source in the future.

In theory, the non-renewable sources of energy could be replenished, but their formation is so slow that on a human time-scale they can be taken as spent. Processes of coal formation, for example, still happen on earth, but it will probably take hundreds of thousands of years for decomposing vegetation to transform into coal. In a review of the energy of the world, Putnam in 1953 used the terms 'energy capital' and 'energy income' to refer to non-renewable and renewable energy sources respectively. These terms are excellent analogues for conceptualizing the problem of depleting energy resources.

At present, about 90 percent of the total energy consumed comes from oil, coal and natural gas, with oil meeting more than a third of the global energy demand. It is estimated (World Resources Institute, 1994) that the contribution of oil and natural gas will be severely diminished in the twenty-first century. Coal, hydro, nuclear and other types of energy (which are currently considered non-conventional) will be used beyond that date to meet the world demand.

Coal

Coal is derived from dead vegetation accumulated in layers, and decomposed in primarily anaerobic conditions in the middle of sedimentary deposits such as sand and mud. Sand is lithified to sandstone, mud to shale, and the dead vegetation to coal. The vegetation is commonly that of extensive tropical to mild-temperate swamps, and after the dead plants are deposited, the anaerobic decomposition increases the concentration of carbon and decreases the proportion of oxygen. The product of the first stage of conversion from decomposing vegetation to coal is known as peat: a moist, brown layer, which comprises at least 50 percent carbon, and is often found in wetlands. Further decomposition, dehydration, and pressure from sedimentary layers above, convert peat to the brown-coloured lignite, in which the carbon content is about 70 percent. Lignite is sometimes referred to as brown coal. Further alteration produces bituminous coal, with 80 percent or more carbon. The rest could consist of various materials including sulphur, phosphorus or ash, which also burns when coal is ignited. Some of these impurities give rise to air pollution. For example, a coal deposit containing sulphur is a potential source of SO_2 in air, which in turn produces deposits of acidic material on land or water or vegetation. With water it produces H_2SO_4, giving rise to the infamous acid rain. Bituminous coal is the most commonly used variety. Anthracite coal is coal with a maximum concentration of carbon (up to 98 percent).

The origin of coal indicates that the coal reserves of the world were formed a long time ago. Most of the coal was formed either in the aptly named Carboniferous (345–280 million years BP) or Cretaceous (140–65 million years BP) time periods. Tertiary (65–2 million years BP) lignites are common, and peat is currently forming in various parts of the world. Coal is mined either by removing the surficial deposits covering its bed (opencast mines) or, where coal beds are found at depth, by subsurface mining using shafts and galleries. Although coal is widely distributed on earth, among the developing countries large reserves are located only in China, India, Mongolia, South Africa, Botswana and Colombia. Although some other countries such as Iran, Mexico, Brazil, Chile, Zimbabwe, Swaziland and Madagascar have some deposits of varying quality, most of these countries are dependent on imported sources of energy. This imported type of energy is almost always oil, not coal.

Oil

Petroleum is formed from minute organisms which accumulate on the floor of seas or inland water bodies, and are converted in oxygen-poor environments to oil. The particles of oil then flow and accumulate in sedimentary rocks with pores, such as limestones or sandstones. The accumulation builds up under certain circumstances:

- the oil is prevented from moving upwards by a layer of impervious rock
- structural features such as folds in sedimentary rocks provide suitable places for the oil to accumulate
- oil flowing through the porous rock is trapped by a fault, or a stratigraphic barrier

In such places oil concentrates, and could be collected by putting in drill holes and pumping the oil out. The demand for oil has led to remarkable advances in petroleum prospecting and extraction. Oil is collected from below sea beds using offshore drilling platforms (e.g. in the North Sea, the Gulf of Florida, and the Arabian Sea near Bombay) and from hostile environments such as the frozen Arctic (e.g. Alaska and Siberia) or harsh deserts (e.g. West Asia and North Africa).

Outside the countries such as the United States, Canada, the Commonwealth of Independent States (CIS), and Britain and Norway (both of which use the North Sea oil), the major fields are located in West Asia, North and West Africa, the Caribbean, and Southeast Asia. The countries that produce oil in large quantities are Algeria, Libya, Nigeria, Mexico, Venezuela, China, Indonesia, Iran, Iraq, Kuwait, United Arab Emirates, and above all, Saudi Arabia. Most of the latter group of countries belong to the Organization of Petroleum Exporting Countries (OPEC), which collectively attempts to control the production and prices of most of the oil which comes to the world market. For example, the price of oil went up dramatically twice in recent years (1973–1974 and 1979–1980), events which are known as oil shocks. Prices have come down since then.

The chief markets for imported commercial petroleum derivatives include USA (in spite of its large oil deposits), countries of Western Europe (excluding Britain and Norway), and Japan. Many countries of the developing world have little alternative but to import oil as a chief source of energy, which accounts for a large proportion of their spending on imports. The combination of oil fields being concentrated in certain regions and yet the demand for energy being global has led not only to economic problems but also to problems of environmental degradation. The latter occurs in oil extraction, refining and transport.

The world supply of both oil and natural gas is expected to be severely depleted towards the end of the twenty-first century, and new sources (extraction of oil from which is not economically feasible at present) are being actively investigated. Two of these potential sources are called oil shales and tar sands. Oil shale is a generic term for a variety of fine-grained laminated sedimentary rocks which include an organic compound called kerogen which yields oil. Oil shales are widely distributed but only a few deposits can be considered economically feasible for oil extraction. Again, such extraction, it is believed, will raise a host of environmental problems. Tar sands, in contrast to oil shales, are porous sands where the pores are filled with a nearly solid hydrocarbon called bitumen.

Natural gas

Natural gas, a mixture of mostly methane (CH_4) with other hydrocarbons such as propane (C_3H_8) and butane (C_4H_{10}), is usually but not always found in the subsurface above oil. From a natural gas field, methane is dehumidified, cleaned of impurities, and pumped under pressure to pipelines. Propane and butane are collected and stored under pressure in tanks as liquefied petroleum gas (LPG). Such gas tanks can be used to supply areas not served with piped gas. To meet long-distance demands, natural gas is shipped at very low temperature in a highly inflammable form, known as liquefied natural gas (LNG).

The occurrence of natural gas is widespread, but the countries with the major gas fields (ranked according to the size of significant reserves) are Iran, the United Arab Emirates, the United States, Qatar, Saudi Arabia, Venezuela, Algeria, Norway, Canada, Nigeria, Mexico, Indonesia, the Netherlands, Malaysia, Australia and Kuwait (World Resources Institute, 1994). It is a cleaner fuel than coal or oil, with fewer environmental problems. It is used mainly for domestic purposes, although there is some demand from industry and also for the production of electricity.

Nuclear power

Nuclear fission is the first step towards the generation of electricity in a nuclear power plant. Nuclei of atoms such as uranium-235 and plutonium-239 are split by bombarding them with neutrons, at which point a small amount of matter is converted into energy, mostly into high-temperature heat. The plant contains, inside a protective shield, a core with rods of fissionable uranium and retractable rods of neutron-absorbing material which controls the rate of fission (Figure 8.1). The velocity of the neutrons used for fission is controlled to an optimal level by graphite packing of the core or by filling the core space with ordinary water, the latter type (light-water reactors) being more common. Almost all electricity-generating nuclear power plants are light-water reactors. The heat from the reactor is used to produce steam, which in turn drives turbines that generate electricity. The uranium-bearing rods are replaced from time to time, and the used rods (spent rods), being radioactive, are stored very carefully. The United States currently produces almost a third of the global production (World Resources Institute, 1994). Countries that are known to have large deposits of the source material, uranium, are Australia, South Africa, Niger, Brazil, Canada, USA, France, India, Spain, Algeria, Gambia, Argentina and the Central African Republic.

Nuclear power is clean so long as there is no leakage or the nuclear waste rods do not need replacement. Disposal of nuclear waste is an expanding problem, as the spent rods have to be kept sealed for hundreds of thousands of years until their radioactivity diminishes significantly. Salt mines, deep underground, have been used to store nuclear wastes, as thick beds of salt provide a good shield against radioactive diffusion, but there is always the threat of contaminated groundwater seeping out into wells or mineshafts. Other storage places planned include putting the wastes in steel drums and depositing them under water in subduction trenches to disappear hopefully into the earth's mantle, or dumping them at the base of the Antarctic ice sheet. None of these guarantees safe storage without leakage or reappearance. Shooting canisters of radioactive wastes into outer space has also been considered. Disposal is not an easy problem to solve. The plan for depositing US nuclear wastes underground at the base of the Yucca Mountains in Nevada, proposed in 1987, is still held up by reports of geological faults and the reluctance of the state of Nevada to accept the waste.

The problem of disposal of nuclear wastes is a gigantic one for the United States, which has to safely dispose of about 30 000 t of spent fuel rods from power plants and a further 380 000 m^3 of high-level radioactive wastes from activities related to nuclear weaponry (Whipple, 1996). As Whipple points out, the term 'safety' is

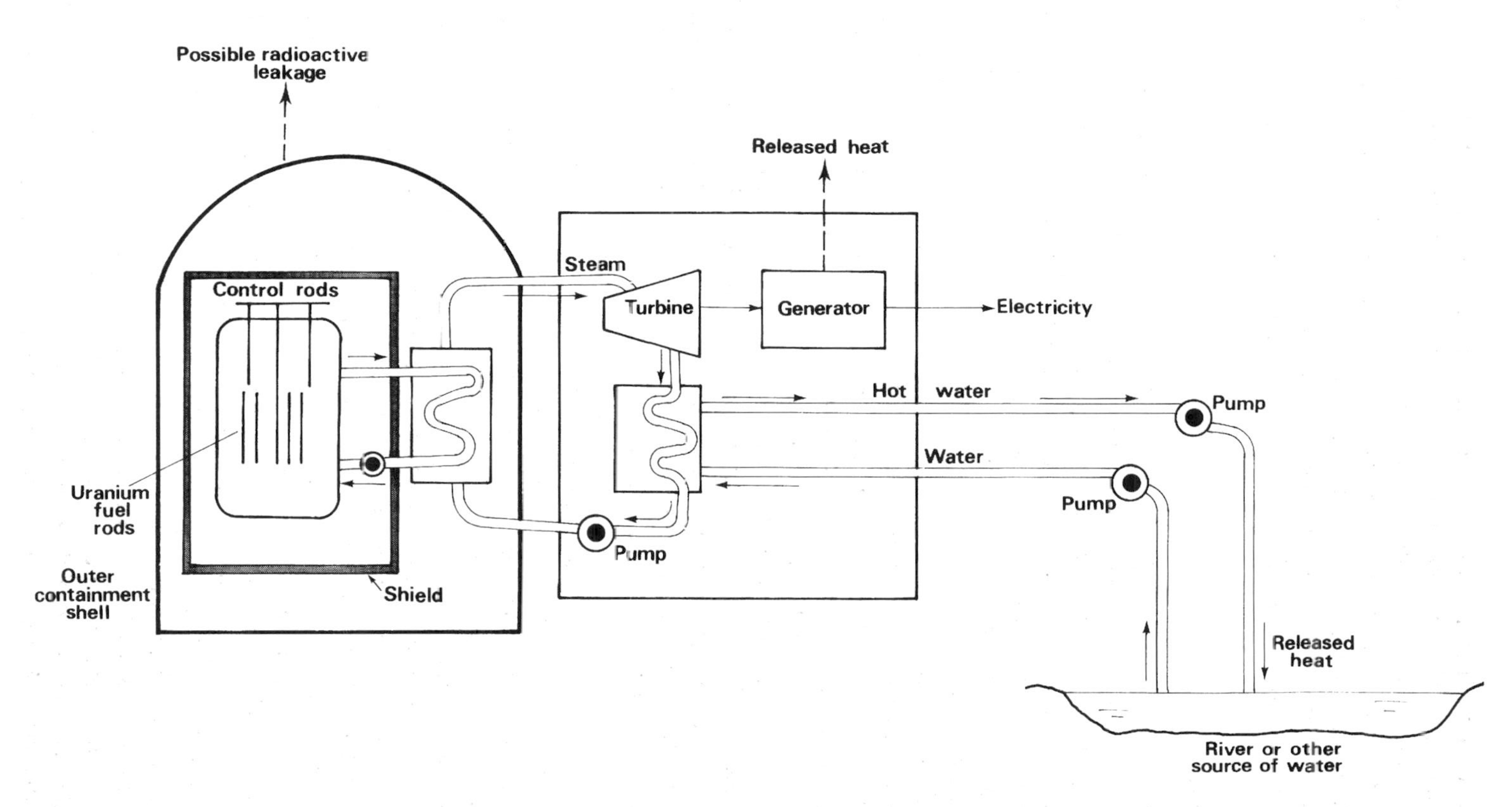

Figure 8.1 *Simplified design of a light-water reactor. Adapted from G. Tyler Miller, (1996)* Living in the Environment, *9th ed., © Wadsworth Publishing Co., Belmont, California. Used by permission*

inadequately defined at least in terms of duration of storage. The Environmental Protection Agency's guidelines demanded 10 000 years of safe storage. The National Academy of Sciences has recommended that the waste be stored until the hazardousness declines. The storage could be very long-term (Whipple, 1996). The original plans were to create a repository of nuclear wastes below the Yucca Mountains in the state of Nevada. There the spent fuel rods were to be stored in canisters arranged horizontally in chambers dug into the rock 300 m below ground surface and between 240 and 370 m above the water table. When full, the repository was to be monitored for 50 years and then sealed.

Certain problems remain to be resolved:

- The state of Nevada is opposing the storage within its territory.
- It is possible that the repository would not have enough space to contain all the fuel rods and the military nuclear waste.
- The repository needs careful planning in view of the old geological faults of this area.
- The canisters have to be corrosion-proof and no leakage to groundwater should be possible. This being a very dry place helps.

The scale of cleaning up the nuclear weapons complexes is even bigger. For example, deactivating and clearing up one storage site located at Hanford in the state of Washington started in 1989. Since then a total of $7.5 billion has been spent on cleaning up Hanford. It is expected that the job may take another 40 years, during which the bill would be around $1 billion every year. Hanford is only a segment of the total programme in the USA to ameliorate the hazards related to the nuclear weaponry since the cold war has ended (Zorpette, 1996). The job is extremely dangerous and there is no alternative.

Some countries still dispose of material that releases small amounts of radioactive wastes as landfills or into the oceans. There have been incidences of radioactive wastes from developed countries turning up in poorer countries of the developing world. For example, the South Pacific, because of its isolation and vastness, has for some time been used as the dumping ground for nuclear wastes of several developed nations.

Certain countries such as France or Japan generate much of their electricity from nuclear plants, but the continuing incidents of nuclear disasters which peaked with the Chernobyl disaster of 1986 have caused a number of countries to revise their options. Siting of new nuclear plants, such as the one occasionally considered by Indonesia for the seismically active island of Java, should therefore be a ground for concern.

Renewable energy sources

The most important renewable energy source is hydroelectricity, which accounts for about a fifth of the total energy produced world-wide. Water from the reservoir behind a dam is taken through large pipes and allowed to fall through a controlled height to turn turbines and produce electricity. Large hydroelectric plants are usually located where the channel slope is steep, such as a river gorge or a waterfall. A combination of rivers and high relief is therefore ideal for generating hydroelectricity. Norway and Switzerland, for example, use hydroelectricity to meet a substantive part of their energy demand. Globally, however, only about 15 percent of the total exploitable hydroelectric potential was in use in 1990 (World Resources Institute, 1994). Large plants are not always necessary. Small and medium sized plants, usually more environmentally friendly, generate electricity next to small rivers. The energy generated is often enough for rural electrification. This is a renewable source of energy and the level of pollution is insignificant. The problems that arise with dams across rivers are discussed in Chapter 7.

Energy from biomass is used for both cooking and space heating. The collection and burning of fuelwood and the associated environmental degradation are discussed in Chapter 5. Crop residues such as bagasse (the residue left after the harvesting of sugar cane), coconut shells and cotton stalks are burnt to produce heat, and in rare cases, electricity. Vegetative matter, organic wastes and sewage are converted inside a large container called a digester, by anaerobic bacteria and other chemical processes, into liquid and gaseous fuels used for cooking or space heating.

Box 8.1
Biogas in the Developing World

The demand for electricity at the village level has led to the use of biogas in the developing world, particularly in China and India. Biogas is produced in a digester where organic material is fermented to produce electricity, and the residue from the digester can be used as fertilizer. For example, cattle dung mixed with water can produce a mix of methane and carbon dioxide. The gas, after passing through a condensation trap, can be mixed with diesel fuel to operate a generator. The electricity produced can be used to pump out underground water, to light up village homes and streets, and in cooking. Smaller family-sized digesters are more common, and in China, about 4.6 million such digesters are estimated to be in use, especially in the south. In India, the number is smaller but growing (World Resources Institute, 1994). Bigger community-sized digesters are more economical but require village level organization which is not always available.

The biogas system is environment-friendly. It is inexpensive, clean, produces hardly any pollution, and provides employment for the local people. If there is enough demand for the electricity, it is also an economically viable project. The number of biogas digesters looks likely to increase steadily in the developing world and biogas has the potential to become a significant source of energy in the villages of the developing countries. This is an example of an innovative measure meeting the demands of development without causing significant environmental degradation and at an affordable cost.

Besides hydroelectricity and biofuels, the type of renewable energy with low pollution that is progressively coming into practice is the harnessing of solar energy. Solar energy can be used for space heating as in a greenhouse where the heat coming in through glass roofs and windows is stored by the walls and the floor. It can be used to heat water in a tank exposed to solar radiation. The more common practice is to use photovoltaic cells to convert solar energy to electricity. Most of these cells (also known as solar cells) are made of silicon, a cheap and abundant earth material. Very small amounts of other substances present in the cells turn these into semiconductors which produce a small amount of electricity from solar radiation. A large number of cells are linked together to build up a reasonable amount of electricity. This technique is becoming common in the tropics and subtropics. Solar cells are used to produce electricity for individual houses (as in Australia) and remote villages (as in India) where the cost of supplying electricity via the usual power lines is prohibitive. The cells have been used for stand-alone phone booths, as in rural Malaysia. The initial cost of installing solar cells in a house is rather high, but such costs are more than made up for by the very low running costs and the lack of pollution. Solar energy is seen more and more as the power source of the future, at least for part of the world.

Certain characteristics in a country's physical geography lead to the generation of renewable energy of special types. Geothermal energy in volcanic regions is generated by trapping heat and steam emerging from underground. This at least is a supplementary source of clean energy for countries located at plate margins such as Indonesia, the Philippines, Japan, New Zealand, Mexico or Iceland, where considerable scope still exists for the utilization of geothermal heat. However, almost half of the geothermal energy produced in the world is generated in the western United States. The cost of electricity production is low, but the pollution from dissolved solids, subsurface gases such as H_2S or NH_4, and toxic substances such as arsenic and mercury needs to be monitored.

Similarly, in windy areas of mountain passes and coastal plains, wind energy can be used to turn turbines for generating electricity. Such turbines are usually placed in groups and described collectively as wind farms. Almost all wind farms are located in the developed countries, chiefly Denmark, USA and Australia, although wind power has established a foothold recently in India. Environmental degradation via pollution is not an issue and the efficiency of the plants is expected to improve, but at the moment wind farms tend to require tax incentives and low-interest loans in order to survive (Miller, 1993).

It is also possible, in areas of very high tidal range, to use the tidal flows to turn turbines to produce electricity. However, only a few sites have the required range. Such plants are located in Canada (Bay of Fundy) and France (La Rance). A few small plants have been built in China. Similarly, it is possible to generate electricity from the pounding energy of coastal waves but the technology has not advanced beyond the experimental stage. Environmental degradation is not an issue but it does not seem likely that either wind farms or tidal flows will be used to produce a significant amount of electric power in the world.

8.3 Energy and the Developing Countries

The developing world suffers from technical inefficiencies, capital constraints, and allotment of subsidies that work against incentives for energy conservation (World

Resources Institute, 1994). About half of the total global commercial energy is produced in the developing world, but only a quarter of it is consumed here. This indicates the presence of a number of oil-exporting nations (listed earlier) whose economy is dependent on oil export. In contrast, a number of developing countries (Gambia, Mauritania, Somalia, etc.) import energy, which negatively impacts their balance of payment and results in their large external debt. The two largest countries, China and India, are nearly self-sufficient in energy because of their coal reserves; India being a modest importer of energy, and China exporting a very small amount of its domestic production (World Resources Institute, 1994).

The rural parts of the developing countries are still dependent on using biomass as fuel in the traditional manner. About a third of the energy needs of these countries are met in this fashion (World Resources Institute, 1994). Biomass, animal and people power, along with a quarter of the total world commercial energy mentioned earlier, provide the developing nations (which make up about three-quarters of the world population) with energy.

Energy consumption in the developing countries has increased rapidly due to demands for power in industry, agriculture, urban settlements, and the general rise in population. The shortage of power in these areas leads to decreased national productivity. Power is also used inefficiently in old industrial establishments, which cannot be updated due to a lack of capital. More power is lost in the developing world between the power station and the customer than in the developed world due to unsatisfactory equipment and pilferage. Subsidies in the developing countries keep the costs of power utilization low and affordable for the poor, especially farmers using electric pumps for irrigation. It has been argued that such subsidies prevent the consumers from being more economical about energy consumption.

In general, the energy problem for the developing nations is not solely in finding a cheap source of energy to meet the demands of economic development. Constructing power plants and transmission lines and transporting coal or oil across the country require huge financial resources and technical expertise which are not always available. The price of environmental degradation (discussed in Section 8.4) drives the costs up even higher. As the energy-related environmental damage is often global, and as most of such global damages has been caused by the huge energy consumption of the industrial nations, it seems logical for the developed countries to transfer clean technologies for energy generation and financial mechanism to the developing world. Even the traditional burning of biomass uses innovative technology (as described in Box 8.1) which can conserve energy significantly. This not only decreases air pollution but also saves the forests.

8.4 Energy-related Environmental Degradation

Serious environmental degradation occurs from the production and consumption of energy for all types of non-renewable energy. Among the renewable sources, the production of hydroelectricity and the burning of biomass could give rise to the kinds of environmental degradation discussed in detail in Chapters 7 and 9. Dams and reservoirs may cause deterioration of the river channel and also of the physical environment in the immediate vicinity of reservoirs and canals. Biomass burning may destroy a significant amount of local vegetation, and release CO_2 and suspended particulate matter. But environmental problems are generated mainly by the mining, preparation, transportation and utilization of solid and petroleum-based fuels. Nuclear power is clean until a leakage occurs, which turns the power plant into an extremely serious environmental hazard. The nature of environmental degradation associated with energy is summarized in Table 8.2.

Large opencast coal mines create environmental problems in the vicinity. The overburden (the layers of material overlying coal seams, which have to be removed before extraction of coal) gives rise to air pollution as fine material is blown off the dumps and remains suspended in the air. So does dust from coal and overburden removed during blasting operations. Pollution of air is further accentuated both by blowing about of fine-grained coal dust and also from the exhausts of the diesel-driven heavy machinery and vehicles that concentrate in the area. Following rainfall, material from the overburden is carried by surface runoff into the local streams, which become overloaded with sediment; the water is often blackened with coal dust. Power stations, which provide electricity by burning coal (thermal power stations), are also sources of pollution. Large (superthermal) power stations are built near the coalfields to utilize large reserves of low-grade coal, as it is easier

Table 8.2 *Energy-related environmental degradation*

Type	Environmental degradation related to		
	accession and processing	transportation	utilization
Petroleum and its derivatives	Seepage, burning of gas in oil fields	Pipe leaks, oil slicks in oceans	Air pollution: global warming from non-CO_2 gases
Solid fuels (coal, etc.)	Mining overburden, quarries	Coal dust	Air pollution: SO_2, CO_2, particulates; contribution to global warming
Nuclear power	Uranium tailings, release of radioactive matter and heat, accidental meltdown, disposal of spent fuel rods		
Hydroelectricity	Dam and reservoir related problems		
Biomass	Deforestation, soil erosion, flooding		Air pollution: release of CO_2 and particulates; contribution to global warming

and more economical to transfer the electricity generated over long distances to the demand areas than it is to transport the low-grade coal. India, for example, has built a number of such superthermal stations with a capacity for generating more than 2000 MW of power.

Thermal power stations pollute the air by discharging the impurities in the coal into the atmosphere via tall stacks. Fly ash, SO_2 and toxic metals are released in this fashion; the extent of the pollution depending on the impurities and the size of the plant. Water in nearby rivers and lakes is polluted by the washing of coal and also by the discharge of warm water after industrial use. Techniques such as electrostatic precipitators and mechanical filters (described in Chapter 9) do exist for reducing air pollution from power plants, but in a developing country are not always in operation. Agarwal and Narain (1986) have discussed the environmental degradation caused by the superthermal power plants of India in the early 1980s. The absence of an account from other countries does not indicate that such pollutive events do not occur. On a positive note, technologies for producing and utilizing clean coal are becoming common in the developing countries. Both China and India, for example, have started to use the fluidized bed combustion (FBC) technique. Small-scale oil spills may occur naturally from permeable rocks that store the oil, but it is the deliberate and accidental discard of oil from storage that causes alarm. Oil tankers flush out their tanks directly into the sea, polluting the sea water. At times, such spills may come ashore. Catastrophic oil spills, extremely destructive to the local environment, are caused either by oil spilling out of damaged tankers at sea or from the accidental blow off of oil wells. The major incidents are publicized, but tens of thousands of small incidents are probably never reported. Details of oil spills and their effects on marine life and coastal ecosystems are discussed in Chapter 11.

Only a small number of nuclear plants have been built in the developing countries. This in a way is fortunate as environmental degradation from nuclear power plants is difficult to control and a leakage has the potential to be disastrous. Nuclear power plants are concentrated in a number of the developed countries. Certain countries (e.g. France and Japan) produce most of their electric power from nuclear sources. The danger of radioactive contamination occurs in every step of the process. It starts with the problem of the safe disposal of tailings from the mining of uranium ores, and also the disposal of waste products from the enrichment plants which concentrate radioactive uranium from the ores for use in the reactors. One well-studied example of the problem of uranium mining comes from the Ranger Uranium Mine of the Northern Territory, Australia. The disposal of the wastes there is unusually complicated due to the location of the mines inside the Kakadu National Park and aboriginal lands.

A nuclear reactor has to be protected inside a specially constructed shell, and a coolant needs to be circulated constantly through the reactor's core to transfer the heat from the fuel rods. This prevents a temperature build-up inside the core, which may lead to meltdown and an explosion. This is basically what happened at Chernobyl in the Ukraine on 26 April 1986, when the explosion blasted off the roof of the reactor building to disseminate radioactive material in the atmosphere which then was carried by the wind over large areas of Europe, reaching distances about 2000 km from the plant.

A number of people died or became seriously ill from the high level radiation at or near the plant. But the effects of the atmospheric diffusion of radioactive material were far-reaching and persisted for a long time. The radiation affected the crops and vegetation of Europe that year, including the milk of grazing animals. The final toll of Chernobyl is still mounting but hundreds of thousands of people have been affected by radioactivity, which leads to cancers, thyroid tumours, eye cataracts, sterility, and finally death. The radiation sickness among the children living around Chernobyl is particularly distressing.

Partial meltdown of a reactor occurred at the Three Mile Island plant in Pennsylvania, USA, on 29 March 1979. If the meltdown had reached completion, its effects would possibly have been of the same order as Chernobyl. On 7 October 1957, a plutonium reactor near Liverpool, England, caught fire. The worst nuclear disaster is suspected to have occurred in 1957 in the southern Ural Mountains of the then USSR, but the details are unavailable.

Using nuclear energy as a source for power also brings in the awesome responsibility of operating proper safety precautions at all times and at high expense. The most demanding part of this is probably the disposal of the radioactive wastes of the reactor such as the spent fuel rods, the filter systems, and the chemical sprayers inside the plant. Disposal and storage of radioactive waste is a very difficult and costly proposition, and parts of such wastes have to be stored for tens of thousands of years. After nearly 40 years of experience, the United States still does not have a comprehensive and integrated policy for storing radioactive wastes and the expenses are becoming impossible. Short-cut methods to meet either budget or technological restrictions, such as dumping low-level radioactive material in steel drums into the sea, is extremely short-sighted and merely postpones paying the price to a future generation.

8.5 Strategies for Environmental Management

The demand for energy in the developing countries is expected to increase rapidly in the future. The direct relationship between energy consumption and environmental degradation via pollution, deforestation, etc., implies that this rise should be controlled. There is a need for higher energy efficiency in order to reduce the demand, and also a need to increase the application of renewable energy to control environmental degradation. Increased energy efficiency can be achieved by using modern technology in industry, as the old industrial technology common in the developing world consumes a lot more energy than that used by factories in the industrial nations for the same amount of production. For example, steel production in India uses at least double the amount of energy that is used in the European countries or Japan (World Resources Institute, 1994). If buildings are properly insulated, far less energy is required for either space heating or air-conditioning. Modern cars need a lot less fuel, but it is hoped that a new generation of vehicles such as electric cars will require even less and that such vehicles would be available in the developing nations as the need for personal transport increases. Use of efficient irrigation pumps and replacing incandescent light bulbs with compact fluorescent lights should reduce electricity consumption remarkably

Box 8.2
The Electric Car

Cars using internal-combustion engines and gasoline are significant polluters of air and users of the petroleum resources of the world. Various techniques have been used to deal with these two major problems:

- curtailment of the demand for vehicles through taxes and levies on ownership and usage
- increases in power efficiency
- reductions in the emission of pollutants
- the use of less-polluting driving systems
- change to a less polluting and renewable fuel
- replacing private transport with public transport

It appears that the introduction of electric cars would go a long way towards solving the environmental problems related to transportation.

Electricity-driven vehicles are those whose wheels are turned by electricity from motors. Such vehicles may use

- batteries, charged from time to time using household or petrol station currents,
- electricity generated in the vehicles themselves,
- electricity stored in devices which are not batteries (Sperling, 1996).

Electric cars are more energy-efficient than conventional cars with internal-combustion drives. Electric cars with fuel cells instead of batteries, which generate energy on board from hydrogen, are even more efficient. Also, air pollution is almost non-existent. There could be a marginal increase in air pollution from plants producing electricity by burning coal or oil as the demand for electricity is expected to rise with the number of electric cars, but it would be a great improvement on present-day transportation-related pollution. Some electric vehicles are known as hybrid vehicles as they combine electric motors and power-storing devices with small internal-combustion engines.

It is now being predicted that electric vehicles of one type or another will significantly replace cars with conventional internal-combustion engines some time in the next century. Driven by pollution control legislation and financial encouragement towards innovations, both provided by governments of industrial countries, a number of major automobile manufacturers are now involved in designing and producing electricity-driven vehicles. Sperling (1996) is of the opinion that the attractiveness of electric vehicles is accentuated in areas with the following characteristics:

- severe air pollution
- vehicles are sought for reliability and low maintenance cost rather than performance
- availability of cheap electricity
- small investment in petroleum distribution

It is even possible that, out of necessity, the manufacturing of electric cars may turn out to be a major industry in a developing country.

Two problems remain with electric cars. First, currently they use lead-acid batteries charged via standard electric sources, i.e. wall outlets. Such battery-driven cars have a limited range of travel before they need recharging. Second, electric cars at the moment are much more expensive than conventional vehicles. Both difficulties appear to be temporary. Future electric vehicles are likely to use devices which have a far better capacity for storing power, or they may produce it on board. Ultracapacitors and flywheels are visualized as efficient storing devices which could be used in electric cars. It is also possible that a hybrid car that is driven by electric power, but which has a storage device charged by a small internal-combustion engine, is the answer. Fuel cells may also turn out to be the optimal solution. Fuel cells produce power by burning hydrogen. The other products of this reaction are water vapour and carbon dioxide (Sperling, 1996).

It is fascinating to wonder what the common cars of our cities will be like several decades from now, how much they will cost, how fast and how far they might be driven, and finally what effect this will have on prolonging global petroleum resources and on city air pollution.

Developing countries would need to improve their capabilities to formulate, implement and monitor a coherent energy policy. The main elements of such a policy should include the following:

- a realistic pricing of energy supplies to provide incentives to control demand
- technology-based regulations
- subsidizing creation and diffusion of clean technologies
- generating public awareness of the need for energy conservation

As these elements may benefit certain income groups and activities and not others, fairness issues also need to be addressed.

The dependence on coal and oil is not likely to disappear but the use of renewable energy should increase in the developing world, as the technology becomes efficient and less expensive. Large-scale hydrothermal plants these days are viewed with apprehension, but relatively easily constructed small-scale plants, called microhydro installations and which can generate up to 100 KW without the need for a reservoir, may increasingly come into operation as a convenient source of power to meet the local needs. Both photovoltaic cells and wind-driven turbines are likely to become common, especially the photovoltaics which at the moment can provide individual households with power and water heating. The initial costs of installing photovoltaics and their restricted use in industry are the main constraints, but the potential is extremely high. Luz International Limited has already supplemented the conventional sources for electricity in California from its solar thermal utility in the Mojave Desert (World Resources Institute, 1990). At the rural end, proper use of biomass (e.g. via gasification) should also be more widespread in the developing world.

9

Changing Air Quality

9.1 Constituents of Air

Air is a mixture of gases that envelops the earth. Almost all (99.999 percent) of its total mass of 5.6×10^{15} t remains within 80 km of the earth's surface. The composition of air within this 80 km is remarkably constant (Table 9.1). This consistency holds even when tiny gas bubbles from deep ice cores are analysed to determine the atmospheric concentration of gases thousands of years ago (Chapter 14).

Table 9.1 *The composition of dry air (selected constituents)*

Constituents	Concentration (percentage by volume)	Approximate total mass (million tonnes)
Major		
Nitrogen (N_2)	78.084	4.22×10^{10}
Oxygen (O_2)	20.948	1.29×10^{10}
Argon (Ar)	0.934	
Carbon dioxide (CO_2)	0.036	
Minor		
Neon (Ne)	0.0018	
Helium (He)	0.00052	
Methane (CH_4)	0.00015+	
Krypton (Kr)	0.0001	
Hydrogen (H_2)	0.00005	
Nitrous oxide (N_2O)	*0.00002*	
Carbon monoxide (CO)	*0.00001*	
Ozone (O_3)	*0.000002*	
Ammonia (NH_3)	0.000001	
Nitrogen dioxide (NO_2)	*0.0000001*	
Nitric oxide (NO)	*0.00000006*	
Sulphur dioxide (SO_2)	*0.00000002*	

Sources: Stoker and Seager (1976) and Berner and Berner (1996). Note: The major pollutants of air included in this list are in italic type. A complete list of such pollutants is given in Section 9.2. A number of the minor constituents vary over space and time. Some constituents, such as carbon dioxide, are increasing.

Nitrogen (78 percent) and oxygen (21 percent) are the two major constituents, which with argon and carbon dioxide make up about 99.99 percent of clean dry air by volume. The rest includes a number of gaseous components, each of which is present in a very small proportion. However, the mass of the atmosphere is so huge that the mass of even a trace element may be measured in millions of tonnes. The proportions of several of these minor components of air vary as they are created or destroyed naturally. For example, from time to time hydrogen sulphide, sulphur oxide and carbon monoxide are added to the atmosphere from volcanic eruptions. Electric discharges during thunderstorms produce oxides of nitrogen. The anaerobic decay of plant and animal remains releases methane, ammonia and hydrogen sulphide.

Table 9.1 represents dry air. Water vapour normally occurs in variable amounts between 0.01 and 5 percent, a figure between 1 and 3 percent being common. The presence of water vapour obviously reduces the concentration of all other constituents, although the proportionality remains constant.

9.2 Sources and Types of Air Pollution

Air is polluted when one or more of its minor constituents are present in sufficient amount to affect the physical well-being of people, animals, vegetation or materials. Such a presence is measured as a concentration, either by volume (ppm) or by the mass of the pollutant present in one unit volume of air ($\mu g\ m^{-3}$). Air pollution is a low level phenomenon, usually confined to elevations below 600 m, and is almost entirely an anthropogenic contribution to the atmosphere. Cities and industrial areas usually release the prime elements of air pollution. Certain physiographic conditions such as mountain-girt basins and cold ocean coasts are favourable conditions for pollution to build up as temperature inversion is common in such areas. Bright sunny days in cities may do the same. The worst scenarios are found when a combination of these conditions is present, as in Mexico City or Los Angeles.

Air pollutants usually occur simultaneously and in various forms (gaseous state, liquid droplets, tiny solid particles), or as a combination of these. Mixing with non-polluted air and dilution of the concentration of pollutants end an episode of air pollution.

Transportation, power plants, and biomass burning are the chief sources of pollution in the developing world. For example, a study of air pollution carried out by the Division of Environment in Malaysia indicated that in 1980 about 82 percent of the polluted material came from transportation, 11.5 percent from burning liquid fuel in boilers and power plants, and the remaining 6.5 percent from the burning of solid and agricultural wastes (Sham Sani, 1987).

Urban air pollution has a long history in temperate countries. Cases of air pollution in London from inefficient burning of coal are on record for AD 852. Several acts of Parliament were passed in the twelfth century to mitigate the hazards of air pollution. In 1661, John Evelyn wrote a scholarly report on the air pollution in London:

> For when in all other places the Aer is most Serene and Pure, it is here Ecclipsed with such a Cloud of Sulphure, as the Sun itself, which gives day to all the World besides, is hardly able to penetrate and impart it here; and the weary Traveller at many Miles distance, sooner smells, than sees the City to which he repairs.
>
> (Evelyn, 1661; quoted in Landsberg, 1956)

Air over Britain and the USA grew seriously polluted in the nineteenth and twentieth centuries. The pollution was essentially due to industrial development and coal burning in winter for domestic heating. In certain years death from air pollution, unusually intensified by meteorological conditions, climbed into four figures, most of the dead being elderly people. As late as 1952, 4000 deaths in four days in London were attributed to air pollution. The SO_2 concentration was more than seven times that which would be admissible today. Apart from London, serious instances of air pollution with multiple deaths were reported from the Meuse Valley (1930), and Donora, Pennsylvania (1948). The air over New York in the 1960s was extremely polluted at times. A distinct difference existed between the polluted city and the countryside:

> At one o'clock yesterday afternoon the fog in the city was as dense as we ever recollect to have known it. Lamps and candles were lighted in all shops and offices, and the carriages in the streets dared not exceed a foot pace. At the same time, five miles from town the atmosphere was clear and unclouded with a brilliant sun.
>
> (Luke Howard on London of 16 January 1826, quoted in Landsberg, 1956)

Paris showed similar variations between the city and the countryside. Besson listed a probability of 0.350 of light and moderate fog over the city in winter (October–March) when for the same period the figure dropped to 0.049 in the country (Landsberg, 1956).

Air quality in the cities of the developed countries has dramatically improved with the implementation of powerful legislation, the use of economic policies and instruments, and technological innovations and control. The spectre of air pollution, however, hangs over many of the urban and industrial centres of the developing world and the transition economies. Air pollution is not necessarily confined to its source. The polluted air is carried by the regional wind pattern across international boundaries to affect the forests and lakes of neighbouring countries. As a result, air pollution requires management and controls at all levels from local to global.

Air pollution affects a number of cities but the types, sources and effects vary. Delhi, Los Angeles, Milan and São Paulo all have their own way of polluting the air. Five types of substances cause more than 90 percent of all air pollution. These five types are known as the *primary* pollutants. These are

- particulates
- sulphur oxides
- carbon monoxide
- nitrogen oxides
- hydrocarbons and photochemical oxidants

A summary of their sources, effects and controlling techniques are given in Table 9.2. The rest of the pollution is caused by *secondary* pollutants, which are formed in the atmosphere following a reaction among the primary pollutants or involving one primary pollutant and one or more minor constituents of air. Lead and ozone, when present beyond a threshold, are also important sources of air pollution. The importance of a pollutant depends simultaneously on its amount and effectiveness.

Particulates

Particulates greatly pollute the air over the cities of developing countries. By definition, particulates are any matter in solid or liquid state which is dispersed through

Table 9.2 *Major pollutants of air – summary observations*

Type	Source	Effect	Control
Particulates	Industry; stationary source of fuel combustion; solid waste incineration; biomass burning; car emissions	Respiratory problems; possibility of lung and stomach cancer, if toxicity is present; loss of visibility	Gas cleaning devices; use of less coal; separation between industry and residential areas; controlled incineration
Sulphur oxides	Fossil fuel combustion, especially of sulphur-bearing coals (mainly for power generation); smelting of sulphide ores especially for copper; petroleum refining	Acid rain damage to vegetation and buildings; corrosion of material; cardio-respiratory problems; impaired visibility	Reduced use and cleaning of coal with sulphur; fluidized-bed combustion
Carbon monoxide	Incomplete combustion of carbon in car exhausts, fossil fuels, and biomass burning; industrial activities	Inhalation leads to loss of oxygen in blood, leading to nerve and brain dysfunction and cardiac problems	Control of vehicle emissions; use of public transport
Nitrogen oxides (chiefly NO and NO_2)	Transportation; stationary source of fuel combustion	Respiratory tract problems; damage to plants; contribution to photochemical smog	Difficult
Hydrocarbons and photochemical oxidants, e.g. ozone, aldehydes and PANS	Car emissions; industrial processes, especially petroleum refining; incineration of solids; organic solvent evaporation	Respiratory problems; eye, nose and throat irritation; plant damage; photochemical smog	Emission control; fuel efficiency
Lead	Car emissions	Behavioural changes and brain damage	Use of unleaded petrol

the air as individual aggregates between 0.0002 and 500 μm in size. They are also referred to as *suspended particulate matter* (SPM) or *total suspended solids* (TSP). Particulates cover a wide range of sizes, chemical compositions, and states. The size of the particles is expressed as the size of spheres with the same settling velocity. This is known as the aerodynamic diameter. Particles with aerodynamic diameters greater than 10 μm settle quickly; those with diameters between 0.1 and 10 μm settle more slowly but are more common. Smaller particles follow a random motion in the air until coagulation increases the size to more than 0.1 μm. The large particles with diameters bigger than 10 μm are produced by mechanical processes such as wind erosion, grinding, pulverizing and spraying. Those between 0.1 and 10 μm originate from fine soil material, industrial combustion and sea salt diffusion in coastal areas. Particles below this size are commonly photochemical and combustion products. Size, composition and concentration determine the effectiveness of particulates as an air pollutant.

The following terms are also used to refer to particulate matter in the atmosphere, the most common of these being aerosols.

Aerosols	Small-sized solid or liquid matter dispersed through the atmosphere.
Dusts	Solid particulates produced by mechanical disintegration as in mining.
Mist	Suspended liquid matter; a dense mist may be referred to as fog.
Fumes	Solid matter of metallic or organic origin produced from the condensation of vapour.
Smoke	Particulate matter consisting mostly of carbon; also called soot.

Particulates are formed both by disintegration of larger bodies and by aggregation of minute units. They can be formed either naturally or by anthropogenic processes. The natural sources are windblown dust, volcanic (particularly pyroclastic) eruptions, forest fires, emission of natural gases, and sea salt diffusion in air. The anthropogenic sources are industrial processes, transportation, and stationary source fuel combustion.

The chemical composition of particulates covers a very wide range, being dependent on the source of the particulate concerned. For example, about a third of the fly ash (the airborne emission from coal-burning power stations) could be carbon, but the rest may include, in varying proportions, aluminium, calcium, carbonates, iron, magnesium, phosphorus, potassium, silicon, sodium, sulphur, and a number of trace elements. The specific percentages are determined by the nature of the coal being burnt. Some of the trace material, in spite of its limited presence, is toxic. High concentrations of sulphur in coal give rise to acid rain damage, discussed in detail later in this chapter.

All airborne particles, given sufficient time, come down to the earth's surface. This can happen in a dry state, such as when the particulates increase in size by compaction; or in the wet state, being brought down by precipitation. The latter process is far more common. In extreme cases, particulates in the air may even reduce the amount of solar radiation reaching the surface of the earth at a location. Urban areas may receive 15–20 percent less solar radiation than the surrounding rural areas (Stoker and Seager, 1976). High concentrations of particulates also reduce visibility by light scattering as it strikes the particulates. A typical rural concentration of 30 μg m^{-3} gives a visibility of about 40 km. In heavily polluted urban areas with a concentration approaching several hundreds of micrograms per cubic metre, that visibility may be reduced to a couple of kilometres (Stoker and Seager, 1976). Particulates settling as dust on vegetation may interfere with photosynthesis in plants. A wide range of damage can occur in materials such as metals and textiles depending on the concentration, chemical constituents and the state of the particulates. Painted surfaces are very susceptible, especially before the paint is dry. However, most studies on air pollution probably deal with its effect on the human body.

It is the combination of the toxicity and size of the particles that determines their effect on people. Particulates enter the body almost entirely through the respiratory system. The degree of penetration depends on the size of the particulate, and the effect on human health depends on its toxicity.

The respiratory system can be divided into upper and lower parts (Figure 9.1). The upper respiratory system consists of the nasal cavity, pharynx and trachea; and the lower part includes the bronchi and lungs. The two bronchi divide several times into the bronchioles in the lungs, which end in millions of air sacs called alveoli. Oxygen and carbon dioxide are interchanged with the bloodstream across the membrane of the alveoli.

Particles larger than 10 μm which enter the nasal cavity tend to be captured by the hair and lining of the cavity, and then expelled by coughing and sneezing. The mucus in the nasal cavity and trachea also captures these particles, which are then transferred to the throat by cilia (tiny hair-like structures), and removed when we expectorate or swallow. Smaller particles may, however, penetrate deeper into the bronchi. It is possible for the tiny particles to come out again in the airstream but those between 0.5 and 5 μm may reach the lungs and be deposited there, although over time a number of these will be removed by cilia. Those at the smaller end of this size range may settle in the alveoli, carrying irritating material into the sensitive areas of the lungs, and even worse, depositing toxic trace metals in the lungs. Such trace elements usually come from high combustion sources (power plants, cars, smelters and furnaces), and may be carcinogenic or cause brain, kidney, liver and nerve damage (Stoker and Seager, 1976).

In the past few decades, particulate emission has dropped remarkably in the developed countries, especially with tighter legislation such as the Clean Air Act of 1970 in the United States. Annual particulate emission over the United States in 1950 was 25×10^{12} g. By the mid-1980s it had dropped to $5–10 \times 10^{12}$ g. The story of the developing countries, however, is different.

Particulates are probably the chief type of air pollutant over the developing countries, irrespective of land use. In *rural areas* biomass burning produces enough particulates to darken the skies and, depending on the wind direction, causes air pollution over the nearby cities. Clearing of forests, both of primary and secondary type, results in large-scale biomass burning. For example, clearing of land by burning off the secondary vegetation during the dry season in south Sumatra has created respiratory problems for dwellers of nearby cities. This is discussed in detail in Box 17.1. Even rural air pollution is not necessarily confined to national boundaries.

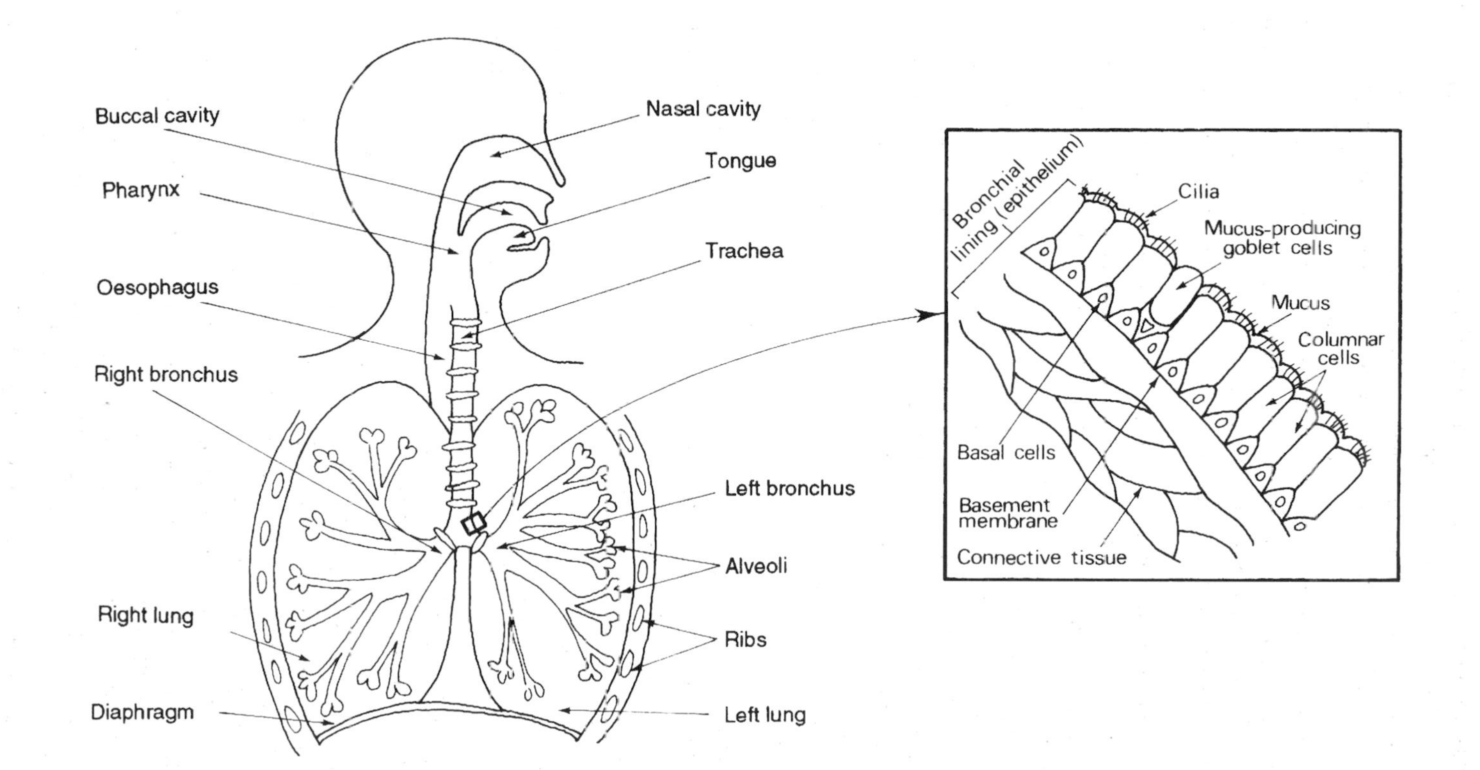

Figure 9.1 *Sketch of the human respiratory system. Based on Williamson, S.J. (1973)* Fundamentals of Air Pollution, *Addison-Wesley Publishing Co., p. 37*

The concentration of particulates over *mining areas* is common in the developing countries. Such air pollution occurs either from blasting practices in mines or due to strong winds blowing across opencast mines, waste heaps, and toxic dumps of mined material such as asbestos. The common need for extraction of primary resources and the neglect of safety regulations together cause respiratory diseases and eye ailments among the miners. Stone quarries and crushers spread a layer of silica dust over the area which may cause silicosis, and as a lesser evil, impair agriculture in the nearby fields. Silicosis starts with an irritant cough, shortness of breath, and chest pains. Agarwal and Narain (1986) have described the grim picture of Mandsaur in the state of Madhya Pradesh in India where industrial silica dust kills workers before they are forty. Asbestos mines produce worse incidents with fine fibres of asbestos being deposited in the lungs, causing pulmonary fibrosis or possible cancer of the lungs or gastro-intestinal tract.

The air is thick with particulates over many of the *urban centres* of developing countries. The common sources – exhausts from old cars and buses, uncontrolled emission from industries often sited next to dense residential areas, stationary fuel burning, ground-level incineration of garbage – are all usually present. The World Resources Institute described the status of pollutants in 20 megacities of the world for 1992 (World Resources Institute, 1994). All except three showed moderate to heavy pollution from suspended particle matter, which in 12 megacities reaches serious proportions. The three exceptions are London, New York and Tokyo. This illustrates (i) the seriousness and commonality of SPM pollution in the cities of the developing countries, and (ii) the existence of technology to clean up this particular type of air pollution. The concentration of SPM may reach several hundreds of micrograms per cubic metre in these cities with high air pollution. This is discussed in detail in Chapter 10 which covers the urban environment.

Oxides of sulphur

Anthropogenic sulphur dioxide emissions come primarily from fossil fuel combustion, and especially from electric power generating plants. A small amount is contributed by the smelting of sulphide ores and automobiles. Sulphur in coal is a relict from the proteins of plants and aquatic organisms that were transformed into coal. The proportion of sulphur in coal varies, as does the SO_2 emission produced by burning coal. A little amount of SO_3 is present after burning, but the SO_2 itself is changed into SO_3 via a series of reactions. Sulphur trioxide rapidly reacts with water to form sulphuric acid. The molecules of H_2SO_4 then become particulates by either condensing on other SPM in the air or mixing with water vapour to form droplets. Such droplets often form a significant proportion of the SPM. The return of sulphur to the surface of the earth could be in either a dry or a wet form. The latter is known as acid rain. The coal that comes from the eastern United States contains significant amounts of sulphur. When the pH of the rainfall is compared across the country, the value drops significantly east of the Mississippi River, reaching the lowest value in western and northern New York and Pennsylvania, and eastern West Virginia. A number of lakes in southwestern Sweden in the 1950s and 1960s showed evidences of acidification, with an ultimate pH of 4.5–5. Although part of this acidification was due to land use changes, the sequence of industrial expansion, increased air pollution and subsequent deposition of acid material in the lakes played an important role. This is clearly shown by the post-1940 spike which emerges when carbonaceous fly-ash particles in the lake sediment are plotted against time (Renberg, Korsman and Anderson, 1993).

Sulphate particles found over cities usually have an effective size between 0.2 and 0.9 μm (Masters, 1991). Suspended material of this size interrupts the visible part of the solar radiation and creates foggy conditions. The particles also penetrate deep into the human respiratory system. There the affinity for water results in most of them being absorbed in the moist upper respiratory tract. Even low concentrations of sulphates (8–10 μm^{-3}) are sufficient to cause respiratory problems, especially to elderly people or people with asthma. Sulphur riding piggyback on an aerosol may even reach the lungs, where the two pollutants act synergistically, increasing the detrimental effect of both. This synergy caused the killer events of air pollution in London mentioned at the beginning of this chapter.

Sulphur oxides may damage plants although the serious damage to forests, especially temperate forests, has been caused by acid rain. For example, the forests in the Sudetes Mountains of Poland have been seriously damaged by air pollution, mainly from the burning of high sulphur coal in Eastern Europe (Mazurski, 1990). Rodhe (1989) identified three areas with high acid deposition problems due to anthropogenic emissions of oxides of sulphur and nitrogen: most of Europe, the eastern United States, and southern China.

A prolonged high concentration of sulphates in the air damages paints, metals and organic fibres. The most striking damage is caused to buildings composed of marble or limestone. This has necessitated special maintenance for some of the world's most famous buildings, such as the Taj Mahal and the Acropolis. Such arrangements, however, are not always forthcoming with the required promptness.

Box 9.1

Taj Mahal and Air Pollution

The threat of air pollution hangs over historic monuments that are located close to urban or industrial areas. Many of these ancient monuments are built of limestone or marble and are therefore particularly susceptible to damage by air containing SO_2. The Acropolis in Greece is a classic example of air pollution damage to a monument that is recognized as part of human heritage. This case study on the Taj Mahal illustrates the vulnerability of these structures, the problem of piecemeal development, and a mechanism for the protection of historical buildings.

The Taj Mahal (Figure 9.2) was built by the Mughal Emperor Shah Jahan as a mausoleum for his wife Mumtaj Mahal. It is lined on the outside by Makrana marble of Precambrian age. Situated at Agra, on the bank of the Yamuna River, the Taj Mahal has survived many acts of vandalism, mostly by marauding armies, first the Jats and then the British. During the early days of the British occupation of India it was used as a place for throwing lavish parties. It was saved by providence (the market demand being non-existent) from being dismantled in order to sell individual bits in England as art objects. The Taj Mahal was finally rescued by Lord Curzon, acting in his capacity as the Governor General of India in the early twentieth century, and since then it has been restored to its proper position as one of the most beautiful buildings in the world.

Figure 9.2 *The Taj Mahal, Agra, India*

The recent threat to the Taj came in the 1970s with the building of a petroleum refinery at Mathura, at an upwind distance of 40 km from Agra. The need for a refinery to serve northwestern India had been recognized for some time, and when Mathura was chosen, environmental issues were not automatically considered as an important part of project evaluation procedure. The Taj Mahal was not perceived to be in danger. The foundation stone of the refinery was laid by Indira Gandhi in 1973 and the construction commenced, but the spectre of irrevocable damage to the marble of the Taj Mahal was raised by a number of distinguished Indians including academics specializing in air pollution, naturalists, lawyers, former administrators, and politicians. Expert committees were set up and after long deliberations it was decided to implement certain safety measures to restrict pollution. The 6 million tonne a year refinery was commissioned in 1983 but it was required to use low-sulphur crude for half of its requirements in order to limit the SO_2 released. Two ancient coal-burning power stations have been removed from Agra and the trains now run on diesel replacing the coal-burning locomotives; two existing causes of air pollution which were found to be more damaging than the release from the refinery. By the mid-1980s, the level of SO_2 has fallen by about 75 percent (Agarwal and Narain, 1986), but the air is still polluted from a number of small iron foundries which have not yet been moved out of Agra as recommended. A green belt of trees is also being planted as a buffer against air pollution. The attempt to move industries out of Agra has not been kindly received by either the local politicians or the population.

The problem of acid rain prompted the Convention on Long-Range Transboundary Air Pollution in 1985 to work on a Sulphur Protocol. The signatories to this protocol were expected to reduce their national emissions by 30 percent of the 1980 level in eight years. This has been achieved to a large extent. In 1980, the total European emission was about 55 million tonnes. By 1991, it had come down to almost 40 million (World Resources Institute, 1994).

Carbon monoxide

Carbon monoxide builds up in the atmosphere when the incomplete combustion of carbon produces CO instead of CO_2. It is therefore associated with automobiles, coal and biomass burning, and certain types of industrial activities. The linkage between anthropogenic CO and transportation is well illustrated when the CO concentration in the urban air is compared to the time of day: on weekdays, both the morning and afternoon rush hours produce peaks of CO in the city air.

The anthropogenic contribution of CO, however, is small compared to the contribution from natural sources. Its importance lies in the intensity of damage to people that CO is capable of inflicting, the localized concentration of anthropogenic CO in the urban areas, and the high rate at which such CO is produced.

Haemoglobin (Hb) in blood acts as a two-way transportation system, carrying oxygen (as oxyhaemoglobin, O_2Hb) from the lungs to the cells of the body, and CO_2 from the cells to the lungs (as CO_2Hb). With CO, it forms carboxyhaemoglobin (COHb), and at a rate about 200 times greater than that with O_2. Therefore the presence of CO leads to COHb being formed rather than O_2Hb, leading to a shortage of oxygen in the bloodstream. An increasing concentration of COHb in blood leads to nerve, psychomotor and brain dysfunction, and high and continuous inhalation may lead to cardiac problems and death. Psychomotor dysfunctions seem to occur at concentrations above 5 percent, dizziness and headaches beyond 10 percent, and death beyond 50 percent. A national survey in the United States, carried out between 1969 and 1972, indicated a high COHb content in the blood of smokers and people in the transportation sector, particularly taxi drivers. Employment in certain industries also seems to be associated with a high COHb content (Stoker and Seager, 1976). Driving on congested highways may result in being exposed to CO concentrations of around 100 ppm. Cigarette smoke carries more than 400 ppm CO, which may raise the COHb content in blood to an unacceptably high level (Masters, 1991). Breathing clean air removes COHb from blood, so for most people adverse effects are usually temporary.

Nitrogen oxides

Of the eight known oxides of nitrogen, only three normally occur in the atmosphere: nitrous oxide (N_2O), nitric oxide (NO) and nitrogen dioxide (NO_2). Anthropogenic emissions produce NO and NO_2, the latter being almost entirely an anthropogenic product. These two oxides are considered as pollutants. The oxides are produced mostly during combustion of fossil fuels when nitrogen and oxygen in the air are heated to a high temperature. Of the two, NO_2 is the more toxic although both can be injurious to plants, and NO_2 may give rise to nitric acid in the atmosphere, thereby causing acid rain.

Our knowledge regarding their adverse effect on human health is still uncertain, but at high concentrations NO_2 is an irritant and prolonged exposure may lead to bronchitis in children. Nitrogen dioxide reacts with hydrocarbons to produce photochemical smog. This is probably the most injurious effect.

Hydrocarbons and photochemical oxidants

Tens of thousands of hydrocarbons exist in all three physical states and are derived from different sources. Hydrocarbons are classified as primary pollutants as they are directly added to the air. Photochemical oxidants occur from reactions among primary pollutants. They are therefore secondary pollutants. The two groups are related as photochemical oxidants result from reactions which involve hydrocarbons.

The number of hydrocarbons that are involved in air pollution is fairly large. Most of these, e.g. methane, originate from natural sources. Those of anthropogenic origin are released from automobile emission, petroleum refining, solid waste incineration, biomass burning, and evaporation of organic solvents. Photochemical oxidants include compounds produced by the process of oxidation in the presence of sunlight. The reactions are many and complicated, and will not be described here. Low-level ozone is the most common member of this group, which also includes peroxyacetyl nitrate (PAN). Ozone is produced by a set of reactions that involve oxides of nitrogen in sunlight. The reactions produce highly reactive hydrocarbon-free radicals which react with a number of materials in the air and produce a mixture which is referred to as photochemical smog. This implies that in a sunlit city with nitrogen oxides in the air, ozone production increases from late morning onwards and as a consequence hydrocarbons are converted into photochemical smog.

Ozone and PAN are both injurious to plants, especially to leaves. Ozone also reduces crop yields and this type of damage to agriculture is serious. It also causes cracks in rubber especially when rubber is under tensile stress, which is how car tyres are damaged by ozone. Exposure to air with more than about 0.3 ppm concentration of ozone causes nose and throat irritation to people (Stoker and Seager, 1976). The oxidants involved in photochemical smog (especially peroxybenzoyl nitrate, PBzN and PAN) may cause eye irritation. Respiratory problems could also be caused by photochemical oxidants. The classic example of this type of air pollution occurs over Los Angeles where car emissions are high, the sun is bright and frequent temperature inversions prevent the diffusion of air which is kept areally confined by the mountain-girt nature of the land.

Lead

The lead content of air has risen in recent years due to car emissions. The dramatic nature of this increase is demonstrated by the sharp rise in lead content since 1940, as shown in samples from ice cores in Greenland or Antarctica. Lead concentrations in urban areas show a significant increase when compared with those in rural air. Lead in the air can be in the form of either gaseous compounds or particulates. Almost the entire amount is from the exhaust products of gasoline combustion. The sources of such lead are the antiknock additives, tetraethyl lead, $(C_2H_5)_4Pb$, and tetramethyl lead, $(CH_3)_4Pb$. In developed countries such as the United States, lead-free petrol is freely available to meet the air quality standards and also to allow the use of catalytic converters which are operable only with unleaded petrol. In the United States this is reflected in the sharp drop (more than 90 percent) in the concentration of lead in the air when compared with data from about 1980.

Airborne lead enters the human body mainly through the respiratory system although other processes of entry, such as the intake of foodstuff over which lead from the air has been previously deposited, are also present. Lead concentration in blood leads to behavioural changes, and in extreme cases, seizure, permanent brain damage, and death. In the developing countries, however, petrol is usually still leaded, and lead is very much a component of urban air. In Calcutta, for example, lead has a concentration in the 0.96–7.42 $\mu g\ m^{-3}$ range in air (Das *et al.*, 1992). The situation may change, however, even in the developing countries. From 1995, lead-free petrol has become available in India and catalytic converters are required on all new cars. It will, however, be some time before the old cars are phased out. It is not unreasonable to expect that similar changes will occur in a number of other developing countries in the near future.

9.3 Physiographic Control of Air Pollution

Air pollution persists until its high concentration is destroyed by diffusion or mixing with other masses of air. Certain environmental conditions enhance the concentration of pollutants in the air and also prevent its dispersion. Such conditions are either topographic or meteorological.

A mountain-girt basin or a deep valley creates a pollution-prone environment. At night, cold air from the upper slopes sinks by gravity to the valley-bottom or the basin floor. This traps the polluted air until sometime during the next day when solar radiation via terrestrial heating destroys the inversion and re-establishes the normal lapse rate and upward motion of air. This pattern is common in winter when the ground is cooler at night. Inversions are also created by cold air blowing inland from a cool ocean which displaces warmer lighter air upwards.

A number of settlements are situated on coastal plains, valley bottoms or basin floors due to the location of mines or industries. In all these locations a number of pollutants are produced in the air which are trapped by inversion. The unhealthy air pollution over a number of tropical cities results from a combination of topography, meteorology, biomass or solid fuel burning, concentration of motor vehicles, and in a number of cases, polluting industries without emission control.

9.4 Air Quality: Concepts, Standards and Acts

Air is considered clean when all the pollutants are below harmful levels. This requires the setting of standards, crossing of which would legally imply that the air is polluted. Three different kinds of standards are in existence:

- a standard for measuring ambient air quality which considers the concentration of major pollutants individually
- a figure which sums up the overall quality of air
- emission limits for industrial plants.

Standards are set by different legislation in different countries (Table 9.3). For example, in the United States a number of laws (collectively referred to as the Clean Air Act) were formulated between 1955 and 1977. The Clean Air Act Amendments of 1970 are probably the most important of these. The National Ambient Air Quality Standards (NAAQS) were set as required from the 1970 legislation by the US Environmental Protection Agency. The standards are of two types, primary and secondary. *Primary standards* are set to protect public health with an adequate margin of safety. The *secondary standards* are set even higher but it is the primary standards which are usually followed. Individual states such as California, where air pollution locally could be a problem, have also set their own standards. The pollutants for which limits have been set include the following:

- particulates not more than 10 μm in diameter (smaller particulates are potentially more damaging)
- sulphur dioxide
- carbon monoxide
- nitrogen dioxide
- ozone
- lead

On the other hand, the *Pollutant Standards Index* (PSI) is an overall index used to indicate air quality over cities. It is the best known of the standards as it is listed in many newspapers for public information. The basic idea is to examine the pollution level of the following: ozone; carbon monoxide; particulate matter (PM) or total suspended particulates (TSP); sulphur dioxide; a combination of particulates and sulphur dioxide; and nitrogen dioxide. The worst of these pollutants is determined and its value converted to a quality scale divided into five categories: good, moderate, unhealthful, very unhealthful or hazardous. Appropriate public warnings are given by the authorities. Such warnings begin when very unhealthy conditions are reached. Details of this index and its calculation are given in Box 9.2. Standard instruments are available for taking air samples over the required period (as listed in Table 9.4) and for analysing the concentration. This is the index most of us are likely to encounter in our daily life.

Table 9.3 *Ambient air quality standards*

Pollutant	Averaging time	US primary ($\mu g\ m^{-3}$)	WHO ($\mu g\ m^{-3}$)
Particulates			
Diameter ≤ 10 μm	Annual	50	
All	Annual	75	60–90
All	24 h, not more than once a year	260	150–230
Sulphur dioxide	Annual	80	40–60
	24 h, not more than once a year	365	100–150
Carbon monoxide	8 h, not more than once a year	10	10
	1 h, not more than once a year	40	30
Nitrogen dioxide	24 h		150
	Annual	100	
Ozone	1 h	235	150–200
Lead	3 months	1.5	
	Annual		0.5–1

Source: Summarised from WHO and UNEP (1992) Urban Air Pollution in Megacities of the World, *Blackwell Reference, Oxford.*

Table 9.4 *Pollution Standards Index (PSI) breakpoints at 25°C and 760 mm Hg*

Index	24-h PM (μgm^{-3})	24-h SO_2 (μgm^{-3})	8-h CO (mgm^{-3})	1-h O_3 (μgm^{-3})	1-h NO_2 (μgm^{-3})
0					
50	50	80	5	120	NA
100	150	365	10	235	NA
200	350	800	17	400	1130
300	420	1600	34	800	2260
400	500	2100	46	1000	3000

Note: The index is interpreted qualitatively according to its values: good (0–50), moderate (51–100), unhealthful (101–199), very unhealthful (200–299), hazardous (more than 300).
Source: 40CFF (Code of Federal Regulations, USA) *58, App. G (7-1-97 edition).*

The last of the three types of standard, the *emission standards* refer to the allowable output of industrial gases. A number of industries are potential sources of air pollution. The list includes power plants using fossil fuel, industrial smelters, petroleum refineries, cement plants, and incinerators. The emission standards are set according to the type of pollutant released. For power plants, control is set in terms of mass of pollutant per one million units of heat output. In other cases the control is set differently.

Box 9.2
Pollutant Standards Index

Imagine you are in a subtropical city. The common air pollutants are particulates from fuel burning, car emissions, and small industrial establishments. On a day in winter with morning inversion the following concentrations of pollutants are found in the air:

8-hour carbon monoxide	1190 $\mu g\ m^{-3}$
24-hour PM	290 $\mu g\ m^{-3}$
24-hour sulphur dioxide	150 $\mu g\ m^{-3}$

What is the PSI for the day and how would you describe the quality of the air? To solve this problem you will need Table 9.4.

From Table 9.4, it can be seen that PM has the highest value in the index column – a PSI value between 100 and 200. By interpolation, the index for PM is 170. The index for CO is 114 and for SO_2 it is below 100. The PSI is therefore 170 and the air of that day will be reported as unhealthful.

9.5 Quality of Air in Tropical Cities

The nature and intensity of air pollution alter with development. In rural areas air is polluted from two main sources: burnt forests and bare farmland. Mining activities cause air pollution by blasting rocks and overburden, and again when wind blows across opencast mines, and waste heaps, particularly if it blows across toxic material such as asbestos. However, the air is also extremely polluted in many cities of the developing countries, most of which are located in the tropics.

Particulates are the most common group of pollutants in these cities. In 1992, the United Nations Environment Programme (UNEP) and WHO published a joint study of air pollution in 20 large cities. The cities were Bangkok, Beijing, Bombay, Buenos Aires, Cairo, Calcutta, Delhi, Jakarta, Karachi, London, Los Angeles, Manila, Mexico City, Moscow, New York, Rio de Janeiro, São Paulo, Seoul, Shanghai and Tokyo. Most of these are in the tropics. At least one pollutant exceeded the WHO guidelines in all the cities. Fourteen had two such cases and seven had three. In this survey, the air over Mexico City was the worst polluted, carrying unacceptably high levels of suspended particulate matter SPM, SO_2, CO, O_3, lead and NO_2. The hazardous combination of SPM and SO_2 was found in Beijing, Mexico City, Rio de Janeiro, Seoul and Shanghai. This was found in spite of inadequate air quality monitoring in 14 of these 20 cities (WHO and UNEP, 1992).

The problems of tropical cities, including the very large ones, are not limited to high concentrations of various pollutants of the air. These cities generally have grown without much zoning and so industrial plants often exist in the middle of residential areas. Also, industrial workers tend to live immediately outside the plant to avoid transportation costs. There is therefore very little escape for the residents of the cities of less affluent countries from the continuous presence of toxic material in the air they breathe. Such unacceptable conditions become the norm when the government is inefficient and lacks political will, access to better technology is unavailable, industry is indifferent, and the people are uninformed.

The concentration of pollutants may vary considerably from the annual average. The average figure for SPM in Calcutta, for example, is 420 μg m^{-3}. This figure is derived mainly from automobile emissions, frequent and long-lasting traffic jams, burning of coal for domestic cooking, and polluted material being released into the air from thousands of small industrial establishments. Air pollution in Calcutta has been studied in detail by the School of Environmental Studies at Jadavpur University, Calcutta (Chakraborti, Van Vaeck and Van Espen, 1988; Chakraborti *et al.*, 1992). The figure for SPM varies widely at different seasons, rising to a winter average of 1100 μg m^{-3} mainly due to temperature inversions, and dropping considerably below the annual average during the summer and the rainy season. The concentration of SPM also varies through the day, the peak concentrations coming between 09.00 and 11.00 hours and between 16.00 and 18.00 hours. The highest concentration recorded is around 2500 μg m^{-3}. The worst pollution is concentrated over the central part of the city, the winter high average of 1100 μg m^{-3} dropping to 200 μg m^{-3} only 3 km towards the east, just beyond the edge of the city. The lead content of the SPM similarly changes spatially from a horrendous 6.0 μg m^{-3} to 0.09 μg m^{-3}. Worryingly, 10 percent of the SPM consists of polynuclear aromatic hydrocarbon (PAH). The high presence of PAH, Benzo-α-pyrene (BaP), and several toxic heavy metals as part of the SPM, make the air carcinogenic. The average conditions in Calcutta from mid-January to mid-February are worse than the recorded maximum polluted conditions of western European cities (Chakraborti, Van Vaeck and Van Espen, 1988). At 6.0 μg m^{-3}, lead in particulates is also high in Calcutta and the situation is comparable to the early 1980s figures from Kuala Lumpur where the maximum total SPM was between 249 and 313 μg m^{-3} (Sham Sani, 1987).

These are frightening figures but we ought to be looking beyond these. At least for Calcutta we now have some data that immediately suggest the urgent implementation of certain solutions, e.g. lead-free petrol, the relocation of small industries, and the availability of clean fuel for cooking. For most cities in the developing world that are recognized as cities with high air pollution, the absence of hard information hinders even that. The implementation of the suggested solutions is very difficult, especially in the developing world, but the first step to problem-solving is measuring the problem. In the case of Calcutta, academic institutions and NGOs are in a position to do that with locally available resources. The findings are sometimes unexpected; for example, while the SPM in Calcutta air is extremely high, the concentration of heavy metals in the sewage is very low compared to that in developed countries (Das *et al.*, 1992). We need to know the patterns of spatial, seasonal and type variations of pollutants in large cities.

The worst case of air pollution in a city occurred in Bhopal on 2 December 1984. Bhopal is the capital of the Indian state of Madhya Pradesh, with a population of 800 000. In the 1970s, Union Carbide built a factory in Bhopal for manufacturing chemicals for pesticides and storing them after production. At that time it was a popular move, bringing jobs to the city and providing pesticides to meet the demands of the Green Revolution. Several instances of leakage from the plant were known but nothing much was done by the plant management or the local government.

Methyl isocyanate (MIC) leaked out of the plant at about 23.30 hours on that day in December in horrendous quantities and without any warning being given. In spite of heroic efforts of individuals, transport operators, the railway staff, and Indian army personnel immediately after the leak, the death toll rose to several thousands. Many more suffered from serious eye and respiratory ailments. Particularly affected were the poor who lived in shacks near the factory. The compensation to the sufferers took about ten years to organize and even then fell short of general expectations. Agarwal and Narain (1986) have provided a harrowing account of the disaster.

This is the nightmarish scenario of an air pollution incident in a developing country where people live in close proximity to a plant producing hazardous chemicals. In comparison to the developed world, the safety measures are less, the environmental legislations are often wanting (even if they exist they are not strictly applied), and an industrial plant run by a multinational has tremendous clout. As industries grow in developing countries, proper environmental legislation and its strict implementation for the sake of the citizens are imperative.

In the developed world incidents of industrial air pollution have happened earlier. On 27 October 1995, the High Court in Britain ordered a multinational engineering company to pay £65 000 in compensatory damages to a woman diagnosed as suffering from mesothelioma, a rare form of cancer of the lung or the abdominal cavity which is incurable. As a child she used to live next to an asbestos factory and play in the asbestos dust. Another claimant was awarded £50 000 for her husband's death from mesothelioma in 1991 (*The Guardian*, 28 October 1995).

9.6 Indoor Air Pollution

The air indoors, under certain circumstances, can be polluted to a hazardous level. This happens either when polluted air leaks into the house or when cooking smoke or evaporation from volatile organic compounds builds up inside a room. The conditions are worsened to hazardous levels if this happens in a confined space with good insulation which prevents dilution of the polluted air. Table 9.5 summarizes the principle sources of indoor air pollution.

Table 9.5 *Major sources of indoor air pollution*

Type	Source
Particulates	Released from cooking or space heating, especially from woodsmoke; tobacco smoke
Volatile organic compounds	Formaldehyde and other compounds released from plywood, particleboard, panelling, insulation material, paints, varnishes, solvents, carpets, sheets, aerosol sprays
Ozone	Photocopying machines, electrostatic air cleaners
Radon	Rocks, soils, building materials
Carbon monoxide	Woodsmoke, gas heaters, stoves

Indoor air pollution from these sources increases in intensity with insulation. For example, radon gas that seeps indoors from the ground builds up in houses with efficient insulation in a cold climate. Radon is released from rocks and soils that carry radium. It is chemically inert, but its decay products (polonium, bismuth and lead) may be inhaled, thereby leading to lung damage, even though all three have a short life. Radon is the suspected cause of thousands of lung cancer deaths each year in the United States. Asbestos fibres if released from insulation may be a potential source of lung cancer. So can be burning cigarettes as tobacco smoke contains numerous known or suspected carcinogens: benzene, hydrazine, benzo-α-anthracene, benzo-α-pyrene (BaP), and nickel. A single burning cigarette releases particles to the order of 10^{12}, and the small size of the particles determines their passage into the lungs of people present near the smoke (Masters, 1991).

In the developing countries it is woodsmoke which causes the most damage indoors. Part of the biomass burning is for heat in winter or on high mountains, but most of the biomass is burnt in the kitchen. The village kitchens are often windowless or the ventilation hole is closed during the heavy rains of the wet season. The amount of particulates released depends on the nature of wood being burnt; fuelwood from *Acacia nilotica,* for example, smokes less but often costs more. Agarwal and Narain (1986) refer to a study of rural kitchens in the Gujarat state of India by Kirk Smith, A. L. Aggarwal and R. M. Dave. The routine concentration of particulates and BaP in the samples were found to be horrific, and as Agarwal and Narain pointed out, much higher than the figures from 1954 London killer fog episodes. The exposure to particulates for women cooking a meal averages 7000 $\mu g\ m^{-3}$, with a range between 1110 and 56 600 $\mu g\ m^{-3}$. The figures refer to cooking with only one burner. A double-burner woodstove releases far more particles.

Such exposures happen several times a day and may continue for hours. The routine combination of a high concentration of particulates with the presence of BaP and CO in smoke is, needless to say, extremely worrying, especially for pregnant women in the high mountains who may also have to do hard physical work outdoors. The problem is lessened if biomass is replaced by other types of fuel, or if better ventilation exists, or if the stove is better designed. Improved versions of cooking stoves are available and such fuel-efficient stoves also demand less firewood.

9.7 Management of Air Quality

Air pollution standards are set to monitor air quality and to ensure that the air remains unpolluted. The justification for setting such standards comes from various clean air regulations and acts. Various techniques (physical, economic and behavioural) are used to attain such standards. For example, pollution from automobile emissions is a common source of urban pollution. This pollution can be controlled by having fewer and cleaner cars on the roads. This has been attempted (1) by physical techniques such as lead-free petrol, catalytic converters, and insistence on periodic car inspections; (2) by the application of economic controls such as high taxation on car ownership and usage, and special licenses for cars entering city centres; and (3) by behavioural changes such as using alternative and comfortable mass transit systems. Singapore, for example, uses all three types of control.

The major sources of air pollution as listed in Table 9.2 are transportation, fuel combustion in power plants, industry including petroleum refining, and incineration including biomass burning. Different techniques, often in combination, have been used to lessen emissions originating from these sources. The physical controls, which are often engineering in character, can be applied in three stages to reduce emissions:

- Precombustion control: changing to cleaner fuels (with fewer impurities) or treating the fuel before combustion, e.g. washing of coal to remove part of the sulphur or nitrogen content.
- Combustion control: improving the combustion process to reduce emissions.
- Postcombustion control: treating the pollutants after combustion and before they are released to the atmosphere (Masters, 1991).

We will discuss in detail emission control from two of the sources listed above, transportation and power plants, as examples of industrial attempts to reduce air pollution.

Transportation

The motor car is the most polluting type of transport. The standards and regulations are therefore designed with personal automobiles in mind. The common pollutants (CO, NO_x, lead and hydrocarbons) are mostly released to the atmosphere via car exhausts. Pollutants also escape from the combustion chamber of the car engine although in the newer cars such gases are brought back to join the exhaust system. The third main source of air pollution is the evaporation of hydrocarbons from the fuel tank and the carburettor.

The common engine is the standard spark-ignited four-stroke piston engine which operates by compressing a mixture of air and petrol. Complete combustion of petrol occurs when the mixture has an air:fuel ratio of 14.5:1 (stoichiometric ratio). A lean mixture is one with more air, whereas a rich one carries more fuel. As different factors such as maximum engine power, minimum fuel consumption, and reductions of CO, NO_x and hydrocarbons happen at different air:fuel ratios, finding the optimal condition is difficult (Figure 9.3). A mixture near the stoichiometric ratio will provide reasonable power, low fuel consumption, and low emissions of hydrocarbons and CO but will produce high concentrations of NO_x. It is therefore necessary to pass the car exhaust through a control system to reduce air pollution.

Catalytic converters are the best known of all car exhaust control systems. Catalytic converters reduce NO_x to N_2, oxidize hydrocarbons, and change CO to CO_2. There are two other significant benefits. First, the use of a catalytic converter requires an air:fuel mixture close to the stoichiometric ratio, which implies that the car will run at near-maximum power and low fuel consumption. Second, as lead corrodes the catalytic converter, concomitant use of lead-free petrol becomes

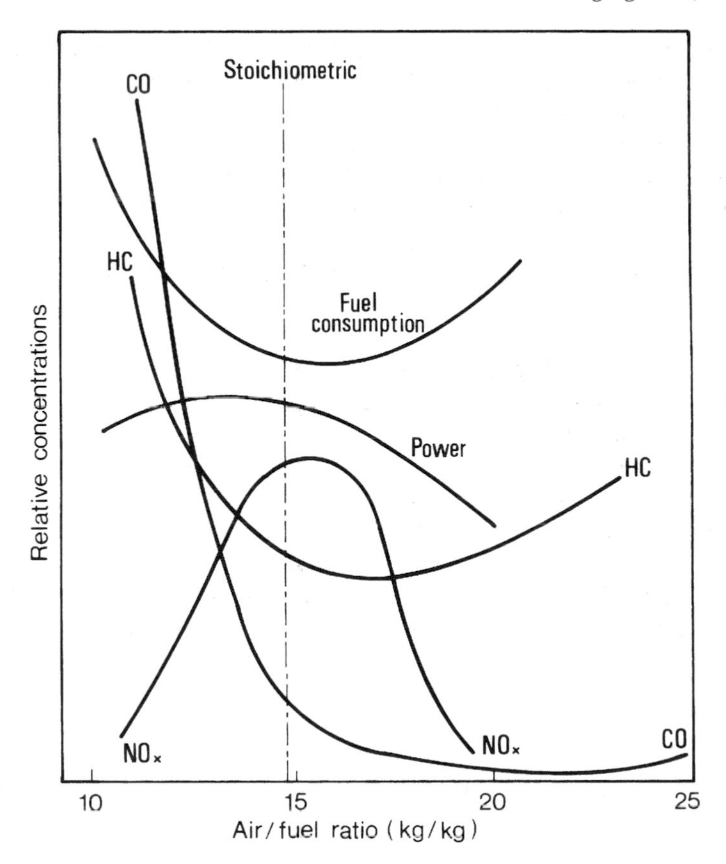

Figure 9.3 *Effect of the air:fuel ratio on emissions, power, and fuel economy. From Masters, G.M. (1998)* Introduction to Environmental Engineering and Science, *2nd edition © 1998. Adapted by permission of Prentice-Hall, Inc., Upper Saddle River, NJ*

automatic. Catalytic converters and lead-free petrol are now common in western countries and the use is spreading. Singapore, for example, started the switch to lead-free petrol and catalytic converters for new cars at the beginning of the 1990s. India started the transition in 1995. The complete change takes years, of course, but meanwhile the city air gets progressively easier to breath.

Other techniques are in practice. Engine designs have been changed to produce less polluting emissions. The Wankel (or rotary) engine and the stratified charge engine with dual combustion chamber are prominent examples. Diesel engines are much more fuel efficient and can run on rather lean mixtures thereby producing low emissions of hydrocarbons and CO. Their exhausts, however, produce considerable volumes of NO_x and carbonaceous particles. Fuels other than petrol or diesel (methanol, ethanol, gasohol, hydrogen, etc.) have also been investigated. The recent interest in reviving the electric car is perhaps the most revolutionary attempt to reduce air pollution. But unlike catalytic converters and small-scale modifications to the standard engine, none of these attempts has yet paid off. The steep rise in oil prices in the early 1970s helped to bring in cars with much better fuel efficiency and therefore less pollution. As discussed in Chapter 8, significant advances have been made with the electric car and it is quite possible that we will see a switch to electric vehicles in the next century.

Power plants

Power plant related air pollution comes mostly from the stations that use coal with impurities. Here precombustion controls are often in evidence, such as changing from using high sulphur coal to a low sulphur variety, or 'coal cleaning' which involves the physical removal of inorganic pyrites (FeS_2) and lowering the ash content. An excellent example of control during combustion is the process known as fluidized bed combustion (FBC). A mixture of powdered coal and limestone is kept suspended by fast rising airstreams in the boiler. This results in the sulphur being precipitated on the floor as calcium sulphate which is then removed. FBC is also a very efficient industrial process that allows the temperature of the boiler to be kept low thereby reducing the volume of NO_x produced.

A number of postcombustion techniques are now in operation in the industry. For example, sulphur is removed by scrubbers which involve spraying a wet slurry of finely powdered limestone (or lime) into the flue gas. This results in precipitating the sulphur as calcium sulphite or calcium sulphate which can then be removed as a sludge. The system is efficient but costly and requires large volumes of water. Particulates can be removed in a cyclone collector. Particle laden gas enters a cylinder tangentially near the top. As the gas spins in the cylinder, particles collide with the walls and slide down to be collected at the bottom. This works well for sizes larger than 5 μm but not for finer particles which are potentially more harmful. Smaller particles are removed in an electrostatic precipitator or a baghouse. In the electrostatic precipitator, ionized gas molecules attach themselves to the particles when the flue gas is passed through the system. The particles are now charged and can be collected electrostatically on a surface and later removed. This is an extremely efficient but expensive technique. In a baghouse, the emission is passed through a number of fabric filter bags hanging upside down in a chamber. The particles are captured by both the dust in the bags and the surface of the bags themselves. This is another efficient but expensive technique.

It is clear that various industrial techniques are now available for reducing air pollution from the standard sources. It is also clear that most of these techniques are costly and not yet appropriate for the developing world. Some of the solutions to the problem of air pollution in the developing world may have to be non-technical and innovative in nature.

Exercise

1. An environmental expert has recommended that the release of CO_2 in the atmosphere can be contained by financially encouraging the use of coal with a

low carbon content. Do you agree with this recommendation? Justify your conclusion.

2. Choose a tropical city with high levels of air pollution. Construct three tables. Table 1 should be a summary of the existing air pollution conditions including the sources, the amounts, and the variable distribution within the city. Table 2 should describe the expected effects of such pollution. Table 3 should list your recommendations for ameliorating the situation. The solutions should be feasible for a developing country.

10

Urban Development and Environmental Modification

10.1 Urbanization of the Developing World

We reviewed the history and the current pattern of global urban transformation in Chapter 3. The current urban transformation of the developing world is distinctly different from that of the developed countries. Although the history of urban settlements in the tropics goes back 5500 years, a number of the major cities date back only to colonial times and are not more than a few centuries old. Unlike the developed countries, the pace of urbanization picked up only in the second half of the twentieth century. In the 40 years between 1950 and 1990, the urban population of the developing world increased fivefold, from 286 million to 1515 million (United Nations, 1994). The growth is continuing and this figure is projected to reach 4 billion by 2025 (United Nations, 1995). The United Nations projections show that by AD 2000, 77 percent of the population of Latin America and the Caribbean will live in urban settlements. For Africa and Asia those figures are 41 percent and 35 percent respectively (United Nations, 1994). However, given the explosive urban growth being experienced by Africa and Asia, the proportion of the population living in the urban areas of both continents could rise to 54 percent by 2025 (United Nations, 1995). This is in contrast to the situation in the developed countries where urbanization has levelled off, the cities currently growing by only 0.8 percent on average (United Nations, 1994).

Urban growth in the developing countries involves settlements of all sizes. In India, for example, towns and cities have been growing for the last 40 years at rates which are comparable irrespective of settlement size (Mohan, 1996). Certain major agglomerations of the world are estimated to grow to gigantic sizes. The projected populations for AD 2000 for Mexico City, Bombay and São Paulo are 18.1, 18.0 and 17.7 million respectively. Of the 22 megacities (defined as cities with a population of at least 10 million) of 1994, 16 were in the tropics. The projected number of megacities for 2015 is 26, and 22 of these are tropical cities, including 18 in Asia (United Nations, forthcoming). The fastest growth rate is projected for cities of between 1 and 10 million (the million cities). In 1990, there were 270 such cities; in 2015, 516 are expected (United Nations, 1995). Labour and industrial production also tend to concentrate in the megacities. In Mexico, an estimated 44 percent of the GDP, 52 percent of the industrial output, and 54 percent of the service sector are found in Mexico City (United Nations, 1994). Such figures ensure the importance and expansion of megacities.

In many cities of the developing world, the necessary services such as housing, transport, the supply of clean drinking water, waste disposal systems, etc., have failed to keep pace with the rising population. Furthermore, urbanization has introduced physical problems such as increased flooding, slope failures and air pollution. In the twenty-first century, the cities of the developing world may very well turn out to be one of the major problem areas of environmental management, along with global warming and the loss of tropical forests.

Urbanization of the developed countries in the 1950s and 1960s and the associated modifications of the physical environment produced a number of case studies. Most of these studies were carried out in the United States, and from such studies we get a basic account of urbanization-related environmental degradation. For example, Helmut Landsberg (1981) carried out extensive observations on climatological changes with urbanization. Urban drainage basins were studied by people from a variety of disciplines (environmental engineering, geography, geology, etc.). Wolman (1967) presented the time-sequence of changes in a drainage basin undergoing progressive urbanization in three stages:

- *Stage 1.* Pre-urbanization: land is under natural vegetation or agriculture, river channels are adjusted to the existing conditions.
- *Stage 2.* Brief period of construction: land is stripped bare, soil and weathered material is disturbed, there is intense erosion, channels receive large quantities of sediment, channels are in disequilibrium.
- *Stage 3.* Post-construction urban landscape: a large proportion of the ground surface is impervious (streets, parking lots, rooftops, etc.), drainage is by concrete drains and sewers, river channels are still in disequilibrium and may continue to be so or adjust over time to new conditions.

Table 10.1 lists the major environmental changes that occur with urbanization. The effect of such environmental modifications are similar to those occurring in non-urban situations discussed earlier in this book, but the intensity and rapidity with which the urban environment is degraded, both physically and socially, necessitates its special treatment. The knowledge that at least half of the population of the developing world will live in urban settlements by the middle of the next century adds urgency to that treatment. The nature of the tropical environment, where most of the cities of the developing countries are located, tends to magnify urbanization-related environmental degradation because of the physical geography. It should be added, however, that only a limited number of urban settlements from the tropics have been studied in detail.

In spite of the small number of case studies, generalizations are possible regarding the nature of tropical rainfall. Tropical rainfall is usually intense irrespective of

Table 10.1 Types of environmental modification arising out of urbanization

Hydrological changes:	increased surface runoff increased flood intensity and magnitude depletion of subsurface water
Geomorphological changes:	accelerated sediment production slope instability modification of natural channels
Climatological changes:	less radiation received increased temperature and the heat island effect increased cloudiness increased precipitation reduced humidity reduced wind speed
Changes in vegetation:	introduction of exotic species increase in species of open habitat increase in pioneer and ruderal species
Air quality changes:	increase of contaminants increase in solid particulates increase in gaseous admixtures
Water quality changes:	problem of water supply waste water from domestic sources waste water from industrial establishments

whether the rain falls in localized thundershowers, or squalls, or as part of the wet monsoon which covers wide areas and lasts for months with a few breaks in between. The amount of rain that falls, however, varies considerably, over both space and time. Griffiths (1972) provides the example of Makindu (Kenya), where the annual rainfall ranges from 67 to 1964 mm. The downpour from a tropical rainstorm is capable of causing considerable erosion when the vegetation cover is removed and the surface exposed. According to Hudson (1971), the erosive power of tropical rain is about 16 times more than that of rainfall in the temperate areas. For most of the tropics, therefore, erosion, sediment transfer, flooding, and channel modification of the three stages of Wolman, achieve greater prominence.

Environmental degradation of urban settlements in the tropics is magnified due to two historical circumstances. First, a number of these cities, especially those that were founded in colonial times, were sited with very little consideration for the local geology or topography. Cities such as Calcutta, Singapore, Kingston (Jamaica) and Hong Kong were originally established as harbours or trading posts in order to propagate the sphere of influence of the colonial power. In many cases the location turned out to be hazardous. Kingston is an excellent example. Studying the seismology of this part of Jamaica, Shepherd (1971) showed that from the viewpoint of a seismologist, Kingston and its suburbs was probably the worst possible place to choose for establishing the capital city of Jamaica. Recurrent seismic disturbances, volcanic activities, and tropical storms at times reaching hurricane force, bring destruction to a number of densely populated cities of the developing world which are located in hazardous zones.

Second, although some of the cities were originally sited in safe locations, their recent spread has extended the urban space over less desirable land such as steep slopes, badly drained floodplains, and coastal swamps. Usually these places are inhabited by the poor squatters and slum dwellers, and these are the places subjected frequently to slope failures, floods and various other kinds of physical disaster. A share of the disasters is at times born by the prosperous inhabitants of the cities who prefer to build their houses on or near the tops of hills and ridges for the view and exclusivity. McGee (1971) has demonstrated a relationship between land price and elevation for Kuala Lumpur. This relationship exists to a great degree in cities that are hazardously situated. The spread of the cities can happen speedily. For example, between 1974 and 1987, São Paulo expanded at a rate of approximately 35 km^2 per year, and mainly over deeply weathered Precambrian rocks. This expansion, often on steep slopes, increased the frequency of mass movements and gully development, with associated siltation in valleys and bottomlands (Coltrinari, 1996). By 1988, São Paulo covered more than 900 km^2 within a metropolitan region of 8000 km^2. The urban expansion has not only reached the steeplands; it has also caused about 1 million people to squat within protected watershed areas. Riverine wetlands are covered with streets and houses. Such is the extension of a megacity, converting agricultural fields and forests to urban areas. Annual estimates suggest that cumulatively nearly 5000 km^2 of arable land is replaced every year by the spread of cities in the developing countries (World Resources Institute, 1996).

Both these circumstances magnify the scale of floods, slope failures, etc., and expose a large number of people living in hazardous places to such dangers. This is well illustrated below by three boxed case studies of degradation of urban environments for different reasons. We end the chapter with discussions on management of the urban environment and also on supplying cities with the necessities of life such as clean drinking water.

10.2 Hydrological Changes

A drainage basin that is at least partially urbanized undergoes three changes in its hydrological character:

- extension of the impervious surface of rooftops, paved areas, streets and drains; the impervious area may vary from 20 percent (low density residential area) to about 90 percent (central business district);
- building of storm drains; the percentage of area served by storm drains varies from 0 to 100 between different parts of the city;
- various types of improved hydrological efficiency imparted to rivers inside cities, such as channel widening and straightening, and lining the channel with concrete, either fully or partially.

Such changes transform the rainwater almost entirely to surface runoff which, aided by storm sewers, rapidly drains off the land. As a result, storm peaks are accentuated and inter-storm flows are very low. Based on data collected by the US Geological Survey, Leopold (1968) showed rapid increases in flood size and frequency for various degrees of urbanization over an area of 1 mile² (2.59 km²). For example, the mean annual flood (defined as a flood that has a recurrence interval of 2.33 years), increased between two and three times its natural size when half the area was impervious and also served by storm drains – a situation commonly found in middle density residential areas. A number of case studies from various parts of the world indicate the same trend towards increased flooding, following urbanization. Not only do the flood discharges increase, but the lag time is also reduced to between one-eighth and one-quarter of its natural value. Lag time is the difference in time between the peak rainfall and the peak in river discharge (Figure 10.1). Hydrological changes following urbanization can be summarized as follows:

- changes in *flow duration*: low water in channels most of the time, but larger and quicker floods occur after rainfall;
- Changes in *flood frequency*: floods of a given size will occur more frequently due to most of the rain being transformed into surface runoff; floods of a given frequency (such as the mean annual flood) will increase in size;
- *modification of runoff* from individual storms: after each rainstorm, water in the streams will rise faster and higher and also travel with higher velocity;
- *the quality of water* in channels will be degraded due to the combination of increased surface runoff and very little dilution between floods;
- diminishing of *aesthetic appearance* of rivers flowing through cities.

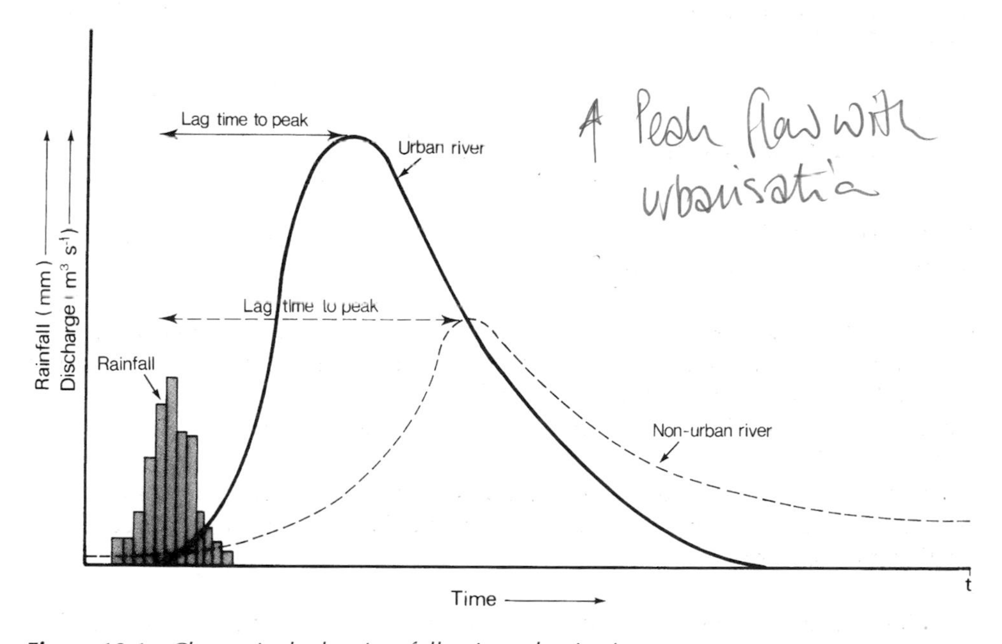

Figure 10.1 *Change in the lag time following urbanization*

The major impact of urbanization is the increase in flood potential which results in eroded streambanks and inundated valley-bottom roads and properties. Such floods are mitigated by two different techniques: structural and non-structural controls. Structural control is essentially engineering in nature, and consists of either speeding the flood discharge out of the city or slowing down the passage of rainwater to a main channel so that the sharply peaked nature of the flood is changed to a lower and flatter peak stretched over a longer time.

The first type of structural control is achieved by altering the drainage network, usually by lining the channels in concrete which, by reducing surface friction, allows high velocity flow. This implies building large channels which carry a considerable amount of water after rain but remain dry at other times. Large lined channels are usually extremely unappealing aesthetically.

The second type is achieved by holding up rainwater at various points on its way to a main drainage channel. This is done in several ways. Water several centimetres deep can be held up on flat rooftops (Figure 10.2) by providing a small opening to the drainpipe, thereby ensuring a slow supply of water to the ground and over a much longer period than would be needed with a conventional pipe. Small depressions in the ground can be used to store water after rain which can then be drained slowly or allowed to enter the subsurface. Blocks of concrete with open gratings (porous pavements) are used to cover car parks and pavements (Figure 10.3). This allows a considerable part of the rainwater to enter the subsurface instead of running off the impervious concrete into a drainage channel. The drainage of Singapore is discussed as an example of flood management in Box 10.1.

Figure 10.2 *Rooftop retention of rainwater*

Figure 10.3 *A porous pavement*

Non-structural controls are basically behavioural adjustments: floodplain zoning to restrict building in floodprone areas, flood insurance, flood forecasting, and evacuation, etc. Another alternative would be to floodproof houses and build small embankments. Some of these innovative techniques are mentioned in the review of the drainage problem of Bangkok later in this chapter.

Box 10.1

Flood Problems in Singapore

Singapore is naturally floodprone. The drainage basins are small, steep hillsides meet valleyflats at sharp contacts, and the rain usually falls in brief but intense showers. The floodprone situation has worsened because of intensive urbanization (often on the slopes) and building of arterial roads along the major valleys. The flood potential, however, has always been high. Even in 1892, an individual swam down Orchard Road, the main thoroughfare of Singapore (also located in a valley), with a three-foot ruler to measure the depth of water (Buckley, 1902). The pace of urbanization since the 1960s has exacerbated the situation so much that more than 100 incidents of waterlogging were reported in the local newspapers in less than 30 years. Flood damage in Singapore is relatively minor when compared to floods in other major cities of the region such as Bangkok or Jakarta. Floods in Singapore primarily inundate roads and cause traffic disruption, which in turn affects the normal activities of this busy city. Damage to roads, vehicles and houses next to the valley-bottom arterial roads still happens but on rare occasions.

Non structural - building restriction in flood plane zones
- Flood insurance ...

The Drainage Department in the Ministry of Environment has the responsibility for alleviating floods and maintaining an efficient drainage system. The regulations required for flood control (the Surface Water Drainage Regulations) were passed in 1976. The 1976 Code of Practice on Surface Water Drainage issued by the Drainage Department provides guidance for planning and alterations of any drainage work in Singapore. Some of the requirements have recently been updated.

The main technique for flood alleviation in Singapore is increasing drainage capacity by widening and lining the channels. Only a few channels of Singapore are still in their natural state, and only in the relatively unurbanized fringe areas. About $130 million was spent between 1980 and 1984 to improve the drainage system including extension of the total length of lined channels from about 300 km to nearly 650 km. The latter figure indicates the existence of more than 1 km of lined channel for each square kilometre of land area. The older drainage channels also had to be enlarged, and all channels are periodically cleaned and desilted. Floodwater from certain basins, where urban development is extensive, also needs to be diverted to a neighbouring watershed. Huge, grey-coloured, lined drainage canals (with an inner channel to carry dry weather flow) have become a conspicuous part of Singapore's urban landscape. In areas that were developed in the 1980s, such large channels were put in at the same time as the houses, but in parts of the older areas of the city finding the space for channel expansion needed some innovative measures. For example, the Stamford Canal and its extension, which drains the prestigious Orchard Road neighbourhood, had to be redesigned to about twice its previous size, and space for it was found by creating a wide and much decorated pedestrian mall on top of the canal. In general the canals are open, usually dry, and aesthetically unappealing. The Drainage Department of Singapore in association with other government bodies is involved in transforming the local environment of these canals. Trees have been planted along their banks and the greenery is expected to attract birds and small animals. Footpaths and bicycle tracks also follow a number of these canals. In an unexpected development, the lined canals are now being designed as surrogate rivers. In an extreme case, a row of mangrove trees has been planted, in between the channel and high-rise residential buildings, on both sides of the lower parts of a canal (which was originally a muddy tidal river).

Singapore is an excellent example of increased flood expectation as a result of urban development. The Bukit Timah Valley, for example, was probably originally drained by a small river flowing through a vegetated floodplain which acted during floods as a temporary water storage area. By the 1960s, the channel was an earth canal with straight banks. In another few years the canal was lined, and in the early 1970s the first flood diversion was built on the Bukit Timah Canal, which carried floodwater to a basin towards the west. It was necessary in the 1980s, following a spell of rapid development in the upper basin, to build a second diversion (this time to the east) and widen the channel considerably (Figure 10.4), for about 2 km before the second diversion (Gupta, 1992).

Urban floods have been controlled in Singapore with three available facilities: financial resources, engineering capability, and an enforced code of practice for surface water drainage. Such resources are not always available for other cities in the tropics. Singapore is also fortunate in not being located in a hazardous area and in not having to cope with seismic disturbances, volcanic hazards or tropical cyclones, which make it nearly impossible to make a city flood-free. The case study for Kingston (Jamaica) illustrates what happens to a city that has to deal with such hazards (Box 10.2).

Figure 10.4 *Expansion of the Bukit Timah Canal*

10.3 Geomorphological Changes

There are basically three geomorphological changes associated with urbanization:

- accelerated erosion and sediment production from construction sites
- increased slope instability and failure
- morphological changes of channels left in the natural state

Accelerated erosion occurs when rain falls on bare soil during construction. In large-scale developments, the entire vegetation cover is removed, the area is graded, and earth piled up in mounds for later removal and adjustments. The duration of the construction phase varies but the ground could be expected to stay open for at least a year, if not more. This is enough time for a considerable amount of erosion to take place.

A few minutes after rain starts to fall on the bare slopes, a sheet of water laden with sediment begins to flow on the surface. After flowing for a few metres, the sheetflow concentrates to form rills or small gullies that then transfer the sediment either to a natural channel or to a perimeter drain. The amount of sediment removed is clearly seen on the bare slopes which, if left ungraded, develop networks of large gullies within a matter of days if the rains continue. Water samples collected from such gullies and perimeter drains routinely have sediment concentration figures as high as 10 000 mg l^{-1} in Southeast Asia. Douglas (1978) reported figures exceeding 80 000 mg l^{-1} from Kuala Lumpur. The sediment, if allowed to reach river channels, forms bars and constrictions and increases the flood potential of such streams. Once the construction phase is over, very little sediment reaches the channels and urban flooding begins. If, however, construction activities persist over a large area, sediment accumulates in rivers to build semi-permanent depositional features. If the urbanization is coastal, plumes of sediment are seen in the sea, as discussed in Chapter 11. Besides taking the extreme step of transforming the river to a concrete canal, some control over channel degradation can be exerted by restricting the amount of sediment reaching the rivers and also by slowing down and spreading the passage of rainwater to the channels over a long time period.

Urbanization also increases the potential for slope failures. This is particularly true for cities where uncontrolled urban expansion spreads over steep slopes, or in areas affected by seismicity or tropical storms. But even in quieter areas, neglect of the underlying geology and operating geomorphological processes may cause slopes to become unstable and fail. Pitts (1992), for example, has attributed all slope failures in Singapore to engineered slopes, i.e. those that have been regraded for landscaping, quarrying, road building, and residential development. Incidents such as aligning roads parallel to strikes in an area of sedimentary rocks, oversteepening cut slopes, or failing to provide subsurface drainage in areas of altered landscape, cause the slopes to fail. The amount of moisture collecting in soils and the soft weathered material is crucially important, especially in the humid tropics. Often relict structures in weathered material or an impervious layer of iron oxide precipitated from the groundwater leads to an accumulation of water above such zones until the upper layers fail. Application of local geology and geomorphology in urban planning, civil engineering or city management would prevent such disasters which could be very expensive both in terms of loss of life and damage to property.

Slope failures and landslides can be avoided by delineation of unstable zones before the beginning of urban expansion. Standard procedures for this exist and have been reviewed by Dunne and Leopold (1978). According to Dunne and Leopold, techniques for identifying areas prone to landslide hazards include the following:

- recognition of past hillslope failures
- recognition of conditions that are conducive to landsliding
- recognition of possible post-construction destabilization of slopes

Such recognition can be rendered by a geomorphologist but, unfortunately, such efforts are not usually required before city expansion.

10.4 Problems of Cities Located in Hazard Zones

Figure 10.5 shows the subduction zones, transform faults, common tracks of tropical cyclones, and a selection of cities in the tropics. Even a generalized map like this illustrates the nature of the environmental problems faced by a number of expanding cities in terms of relief, tectonic disturbances, volcanic eruptions and intense rainfall events. Seismic disturbances are associated with all subduction zones and transform faults. Subduction zones will also be areas of high relief, and, if one of the converging plates is continental and the other oceanic in type, the resulting mountain chain will be volcanic, as in the Andes in South America or the Barisan Mountains of Sumatra. Cities which grow on such slopes will be subject to three types of hazards:

- mass movements due to the presence of steep slopes
- seismic disturbances
- volcanic eruptions, usually of a pyroclastic nature

Where two continental plates have collided, as in the case of the Himalaya, mass movements and seismic disturbances will be frequent. As rainfall usually increases with elevation, rainfall on steep slopes activates slope instability and channel erosion to add to the hazardousness of the place. Even relatively drier slopes are subjected to tremendous erosion by slopewash and debris flow following rare cloudbursts.

Volcanic slopes are especially hazardous as high rainfall drains into stream channels radiating out from the upper slopes of the volcano. In such channels antecedent pyroclastic flows leave large deposits of volcanic material. This, when saturated with water following a rainstorm, gives rise to a type of mudflow called a *lahar*, the term originating from Indonesia. Both pyroclastic flows and lahars are capable of travelling long distances and engulfing parts of settlements with significant loss of life and property (Figure 10.6). A number of cities in the Philippines, Indonesia, and the Andean countries of South America are in areas of great potential danger due to expanding across steep volcanic slopes or simply due to their location near a volcano. The hazard is accentuated by the destruction of vegetation and by uncontrolled settlement. In western Java, river channels that originate on the slopes of the volcanic Gunung Merapi, pass through the historical city of Yogyakarta. It is eminently possible that pyroclastic flows and lahars may one day reach the city. Twenty rainfall-triggered lahars travelled down the Boyong River in the rainy season, which followed the collapse of a lava dome on Merapi on 22 November 1994 (Lavigne and Thouret, 1995).

In Box 10.2, we use the example of Kingston (Jamaica) to illustrate the problem of managing the urban environment in hazardous areas with limited resources. The conditions at Kingston are similar to those of a number of cities located on tropical steeplands.

Box 10.2

Kingston, Jamaica: A City in a Hazardous Environment

The city of Kingston was founded in 1692 on the coastal plains of Liguanea when a large earthquake destroyed the thriving buccaneer centre of Port Royal on the Palisadoes spit.

Over time, Kingston spread out of the coastal plains to the large gravel fan of Liguanea and then up the slopes of the surrounding limestone and highly faulted volcaniclastic mountains (Figure 10.7). The location of the island of Jamaica within a 200 km wide, seismically active zone between the Caribbean and the North American plates has put Kingston in a particularly susceptible earthquake-prone area. Hurricanes are also fairly common and rainfall from smaller tropical storms and depressions are annual events. About 200–300 mm of rain in 24 hours is expected to fall once in less than 5 years. Part of the cityscape is an environment of steep slopes, mass movements and flooded valleys.

The natural vulnerability of Greater Kingston from earthquakes, slopewash, mass movements and floods has been accentuated by deforestation and anthropogenic slope modifications over much of its environment. A cyclic pattern of destabilization of slopes is quite common. Sites of former landslides and the related downslope deposits are often developed for housing after these have been vegetated and to some extent stabilized, presumably because of the shortage of land on which to build, especially in the narrow valleys of the surrounding hills. Large-scale rainfall reactivates the slides, and the slopes are again destroyed along with any development on them. This process also transfers huge quantities of sediment to the valley-bottom streams. Slope failures are so common that about ten slides are expected along each kilometre of road in the hills after a rainstorm.

Given the scale of the problem (and therefore unlike Singapore), engineered slope stability measures are too expensive and impermanent. They will need to be redone after several years. The problem lies in a combination of several factors: bad siting of the city; spread of the city up steep slopes; floodprone lands; the need for cheap land for the poor and wide vistas and exclusivity for the rich (Gupta and Ahmad, 1998). These are common problems for steepland cities in the tropics. To this, volcanic hazards are added for cities located near an active continental–oceanic convergence boundary.

10.5 Climatological Changes

Urbanization creates its own climate, the changes usually being for the worse. Bryson and Ross (1972) list the types of environmental change that alter the local climate:

1. Changes in the physical surface:
   - increased impervious nature of the ground;
   - increased thermal admittance of the ground due to the removal of vegetation and the laying down of the concrete cover;
   - increased aerodynamic roughness due to the construction of buildings.
2. Production of extra heat due to fossil fuel combustion, transportation, industrial activities and human metabolism.
3. Increased turbidity and air pollution.

Landsberg (1981) has discussed such changes. In general, an urban settlement receives less radiation but has a higher temperature and greater cloudiness. The surface relative humidity is lowered but precipitation is increased. The wind speed is less, calm periods are more common, and the city air contains more

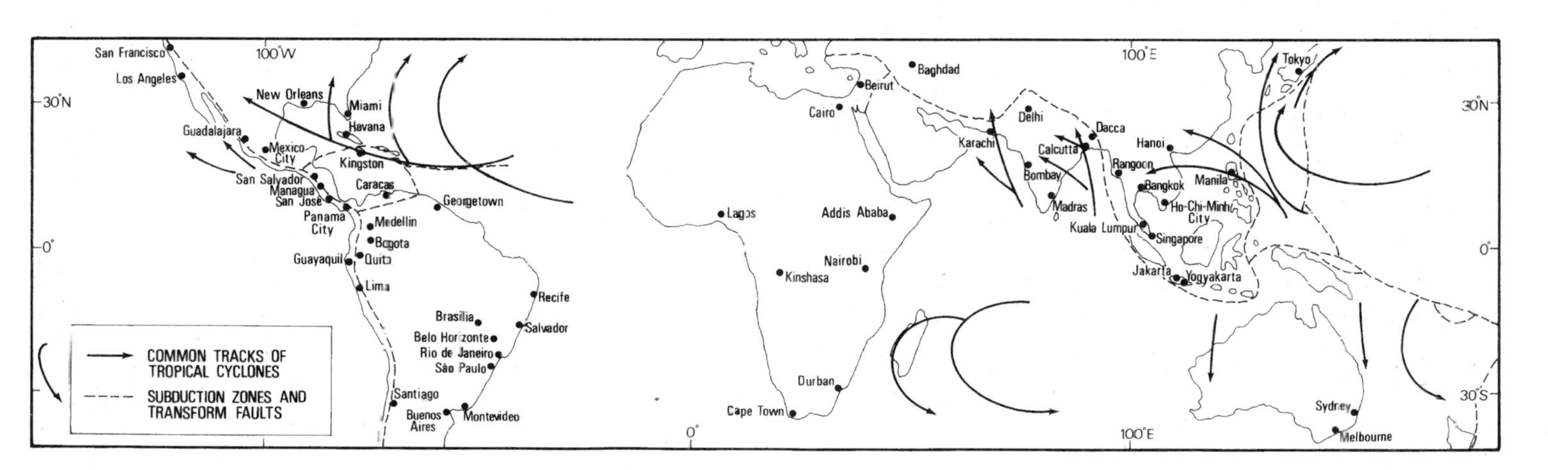

Figure 10.5 *Distribution of large cities in the tropics and hazardous areas*

Figure 10.6 A lahar filled river valley near Yogyakarta, Java

contaminants. As a result of these changes, the urban settlement becomes an isolated area of higher temperature surrounded by cooler countryside, a phenomenon which is described as the urban heat island. The heat island is a constant phenomenon, although it becomes stronger at night. Within the city, however, local variations in temperature occur due to both topographic variations and the distribution of roads and buildings. As a result of the city being hotter than the countryside, an increased buoyancy of the air occurs over the city and winds from the surrounding country blow towards the centre and circulate over the city to build up a climatic dome (Figure 10.8). The climatic dome is strongest when the regional weather patterns are too weak to interfere with its development and persistence. This is a visible dome as it also functions as a dust dome. Dust and smoke in the city are caught in this domal circulation and the haziness over the city is seen from a distance. This lack of mixing of air between the city and the countryside explains why air pollution in cities tends to persist. The exception occurs when a strong regional wind blows. Then a dusty plume from the city extends down-wind across the countryside. The nature and type of urban air pollution is discussed in Section 10.7, the section on pollution in the urban environment.

10.6 Urban Vegetation

The building of a city implies the destruction of natural vegetation; the creation of vacant plots; and the planting of selected species in parks, streets and gardens. The

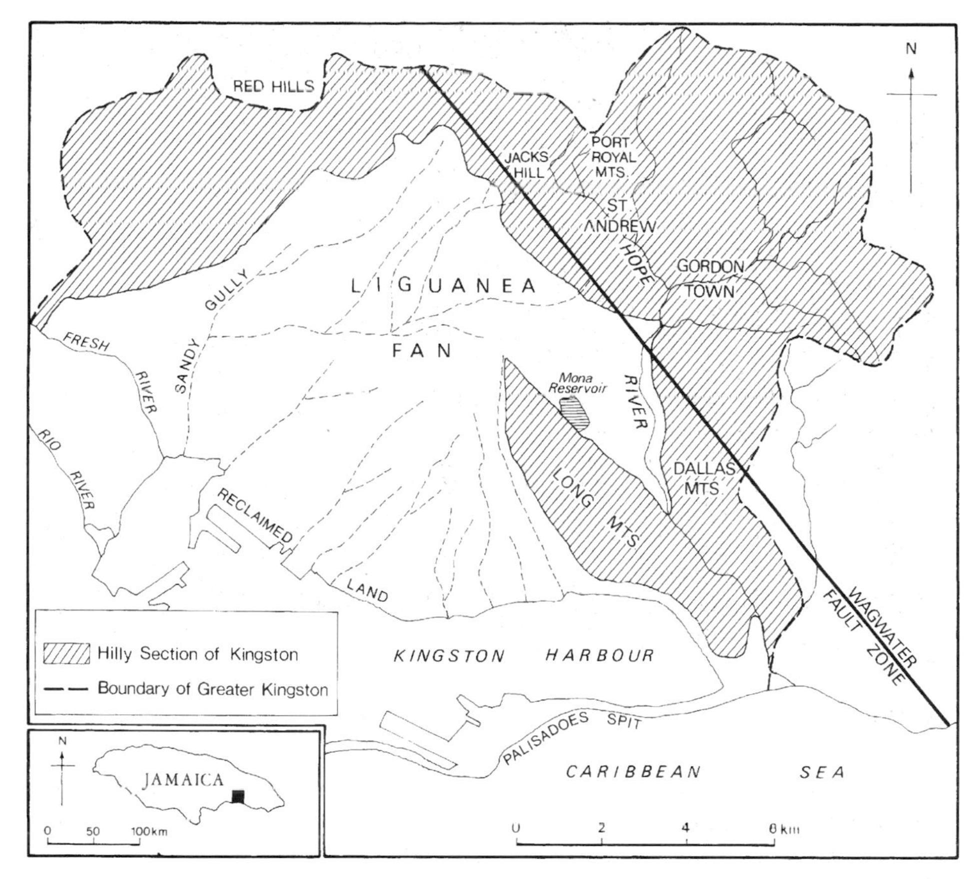

Figure 10.7 Kingston, Jamaica. The hills are prone to landslides and sheet erosion while part of the Liguanea Fan area is subject to flooding. Other problems are flooding on the coast and wind damage during hurricanes and earthquakes

ultimate urban vegetation therefore is a mosaic of different types, the principal components of this mosaic being

- relicts of natural vegetation
- secondary vegetation
- ruderal vegetation
- planned introduced vegetation

Relicts of the natural vegetation survive as patches of forests or coastal vegetation inside the city boundary either in areas undesirable for development or as a deliberate policy to maintain a natural area for the recreation and education of citizens. Secondary vegetation is found in areas which have once been cleared of their natural cover but then left unutilized for a number of years. This allows a number of species with adaptive and quick growing characteristics to establish themselves and grow to substantial dimensions. Such species could be different from those found in

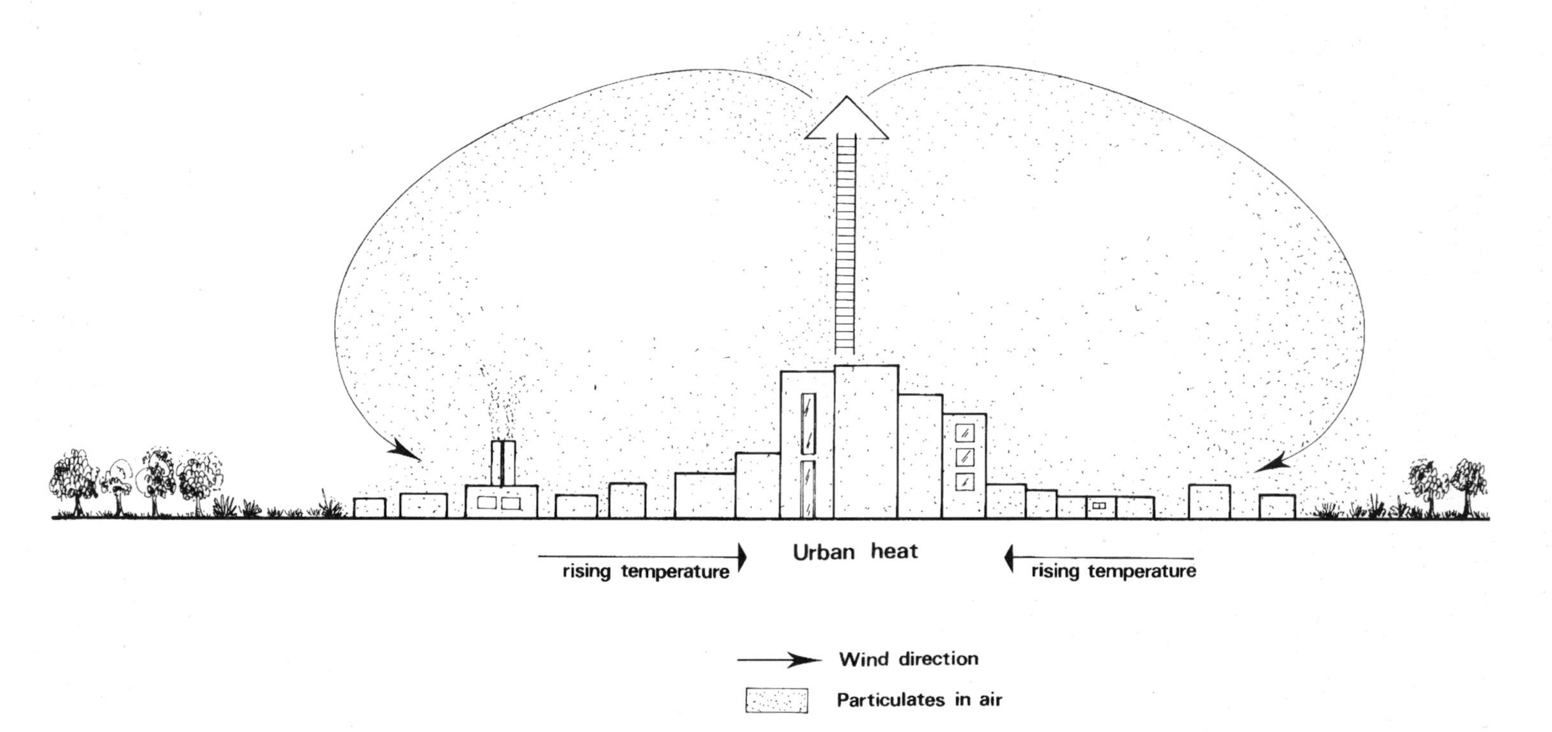

Figure 10.8 Climatic dome over a city

the relict areas due to the specialized traits required to invade a bare area, and also because such vegetation often includes species introduced by people from outside the region. Areas that are constantly disturbed do not achieve such a vegetative cover, but are usually under grasses, shrubs and small trees. This is the ruderal vegetation, usually found in urban wastelands. A large number of the species of the ruderal vegetation are popularly identified as weeds. The most significant change is the deliberate selection and introduction of species in parks, gardens and roadsides. The selection can be based on several characteristics. Roadside trees are planted for their shadiness, beauty or ability to act as fire breaks. Such a practice has resulted in the diffusion of a number of species across the world, climate permitting. For example, a very common roadside tree of the tropics, the rain tree (*Samanea saman*), is originally from seasonal South America. It is now very common in all parts of the tropics after being introduced as a shade tree. Such introduced species may replace local varieties by being better competitors. But most of this managed vegetation is maintained only by regular intervention to keep out the spontaneous varieties found in urban wastelands.

Vegetation that survives in a city (of either native or introduced variety) is often a vegetation of open, warm and windy environments, which are the characteristics of the urban climate. An ability to handle air pollution also helps. Characteristics suitable or desirable in trees of the urban tropics are as follows:

- adaptable to the urban climate
- resistant to air pollution
- fast growing
- provider of privacy and protection (shade giving, noise reducing)
- resistant to most biotic diseases and insects
- flowers, especially brightly coloured flowers
- flowers with sweet scents
- fruits that attract birds

Certain characteristics, on the other hand, are not desirable. These are found in trees which are

- weak wooded
- prone to drip after rain
- prone to clog drains and sewers by dropped branches, leaves and fruits
- prone to breaking pavements
- likely to attract undesirable insects
- burnt easily

(Detwyler, 1972, summarizing American Society of Planning Officials, 1968; Corlett, 1992).

Urbanization brings in a pattern of changed and managed vegetation. Within its common characteristics, as described above, the vegetation is managed differently between cities. In planned tropical cities such as Singapore, the managed vegetation component is rather high, whereas in other cities, apart from private gardens and roadside trees, the spontaneous component may dominate. In developing countries, part of the fuelwood demand of the city population perhaps could be met by social forestry by planting and maintaining suitable trees within the city limits. Rapid urbanization, however, leads to the disappearance of agricultural fields growing grain crops or vegetables within the city. The price of the land is too high.

Grasses also cover parts of urban settlements. In less prosperous cities such grasslands are spontaneous, but the managed grasslands in others could be well-maintained parks and lawns. With increasing prosperity, another type of managed grassland has started to take over sizable parts of the urban area: the golf course. This has immediate environmental impact of several kinds. First, such courses are often located on land which should be better used for housing or providing cover around water reservoirs. Second, the maintenance of golf courses requires regular application of fertilizers and pesticides which may be carried into water courses by runoff after heavy rain. The combination of trees and grasses on golf courses, however, tends to attract birds and animals. The urban fauna in cities (including birds) could be quite impressive. The remaining green space in urban settlements, such as vegetated narrow corridors along pipelines, is well populated and may connect two vegetated areas. The character of the fauna changes as the city is established, even for the avifauna. The cities may carry a large population of birds from the open country such as coastal areas or grasslands, rather than from the forests of the pre-urban stage.

10.7 Resources and Metabolism

An urban settlement exists in conjunction with the countryside. It is never sustainable by itself. People and resources move to cities; and industrial products, services and cultural creations diffuse from cities to the countryside. Cities need resources but one type of output from resource use is waste, which must be properly handled. A city produces huge quantities of wastes in solid, liquid and gaseous form every day, of various types, and from different sources (domestic, industrial, transportation). This input of resources and output of wastes has been described as the metabolism of cities (Wolman, 1965). Management of the urban environment includes the management of resource utilization and of the metabolised waste.

Some of the issues and techniques concerned have been discussed in earlier chapters which dealt with the management of land, water and air. Cities, however, are a special case. In cities, the demand for resources, the ability to transform material, and the production of wastes reach very high amounts within a space of several hundred square kilometres. A high number of people are directly, and immediately, affected by environmental mismanagement. The 1984 disastrous leakage of poisonous gas from a Union Carbide factory in Bhopal, which killed thousands, is perhaps the starkest example. Herein lies the urgency for managing the urban environment properly. The rapid urbanization of the developing countries therefore may turn out to be as demanding from an environmental point of view as saving the rain forest or living on a warmer planet.

Utilizing land resources for city expansion leads to floods, slope failures, and channels in disequilibrium, as discussed earlier. The need for water may also give rise to other environmental problems.

Water for a city comes mainly from rivers, lakes and groundwater, as discussed in Chapter 7. The rising demand with the expansion of cities, and the falling supplies at source, both create difficulties. For example, the degradation of water quality around Shanghai forced the city to move its intakes 40 km upstream. It cost $300 million (World Resources Institute, 1996). The extraction of groundwater often gives rise to land subsidence which augments the problem of inundation and water pollution of low-lying areas. We discuss the case for Bangkok (Box 10.3) to illustrate this, but the subsidence at Venice, Mexico City and Jakarta is due to similar causes. More than 70 percent of Mexico City's water supply is drawn out of its aquifer. Every year, the groundwater is lowered another metre; and in the 50 years between 1935 and 1985, the surface of the central area of the city (with the highest demand for water) has subsided 7 m. In many cities, such as Calcutta (Biswas and Saha, 1985), groundwater-related subsidence is not yet conspicuous but the warnings are in place. Cities located on deltas, coastal plains, and intermontane basin floors which extract water from sand aquifers situated within a thick clay deposit, are at risk from possible subsidence.

Box 10.3

Ground Subsidence in Bangkok

Located about 25 km inland from the Gulf of Thailand, Bangkok has a current population of more than 7 million, which is projected to surpass 10 million by the end of the century. This has been an extremely rapid rise: the 1950 population of Bangkok was below 1.5 million. The city is located in the flat Chao Phraya Delta with the ground elevation varying between 0.5 and 2 m above sea level. The old city was located next to the river, partly on the levees, but urban expansion has caused Bangkok to spread in all directions, including the swampy low grounds of the east and the south. The Lower Central Plain of Thailand (through which the Chao Phraya flows) and the Gulf of Thailand are located within a north–south trending structural depression filled with at least 500 m of poorly consolidated Quaternary sediments over a basement of igneous and metamorphic rocks (Khantaprab and Boonop, 1988). The Quaternary sediments are capped by a thin veneer of fluviatile mud on top, but most of it consists of deltaic deposits, predominantly clay which includes several beds of sand and gravel. The sand and gravel layers behave as aquifers and Bangkok's groundwater is pumped out of these beds.

The annual average rainfall for Bangkok is nearly 1500 mm, almost all of which arrives during the southwestern monsoon which may last from late May to early October. The Chao Phraya River drains the city along with a large number of canals which were constructed for drainage and transportation. During the rainy season, heavy rain, the high level of the Chao Phraya, and high tides may combine to make the city difficult to drain and widespread inundation occurs from time to time. This problem has been tremendously accentuated by land subsidence.

About a third of Bangkok's water is pumped out of the sand and gravel aquifers. The amount thus acquired daily was nearly 1.5 million m^3 towards the end of the 1980s. The area supplied by the city's water authority was about 300 km^2 in 1985, this figure being expected to rise to 815 km^2 by the year 2000 (ESCAP Secretariat, 1988). As the city expanded in the 1960s and 1970s, the use of groundwater grew rapidly, supplying mostly private housing estates and industrial establishments. The quality of the water was good and as the city water rates were not levied, its use was economical, especially for industries such as breweries or paper manufacturers. By 1985, the total amount pumped out by the private and public sectors was about 1.3 million m^3 each day. The piezometric level in the wells fell from the surface to about 9 m depth by 1959, and to nearly 50 m in places by 1983. Saline water also started to leak out of the marine clays in between the sandy aquifer beds into the groundwater (ESCAP Secretariat, 1988).

However, the main effect of groundwater withdrawal on Bangkok was subsidence at a rate which locally approached 10 cm per year. The cumulative subsidence resulted in a bowl-shaped depression over eastern Bangkok (Figure 10.9). The total area affected by subsidence has been estimated to be about 4550 km^2 (Nutalaya *et al.*, 1996). The bowl is very shallow, but during the wet monsoon it is easily flooded and is very difficult to drain as the water has to be pumped out to *klongs* (canals), and then again pumped over the levee to the already high Chao Phraya. Several days of waterlogging is quite common, and if the rains are as heavy as they were in 1983, when 575 mm and 454 mm of rain fell in August and September respectively, water may cover substantial parts of the city for months. The total damage for the 1983 inundation has been estimated to be $241 million (ESCAP Secretariat, 1988). The drainage system is also affected by subsidence as the drains and canals tend to sink with the ground. Structural damage to roads, buildings and pavements is common. Buildings on piles, which do not sink like the road on which they stand, require the addition of an extra step at the entrance; sidewalks develop cracks and scarps; small shophouses subside so much that the road surface could be at the level of their windows; pipes crack and bend; well casings protrude out of the ground surface; and a host of emergency and innovative flood-protection devices (such as small embankments) appear in front of shops (Figure 10.10). Septic tanks which take care of the domestic waste for large areas of Bangkok become a health hazard.

The solution lies in stopping the subsidence and, if possible, raising the level of the land. Given the scale of the problem, the remedial measures need to be incorporated into a long-term and holistic groundwater management system (Nutalaya *et al.*, 1996). At present, stricter controls are being enforced on groundwater pumping and the private consumer is charged a fee compatible to the city water rates. It has been noted that the piezometric level in central Bangkok has started to rise. The problem, however, will continue for years due to the slow rate of natural recharging of the aquifers. Artificial recharging has been suggested which may accelerate recovery, but the technique and its possible consequences require careful study. For example, any water pumped underground needs to be of a very high quality. The flood problem, however, can also be tackled at least partially by land zoning and by controlling the growth of Bangkok and its suburbs.

Urban water coverage ranges from just below 70 percent of the population for Africa to over 90 percent for the Latin American countries. Most of it is supplied by house connection. Sanitary coverage is much lower, ranging from about 53 percent for Africa to almost 80 per cent for the Latin American countries. The majority are served by house connection to a sewer or a septic system, but pit latrines are common in Africa and pour-flush latrines are common in Asia and the Pacific (World Resources Institute, 1996). A large sector still has no proper water supply,

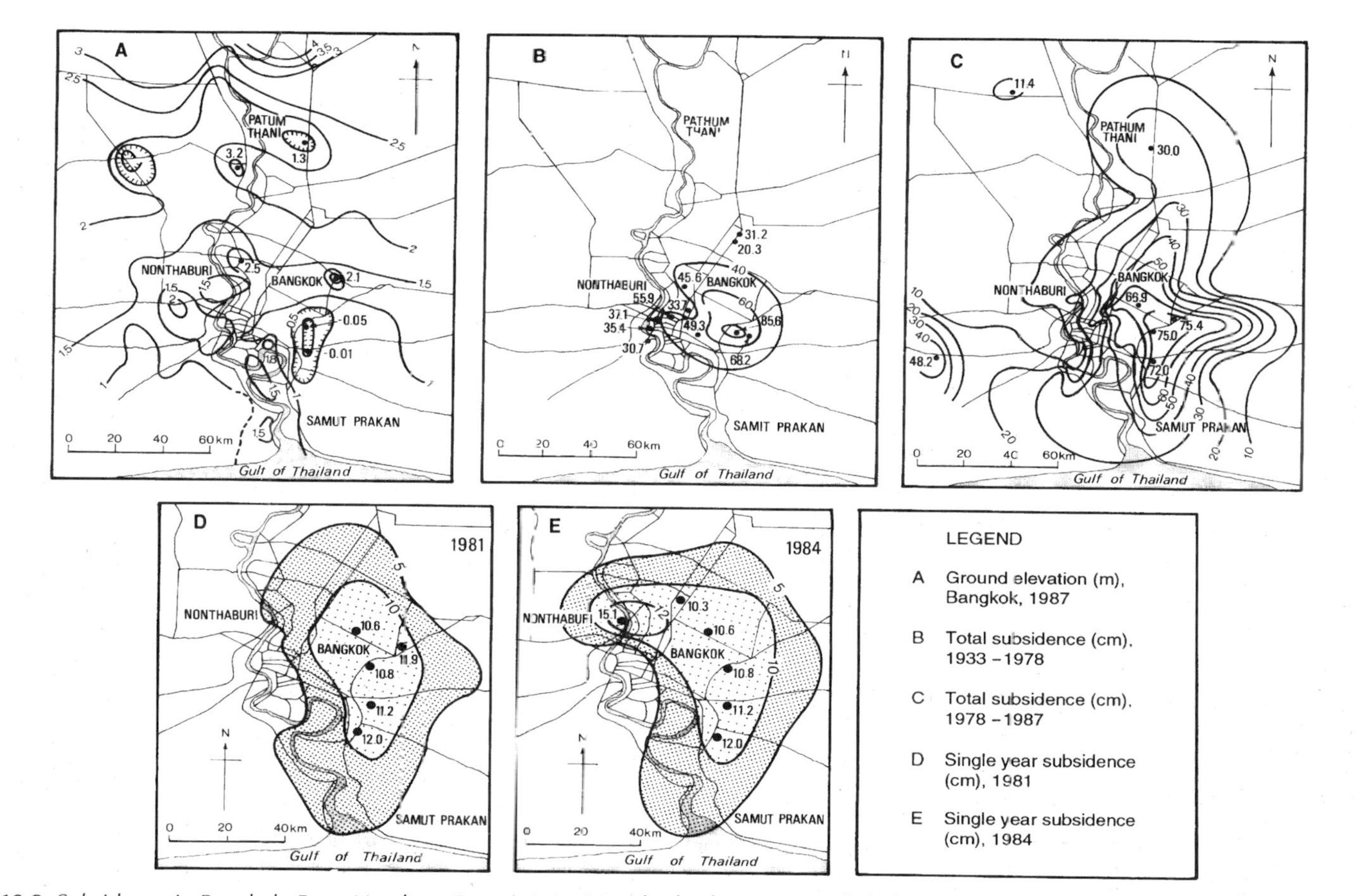

Figure 10.9 *Subsidence in Bangkok. From Nutalaya, P. et al. (1996) Land subsidence in Bangkok during 1978–1988, in (J.D. Milliman and B.U. Haq, Eds),* Sea-Level Rise and Coastal Subsidence, *Kluwer Academic Publishers, Dordrecht, 105–130. Reproduced with kind permission from Kluwer Academic Publishers and Prinya Nutalaya*

Figure 10.10 *Extra step below large buildings due to street subsidence, Bangkok*

and is dependent on water from open drains or canals. In worst case scenarios, as described in Chapter 7, the poor occupants of slums and squatter settlements buy water from street vendors. The unit price of water is several times higher than that of the municipal supply, and of course the water is of dubious quality. The more established residents have cleaner water at cheaper rates.

The metabolism of cities requires disposal of wastes in different states: solid, liquid and gaseous. In general, the disposal of wastes is not a success story in the developing countries.

Disposal of solid wastes

The disposal of domestic urban waste is well organized in the cities of the developed countries. Both putrescible (mainly food products) and non-putrescible wastes (paper, plastic, glass, rags, etc.) are collected regularly in large manually rear- or side-loaded trucks which compact garbage (packer trucks). Since most of the developed countries are located in the temperate zone, garbage may be collected only once or twice a week. In a number of cities, the garbage is classified by the householder and separated into different bins; one for glass, one for combustibles, and so on. The collected garbage is processed and disposed off either in landfills or by incineration. An efficient administrative structure, a fleet of large vehicles for garbage collection and street sweeping, and public participation come together to keep the city clean. The disposal of solid wastes is done differently in cities of the developing countries.

To start with, the garbage has a different composition. There is usually a lot less plastic, glass and paper, and more organic wastes. Plastic containers and glass bottles are reused as much as possible, and waste papers and rags are sold. Given the general location in the tropics, wastes have to be collected every day to prevent decomposition, foul odours, and disease transmission by vectors such as rats and flies. The streets, at least in parts of the city, are too narrow to allow collection by vehicles and the garbage has to be collected manually in wheelbarrows and brought out to the main roads to be picked up by vehicles. The vehicles are usually container trucks rather than packer trucks. Incinerators are costly, and although landfills are common, much of the garbage may get dumped in canals or rivers flowing through the city. Bangkok, for example, had to take steps to ensure a proper garbage collection which included more efficient city collection routes, the installation of a number of garbage bins in public sectors and markets, the building of incinerators, and converting part of the solid wastes into composts (ESCAP Secretariat, 1988). Garbage in many cities continues to be dumped in rivers or the sea depending on the location of the city. Apparently between 20 and 50 percent of the total solid waste of the cities of the developing countries is not collected officially and is disposed off in waterways, streets and unofficial dumps. In Guatemala City, for example, only 65 percent of the total solid waste is collected (World Resources Institute, 1996).

Even after collection, the problem of correct disposal remains. A proper landfill not only requires space, which may be difficult to find, but ideally also has to be correctly constructed with an impervious layer at the base to prevent leaching of polluted water through the fill into the groundwater or nearby surface drainage. The sides of the landfill also need to be sealed and decomposing gases allowed to escape from the pile of garbage. A soil cover should be spread over the day's dumping. Such precautions are not always taken, and often the garbage is tipped over a large area on the outskirts of the city, leaving the poor to scrounge through the pile for anything that has monetary value, e.g. pieces of metal. If it is an industrial city, the proper disposal of hazardous wastes (which may include toxic metals or chemicals) becomes a problem. The dump sites could be large for the megacities. The biggest solid waste disposal site in Manila, dealing with about 650 t of waste each day, has filled in 34 ha of Manila Bay and created a large hill of rubbish that reaches 40 m above sea level.

The disposal of solid wastes is often unsatisfactory in the cities of the developing countries. However, certain cities have tried to deal with it in innovative ways. For example, the solid wastes of Calcutta are dumped in the swamplands to the east of the city. The older parts of the landfill have been converted to market gardening and now supply a significant part of the city's vegetable requirements. Curitiba (Brazil) has been converted over the last few decades into something of an environmental showpiece. There, solid wastes are successfully collected even from the poorer sections of the city by involving the local community, as described in Box 10.4.

Disposal of waste water

Techniques for the treatment of waste water before its discharge into nearby streams or the sea have been discussed in Chapter 7. Again, a considerable disparity exists between developed and developing countries. Only a small number of cities in developing countries are equipped with waste water treatment plants which release water below a certain measure of BOD_5. Most cities discharge their domestic waste water without treatment to rivers, lakes or seas with expected consequences. The drive to clean up the Ganga in India arose in the 1980s primarily because of the domestic and industrial discharge into the river by a number of cities often situated close to each other in a conurbation. Malaysian urbanization in the 1970s and 1980s polluted the Strait of Malacca (Sharifa Mastura, 1987). This is in contrast with the polluted waters of the Great Lakes of North America or the River Rhine of Western Europe for which industrial pollution was responsible.

Industrial waste water is also a problem in the developing countries because of a lack of information regarding the nature of pollution and enforcement of the existing regulations on the errant industry. Not only is surface water polluted, but industries have also been known to pollute subsurface water by careless discharge of wastes, thereby putting at risk the local residents dependent on shallow groundwater (Guha Mazumdar *et al.*, 1992). This case study reported cases of chronic arsenic toxicity resulting from the use of arsenic-contaminated water from shallow tubewells close to a factory which manufactured the insecticide Paris-green (copper acetoarsenite). Uncontrolled discharge of industrial waste water in Jakarta Bay is indicated by the alarming high level of heavy metal accumulation. Prawns from the bay have extremely high levels of mercury contamination (World Resources Institute, 1996). In Hong Kong, the source of a 1988 hepatitis epidemic was traced to contaminated shellfish.

Most cities of the developing world lack the financial resources to build proper purifying plants for waste water for all their cities. Sometimes, however, innovative techniques can be used to get round this problem. Calcutta uses a technique for improving the quality of municipal waste water which does not use a treatment plant but requires sufficient land area and abundant sunshine.

The city of Calcutta evolved on the levee of the Hooghly River. To the east lie the backswamps of the old delta, drained by a network of tidal creeks which ultimately move the water eastward to the large tidal distributaries of the active delta of the Ganga. The municipal waste water, carrying mostly organic wastes, drains eastward towards this swampland. The waste water transported to this area via canals spreads into a set of fish ponds through which the water flows sluggishly. The organic wastes are decomposed in the presence of abundant solar energy by the microfauna in the water, fish, and aquatic vegetation. The semi-purified water then drains through a set of paddy-fields further east and ultimately into the large tidal distributaries flowing towards the Bay of Bengal (Figure 10.11). The final discharge is of reasonable quality, no treatment plants are used, and the city's fish and rice supply is augmented in the process. Based on Calcutta's experience, Ghosh (1995) has prepared a manual for utilizing a wetland system for waste water treatment and recycling in the tropics.

Changing air quality in cities

Deteriorating quality of air is another basic environmental problem of rapidly growing cities. Air pollution created by many sources (Chapter 9) such as industry,

Figure 10.11 *Swamplands east of Calcutta used to treat domestic waste water*

power generation, transportation, and biomass burning leads to increased mortality and problems of pulmonary, cardiovascular and neurobehavioural functions. The cities of the developing countries are particularly vulnerable as sources of pollution and residential areas are usually not far apart and could occur in the same neighbourhood.

Not much hard information, however, exists for the air over cities of the developing countries. The available information is a combination of detailed case studies on certain cities by individual researchers and government-monitored measurements of air pollution. The latter also covers only a small number of cities and only for several years. A summary (WHO/UNEP, 1992) of the quality of air over 20 megacities was published by the World Health Organisation and the United Nations Environmental Programme as part of the Global Environment Monitoring System (GEMS). Of these, most are located in the developing countries, with 13 being in the tropics and subtropics. The major sources of air pollution remain as described in Chapter 9:

- the combustion of fossil fuels for power generation
- domestic heating and biomass burning
- transportation
- industrial processes
- incineration of solid wastes

The air over almost all cities in the developing countries contains suspended particulate matter (SPM) well over the accepted WHO limit. Carbon monoxide, lead, and oxides of nitrogen are the other major pollutants which commonly occur. Sulphur dioxide and ozone have variable distributions, being high in some but not all cities. Ozone-related forest damage has been reported near Mexico City. This variability in the composition of the polluted air is due to the regional differences in the production of different types of pollution, and separates out the cities of the developing countries from those in the developed ones where the sources and patterns of pollution are common. For example, the presence of SO_2 in Bombay is from industrial sources (chemicals and textiles) and power plants. SPM is high in Cairo due to the presence of limestone quarries and associated cement plants. It has been estimated that about 5 percent of all the cement produced in Cairo ends up as SPM in air over Cairo. Cities in Central and South America have to contend with pollution from vehicles, which is not the crucial source for air pollution for many African cities. In general, the large number of older and poorly maintained cars, diesel engines of trucks, buses and taxis, and two-stroke engines of motor-cycles and autorickshaws which ply in large numbers, all spew out SPM, NO_x and lead in the atmosphere. It should be noted, however, that several developing countries have taken steps to phase out lead in petrol and to require catalytic converters for new cars. The presence of toxic and carcinogenic chemicals in the city air includes heavy metals (e.g. berellium, cadmium and mercury), organics (e.g. benzene, polychlorodibenzodioxins, formaldehyde, vinyl chloride, and polyaromatic hydrocarbons or PAHs) and fibres (chiefly asbestos). The sources for these include waste incineration, industrial processes, solvent use, building materials, automobiles, and sewage-treatment plants. The concentration of these is still low but with greater urbanization and industrialization the danger may increase, especially if land zoning is not practised, as often is the case in these cities.

Where records do exist, the trend over the last ten years or so, is one of increasing air pollution. The increase is due to three factors:

- rapid population growth
- industrialization
- increased energy use (WHO/UNEP, 1992)

As the importance of these three factors is not likely to decrease, it is imperative that the pollution of air in the large cities of the developing countries be carefully monitored and contained. In Calcutta and its neighbouring town of Howrah, the number of buses and trucks which use diesel increased by 78 percent between 1980 and 1989. About 60 percent of Calcutta's population already suffers from respiratory disorders due to polluted air, chiefly SPM (WHO/UNEP, 1992). In Mexico City, SPM contributes to the death of over 6000 people each year (World Resources Institute, 1996). Comparable situations occur in many cities which may or may not be monitored for air pollution. A World Bank estimate suggests that if SPM concentration is brought down to the level accepted in the WHO guidelines for air pollution, between 300 000 and 700 000 premature deaths would be averted each year in the developing countries (World Bank, 1992). The risk is particularly high where physiography and climate from time to time prevent pollutants from dispersing. A good example of this is Mexico City which is located in a mountain-girt basin where inversion of temperature occurs from time to time.

Indoor air pollution from the burning of charcoal, biomass or animal dung is not limited to the villages; it is also a problem in a number of cities of the developing countries. As in rural areas, it is the women and children who are exposed to this type of pollution. Hard data are difficult to come by, but based on what is available, the World Bank considers indoor air pollution (both rural and urban) one of the most critical environmental problems, contributing to respiratory infections in the young and cancer and lung diseases in adults (World Bank, 1992; World Resources Institute, 1996).

10.8 Urban Environment in the Developing Countries: the Present and the Future

Urbanization modifies the local environment, mostly for the worse. Although the supply of clean drinking water and proper sanitation are much better in the cities of the developing countries than in the villages, environmental problems such as land degradation, air pollution or disposal of hazardous wastes become extensive. Where it has been possible to calculate the costs in monetary terms of a large-scale urban environmental problem, the figures have been disturbingly high. For example, the cost of traffic congestion in Bangkok has been estimated to be between $272 million and $1 billion, i.e. about 2.1 percent of Thailand's GNP (World Resources Institute, 1996). Even then this takes into account only time lost from work, and does not include the cost of health problems associated with transportation-related pollution of the air. The WHO has estimated that the majority of the five million children that die from diarrheal diseases in the developing world are from poor urban families (World Resources Institute, 1996).

The environmental problems of the developing countries which are due to urbanization are worsening because of two factors: the astounding rate of city expansion, and the lack of financial, technical and legal resources available for environmental management. The existing political systems generally do not give high priority to environmental issues. Certain types of environmental degradation can perhaps be dealt with by innovative measures, as done in Curitiba, Brazil (Box 10.4). The sheer scale of the ecological degradation requires that urban problems are recognized as one of the biggest environmental challenges of the next century.

Box 10.4

Urban Problems and their Solutions at Curitiba, Brazil

Curitiba, the capital of the Paraná State of Brazil, has become an environmental management showpiece by adapting low-cost, innovative and community-based techniques for solving the standard environmental problems that accompany rapid urbanization. Between 1950 and 1990, the population of Curitiba jumped from 300 000 to 2.1 million. Its economic function also changed significantly from being a centre for processing agricultural products to an important industrial and commercial centre. The usual litany of urban degradation was present: in-migration, unemployment, squatter settlements, traffic congestion, waterways blocked with solid wastes, and flooding. Curitiba, however, evolved into a city with significantly less pollution and a better social environment than the regional norm.

This amazing transformation was carried out by innovative and dynamic planning under the leadership of Jaime Lerner, a three-term mayor of Curitiba who subsequently became the state governor. Jaime Lerner has been described as a visionary but professionally he is an architect and planner. His policies continue under the present mayor, Rafael Greca, who has also built on them. Curitiba's success has been attributed to the building of a new large-scale surface public transportation network; working within the constraints of the local environment and resources; using appropriate but low-cost solutions; and encouraging citizen participation, instead of confronting citizens with a static master plan (Rabinovitch and Leitman, 1996). Officials were encouraged to inspect all problems and discuss them with the local people before taking decisions.

It has been said that of all the innovative techniques used in Curitiba the most striking is the bus service, which has been described as a surface metro. In Curitiba, surface transportation and land use planning were tied together in an attempt to motivate the expansion of the city away from the centre along five corridors, thereby preventing the usual concentric growth of a congested city. Along these corridors, three parallel roadways were established. The road in the middle carries two express bus lanes flanked by local roads. Two high-capacity one-way streets run in opposite directions parallel to the busway and separated from it by a block. Land use regulations encouraged high-density services and commercial occupations along the corridors. In contrast, the areas between corridors are zoned for low-density residential use. The central city, with greatly reduced traffic, is traversed by pedestrians and bicyclists.

The express buses are supplemented at various points and at all termini by interdistrict and feeder bus services. A single fare allows passengers to travel on any bus, and as this is collected at specially raised tubular bus stops before boarding, the buses need to stop for very little time. The express buses are double- and triple-length articulated vehicles, the biggest of which can transport 270 people. This is a highly successful surface metro service and was constructed at a cost of $200 000 per km instead of the estimated subway costs of $60 to $70 million per km.

The success of this system is seen in the number of users: about 1.3 million a day, including a high number of car owners. Both per capita fuel consumption and ambient air pollution for Curitiba is extremely low for Brazil. The average low-income residents need to spend only about 10 percent of their income on transport. The city pays one percent of the value of a bus to the bus owners each month, takes delivery of the vehicles at the end of ten years of use, and refurbishes them as free buses in city parks or as mobile schools. The average Curitiba bus is only three years old. The bus companies are paid according to the length of the routes they serve, instead of the number of passengers, thereby destroying the interest to serve only the busy roads.

Before the public transport system was developed, land was acquired and set aside for 40 000 low-income houses near the new industrial centre founded in 1972, about 8 km away from the centre of Curitiba. This allowed the urban poor to have jobs within a short distance of their homes.

Curitiba used land zoning to control urban flooding. During the 1950s and 1960s, dumping of solid wastes, urban development along the channels, and engineering modifications exacerbated the flood problem. The problem was solved by setting apart riverbank land for drainage and preventing building on low-lying areas. Riverbanks have been converted to parklands, and artificial lakes have been used to contain floodwaters. Curitiba thereby developed recreational areas at the same time as the floods were controlled at a low cost. No substantial new investment was required. Surprisingly, this green space next to the rivers was created during a period of rapid population growth.

Providing necessary information and incentives is behind the success story of Curitiba. The City Hall provides information instantly to any citizens about the building potential of any area inside Curitiba. Incentives and encouragement are also supported by Curitiba's Free University for the Environment, which offers practical short courses free of charge to homemakers, building superintendents, shopkeepers and others, regarding the environmental implications of ordinary jobs. Such training is a prerequisite for certain jobs, but the courses are also taken voluntarily. Programmes also exist for supporting school children from low-income families.

Perhaps the most striking example of Curitiba's innovativeness is the way in which solid wastes are collected. Garbage collection being difficult in the narrow streets of the poorer sections of the city, people are encouraged to classify their solid wastes and bring it out to nearby collecting points where the city, in return, supplies them with fresh vegetables and other food items. The city has been known to exchange garbage for bus tokens, special arrangements at Christmas, or tickets to cultural functions. Students can exchange garbage for notebooks. This exchange programme costs no more than hiring trash collectors for the slum areas but provides nutrition and transport facilities for the poor as well as involving them in the smooth functioning of the city (Rabinovitch and Leitman, 1996). About two-thirds of the city's solid waste is recycled due to prior sorting by the residents. Innovative measures, public involvement and labour-intensive approaches have made the waste collection and processing system effective at a low cost and without expensive engineering techniques. Curitiba has demonstrated that it is possible to make city management and planning efficient by (1) taking correct decisions which match the environment and (2) avoiding a fixed top-down approach. Rabinovitch (World Resources Institute, 1996) has stated that Curitiba illustrates how creativity can substitute for financial resources, but the final success also requires a will to change, political commitment, and leadership, which are the hallmarks of Curitiba.

A number of unanswered questions remain. Will the expansion of the cities replacing farmlands and forests accumulate to create a global problem of resource depletion? As the cities expand, would this cause greater erosion and sedimentation leading on to river modifications and, in coastal areas, serious offshore sedimentation affecting marine fauna and flora, including coral reefs and mangroves? If global warming occurs, with the associated rise in sea level, changing climate, etc. (Chapter 14), how would the cities of the future cope? Urbanization in the tropics therefore has a large but hidden future agenda so far as environmental management is concerned. The sheer size of urbanization of the developing countries could make it a global problem, the solution to which may require as much resourcefulness as needed for managing the disappearing rain forests or global warming.

Exercise

1. The government of Wonderland has decided to build a new capital city. It is planned as a showpiece and no 'dirty' industry will be allowed inside the city. The site is an alluvial fan at the base of an escarpment. A major river flows through the proposed site. The annual rainfall is 1200 mm and seasonal. The area is under secondary tropical forest at present. Describe the environmental problems the city may encounter as it grows.
2. Select a city in the humid tropics which is growing rapidly and has at least a million inhabitants. This city should be located on either a coastal plain or a delta. It must be a real city and not an imaginary one.
   (a) Tabulate its environmental problems. Given the business-as-usual scenario, what would be your predictions for the city 50 years in the future?
   (b) Suggest feasible methods for controlling the problems. If your suggestions are accepted and used by the city authorities, how would your predictions change?

11

The Coastal Waters

11.1 The Land–Sea Interface

The coastal environment is a special case, as it interfaces between the land and the sea. The physical processes that work here are a combination of land, marine and wind processes which together accumulate a mixture of material derived from both land and sea. Environmental changes happen rapidly. A rich biological diversity is found in many coastal areas, as well as special types of environment, e.g. mangroves and coral reefs, which are extremely vulnerable to environmental degradation. Not only is the coastal environment fragile, but many parts of it are also highly populated. Such a combination makes the impact of economic development both extensive and highly degrading. The two major types of such degradation are the uncontrolled use of coastal resources and an accumulation of polluted material from both land and sea. In this chapter we examine the nature of degradation of the coastal environment, and the possible strategies for environmental management.

11.2 Coastal Features and Processes

The level of the sea fluctuates according to the tidal movement. The zone between the mean high water level and the mean low water level is known as the intertidal zone or foreshore (Figure 11.1). Higher than the foreshore zone is the backshore, which usually consists of beaches backed by sand dunes or rock cliffs. Lower than the foreshore is the inshore zone, which is usually under water. The far part of the coast beyond the range of breaking waves is known as the offshore zone.

The sea moves over the backshore only during very high tides or storms or when a naturally protective barrier has been removed. Then the beach and even the dune can be severely eroded. At other times, beaches are stable depositional features. The rock cliffs, where they exist, are eroded only by high-energy waves which can reach the cliffs either during storms or at headlands. When eroded, the rock cliffs retreat slowly in the face of the sea, leaving in their former location a rock platform

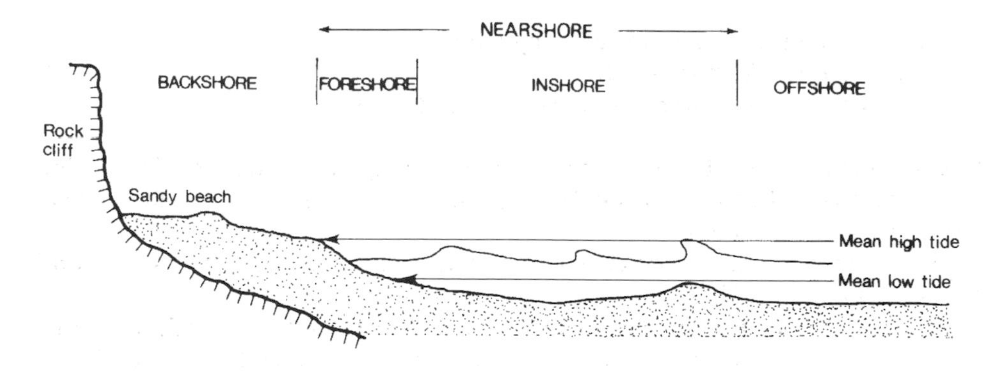

Figure 11.1 *Section through a typical coast, illustrating various terms used in the text*

on which remnants of old cliffs may stand out as pinnacles of various shapes and sizes. Material eroded from a headland against which large waves break is swept parallel to the coast and is deposited at the head of a sheltered bay between two headlands to form a beach (Figure 11.2). Over the nearshore, breaking surf and waves often build a set of bars. In exceptional cases, large islands are formed parallel to the coast known as barrier islands. The bigger the tides and the more plentiful the sediment, the wider the shore zone. The major sources of coastal sediment are three:

Figure 11.2 *A headland–bay coast with erosion on the headlands and deposition in the bays*

- sediment brought down to the coast by rivers
- sediment eroded from rock cliffs
- sediment moved shoreward from the bottom of the sea

All coastal depositional features, unlike the rock cliffs, are prone to rapid alteration by either erosion of existing material or deposition of new sediment. Changes affect both forms and dimensions.

Several special cases of coastal features – estuaries, deltas, mangroves and coral reefs – are highlighted below.

Estuaries are wide river mouths emptying into the sea, with no islands or bars or only a few. However, estuaries, at least those of big rivers, often have depths in the order of tens of metres and the tidal influence can extend up and down the channel. This extends the zone of contact between the fresh river water and saline sea water which leads to flocculation. Flocculation is a process that happens when river water carrying suspended fine particles of clay meets saline water. In the ensuing reaction, the fine clay particles combine with each other to form bigger units which then settle down to the bed of the river. Thus deposition of fine-grained material occurs in river mouths over a sandy or clayey bed. The extent of this zone of sedimentation is directly related to the tidal range.

Deltas are formed where riverine sediment is deposited at or near river mouths by (1) the slowing down of the ambient river flow by sea water moving towards the coast and (2) flocculation in the tidal zone. The character of a delta depends on various geomorphic processes that operate in the area: rivers, waves, currents, winds. Where the river brings in a large amount of sediment and both wave energy and tidal range are limited, the ambient river water pushes into the sea and deposition of sediment occurs in strings of bars and islands radiating out from the river mouth. The final appearance of this type of delta resembles the foot of a bird and a good example is the delta of the Mississippi. In comparison, where tides are strong, the riverborne sediment moves back and forth over a wide zone over which flocculation occurs. The tide-dominated deltas are characterized by funnel-shaped river mouths, elongated bars and islands parallel to the tidal movement, and a number of small tidal creeks separating these islands. Over many tidal deltas, especially those in the humid tropics, salt-resistant vegetation (mangroves) grows, anchoring the muddy bars and islands. The Ganga-Brahmaputra and the Mekong deltas are excellent examples. Where wave energy or wind action is strong and the flow of the river is relatively sluggish, the river-borne sediment is pushed back on to the coast to form a delta with a straight coastal edge and good beaches. The delta of the Senegal River, for example, illustrates this type.

Mangroves form a vegetation community (Figure 11.3) on the coastal wetlands of the tropics, especially on large tidal deltas. A dense community of mangroves is also called a *mangrove forest* or a *mangal*. Individual plants may be shrub-like but most mangroves are communities of trees which may reach tens of metres in height. Mangroves extend from saline conditions next to the sea to brackish waters of the estuaries away from the open sea. They develop best in a sheltered sediment-rich fine-grained environment but occur on sandy and rocky coasts as well.

Figure 11.3 *Mangroves, Malaysia*

Mangroves include a number of species of plants, most of which are found in the tropical parts of the Old World. The mangrove communities of the Americas have relatively fewer species. The plants anchor themselves in the mobile fine-grained coastal environment by specialized root networks which may protrude from the mud in spikes (e.g. *Avicennia*), form a basket of prop roots (e.g. *Rhizophora* or *Brugeria*), or develop buttresses (e.g. *Lagunclaria*). The plants breathe through the exposed parts of their roots and also have physiological arrangements for adapting to a saline environment. The plant community also includes palms and ferns. Mangroves have three important environmental inputs:

- their productivity is high and detached parts of the plants, leaves, etc., after falling into the water, move back and forth in tides providing nutrition to coastal organisms over a wide zone;
- they provide coastal protection, especially from storm waves and high winds;
- they provide a suitable habitat for birds, fish, crabs, crocodiles, insects, etc.

Coral reefs are biological structures built by a number of organisms dependent on a combination of favourable environments. They grow within a narrow temperature range (optimal range 25–29°C), and in the presence of light. As a result they are formed only in tropical seas, both near the coast and in the middle of the ocean. Three types of reefs occur:

- fringing reefs, growing next to the coast;

- barrier reefs, found within a short distance of the coast and separated from it by a shallow body of water, known as a lagoon;
- near-circular or elliptical reefs called atolls found in mid-ocean which enclose a lagoon and form upwards from a geological basement such as a submerged volcano.

The corals of the Indian and Pacific Oceans are rich in biological diversity (about 500 species) compared to those of the tropical Atlantic (about 65 species). The basic unit of the reefs is the coral polyp which is enclosed inside a calcareous skeleton. New calcareous material is added to the outside of the skeleton as the coral grows. The nature of their growth gives the coral reefs a basal plate of calcium carbonate with vertical ribs growing upwards from it. An established completed reef is a combination of various types of corals of different forms and calcareous algae. A complete coral reef has a flat top with two steep sides, one going down to a shallow lagoon and the other descending much deeper towards the bottom of the open sea. The submerged structure of the reef is characterized by flat platforms at different levels, vertical walls, canyon-like openings in the reef, and hemispherical surfaces or open-structured branching on top. Rapid and efficient cycling of nutrients takes place within the coral reefs, which are species-rich ecosystems, including besides corals and algae a host of marine organisms such as varieties of fish, clams, sponges and foraminifera. Different species of coral tend to inhabit different levels and positions in the reef. Often a modern reef has a Pleistocene base (Figure 11.4). All this together with the colourful fish that colonize the reef make coral reefs arguably the most spectacular of ecosystems and capable of attracting a large number of tourists, especially where the reefs are large and wonderful such as the Great Barrier Reef off northeastern Australia or the reefs off Mauritius.

Coral reefs, however, are strictly dependent on environmental attributes such as light, exposure, powerful storms, salinity and turbidity. They survive best at shallow levels with some light penetrating below the water surface and in nutrient-poor conditions. Therefore, a falling sea level, an exceptionally low tide, a change in salinity, or increased turbidity could be lethal for corals. They are also destroyed by reef grazers (particularly the Crown-of-Thorns starfish) and burrowers (echinoderms, bivalves, sponges and worms). Under such circumstances a substantial part of the reef may be physically destroyed or killed or lose out in competition with macroalgae, sponges, bryozoans, etc. Such events happen under natural circumstances but, more frequently, anthropogenic degradation of the coastal

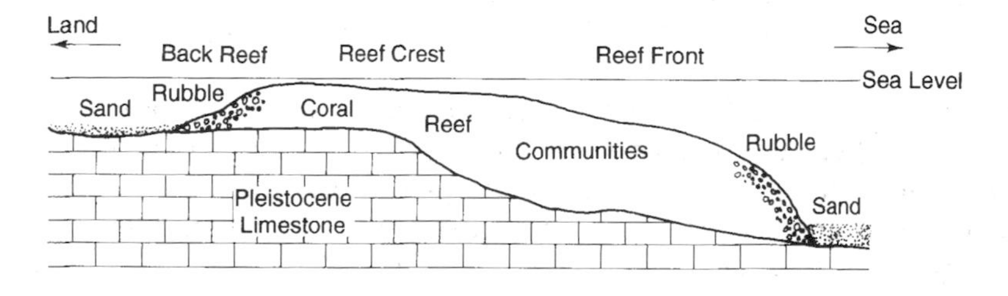

Figure 11.4 *Diagrammatic cross-section of a coral reef*

environments such as increased sediment load or other types of pollution and even physical destruction (for excavating boat channels or by using dynamite to stun fish) is removing large sections of coral reefs. Large parts of the reef are removed as building materials or for lime for making cement. Corals subjected to thermal stress by warm outflows from power stations turn white, a phenomenon known as coral bleaching. Corals are at risk from various types of environmental impacts. About 240 000 billion barrels of crude oil escaped from a ruptured storage tank on the coast of Panama in 1986, causing extensive damage to nearby coral reefs (Guzmán and Jarvis, 1996). Kühlmann (1988) mentioned the destruction of excellent coral reefs between 1978 and 1985 off the volcanic island of Moorea (French Polynesia), apparently due to excessive sediment accumulation following the removal of a belt of coconut palms on the island for the construction of hotels. Dredging and levelling of the sea bed also destroyed coral reefs in French Polynesia. Destruction of coral reefs due to increased sedimentation in coastal waters has been reported from Panganiban Bay in the Philippines. Dumping of mine tailings is also very destructive for coral reefs as has happened regarding copper mines in the Philippines and tin mines in Thailand (McManus, 1988). In Thailand, damage to coral reefs was also done by bucket dredges operating in Phuket Bay where they destroyed reefs near the beautiful tourist island of Phuket (Chansang, 1988).

Coral reefs are apparently disappearing at a rapid pace (Lundin and Lindén, 1993). For example, about 60 percent of the coral reefs of Southeast Asia have either disappeared or are on the verge of doing so (Wilkinson *et al.*, 1993). Increased tourism has also led to the disappearance of coral reefs due to a variety of factors: the collection of pieces of coral as souvenirs; increased pollution, including sediment accumulation in the coastal waters resulting from resort building; and careless dropping of boat anchors.

Three factors make the coastal waters particularly vulnerable to environmental degradation. First, as the interface between land and water they tend to accumulate pollution and other degrading material from both sides. Second, the coastal lands are heavily settled (including a number of major cities) which increases the level of environmental degradation. Third, sections of coasts carry sensitive ecosystems such as mangroves and corals which do not cope well with certain types of anthropogenic degradation. Their disappearance in turn brings in other problems. For example, the disappearance of a mangrove community often leaves the coast with little protection against erosion from waves and tidal currents. In the rest of this chapter we examine the nature and effects of marine pollution and overuse of resources, as well as strategies for coastal zone management.

11.3 Marine Pollution

Inputs of material which alter the nature of sea water can come from both natural and anthropogenic sources. Heat, for example, is added to coastal waters by volcanic eruptions as well as industrial discharges. Such inputs are considered pollution when they degrade the quality of the water enough to cause a negative impact on organisms (including people) or to create an aesthetically displeasing environment. Marine pollution usually has an anthropogenic source.

Pollution has to be assessed with reference to its type, source and effect. By volume, most of the marine pollution is organic material (Clark, 1992), which is subject to biological degradation generally in the presence of oxygen, and therefore its effect is usually restricted over time and area. Certain types of pollution are quickly dissipated in or buffered by sea water and the effect is extremely localized. Examples include heat released from cooling water discharged from coastal power stations and industrial establishments, as well as acids, alkalis and cyanide. In contrast, heavy metals (such as mercury or lead) or halogenated hydrocarbons (such as DDT or PCB) do not break down or dissipate easily, and their effects on various organisms could be persistent and harmful. A particular source of pollution may release multiple types of pollution. Power stations may release water not only at high temperature but also carrying some chlorine and metals. Untreated water discharges from the sewage pipes of cities carry a range of contaminants along with organic wastes. The inputs can be

- off land, via estuaries or outfalls from coastal towns or industries;
- from the sea, usually as deliberate or accidental discharge from ships;
- from the atmosphere, when pollutants are returned in rain or in fallouts, or as directly dissolving gaseous wastes at the water–atmosphere interface (Clark, 1992).

The different types of marine pollution can be listed as

- oxygen-demanding wastes
- oil-related wastes
- conservative wastes
- solid wastes
- heat
- radioactivity
- sediment

Oxygen-demanding wastes

Sewage is commonly released (often untreated) into coastal waters. Organic sewage is broken down by aerobic bacteria in rivers in the presence of oxygen, as described in Chapter 7. Tidal waters are a special case. Flocculation usually occurs in tidal river mouths, and clay particles settle to the bottom, carrying with them contaminants. River beds and estuarine mudflats are places where organic material, pesticides, metals, etc., concentrate. The wastes stay for a long period in the tidal reach, moving back and forth, which leads to the concentration of organic wastes in the reach and a resultant drop in the oxygen content of the water. The sag in oxygen may reach such a low level that life forms disappear from the river reach, except the anaerobic varieties of bacteria and protozoa. Such a condition is easily recognized by the foul odour that emanates from a section of the river. If the estuarine water has a density stratification with the denser saline flow from the sea underneath, the waste may not get flushed out of the river at all. Coastal settlements usually discharge their sewage directly into the sea by pipelines. Where the sewage is untreated or the outfall is badly sited with little mixing, coastal waters and beaches also become contaminated with sewage, which not only interferes with the recreation industry but could also be a health hazard. The current rapid urbanization of the developing countries (Chapter 10) requires that this type of environmental degradation should be taken extremely seriously. The problem is extended when the solid extraction from the sewage in treatment plants, the sewage sludge, is dumped offshore. Continued sewage sludge dumping offshore over the same area leads to an enrichment of metals which may affect the benthic (bottom-dwelling) fauna.

The accumulation of organic wastes, nitrates and phosphates (derived from sewage and agricultural fertilizers carried by runoff) leads to an enrichment of coastal waters. This may cause an increase in fish population (a benefit) or eutrophication and the production of green algae and even red tides (a cost). The latter is an intense bloom of phytoplankton which colours the sea water. This leads to mass kills of many aquatic lifeforms including fish by clogging of parts of their bodies (such as gills) or by toxicity. The enriched plant material finally decays and falls to the seabed causing a serious reduction in oxygen there, with disastrous effects on most benthic life. This, for example, happens periodically in Manila Bay. High phosphate levels in discharged water from Zanzibar Town have been reported to negatively affect coralline algae (Björk *et al.*, 1995).

The presence of bacteria and parasites in untreated sewage is a serious risk for bathers and swimmers, but the main health hazard occurs through the consumption of contaminated seafood. The contamination particularly affects filter feeders such as bivalve molluscs. Certain types of red tides transfer toxins to bivalves and fish that eat plankton. This raises the risk of shellfish poisoning among humans.

Oil-related wastes

This is probably the most publicized of all types of marine pollution, and is usually associated with offshore blow-outs or leakages from tankers. Petroleum-related hydrocarbons, however, may reach the sea by various routes (Clarke, 1992):

- tanker accidents
- tanker operations
- bilge and fuel oil
- dry docking
- offshore oil production
- coastal oil refineries
- miscellaneous sources

A number of shipping accidents involve tankers every year but most of these lead to only minor spillage of oil. The spillage, however, may reach catastrophic proportions when a large tanker runs aground or is badly damaged in a collision, and huge quantities of oil are lost. The biggest of these was the wrecking of the *Amoco Cadiz* in 1978 off Brittany, when 223 000 t of crude oil was lost. Environmental degradation from such spills is very difficult to ameliorate. The probability of such accidents is high along the main tanker routes (Figure 11.5), especially where they pass through narrow straits or along busy shipping lanes.

Even normal daily tanker operations may lead to oil pollution. Carrying water as ballast on the return journey after delivering oil, and the cleaning of cargo compartments both lead to some oil being leaked to the sea. Oil could also be leaked during the period when tankers are at dry dock. Precautionary steps, if taken, reduce oil leaks during such operations. Ships also leak oil into the sea when they discharge ballast water or pump out water from the bilges which are contaminated with oil from the engines. Steps should be taken not to discharge such oil, but illegal leakages are common, and a considerable amount of oil pollutes the waters of the oceans in this fashion.

In offshore oil extraction, drilling mud is pumped down the oil well under pressure. One of the objectives is to prevent a blow-out which leads to a pressure-driven rapid release of oil from the well. These incidents are rare but when they do happen, catastrophic amounts of oil reach the sea. The classic example is the 1979 blow-out in a well in the Ixtoc field, off the Mexican coast. Not only did it catch fire but it also released 350 000 t of crude oil into the Gulf of Mexico over the nine months needed to bring the blow-out under control.

Petroleum derivatives also reach the sea in other ways. Effluent from coastal oil refineries contains some oil, as do domestic and industrial wastes. Oil often leaks out from near-surface oil deposits. Individually, these incidents are not usually large-scale, but the accumulated amount could be quite impressive. Regulations exist for discharging ballast properly but the lack of surveillance in many parts of the world means that such restrictions are not necessarily followed. Ngoile and Horrill (1993), for example, mention tankers discharging ballast and cleaning their tanks in East African waters. Instances of oil pollution of this type also have been reported from Southeast Asian seas and even seen on satellite images.

When an oil spill occurs, it spreads over the surface of the sea to form a thin layer: an oil slick. The lighter the density of oil, the thinner the layer and the further it spreads. Following an oil spill various changes take place:

- the light fraction evaporates into the atmosphere
- the soluble component is dissolved in the sea
- the non-soluble part becomes emulsified and is scattered around in droplets
- the heavy parts form balls of tar

The droplets could be degraded biologically but the heavy fraction is difficult to clean, especially when the oil comes ashore. The components of crude oil that dissolve in water contain various compounds toxic to marine organisms. By the time spilled oil reaches the beaches its toxic properties have often been reduced by the passage of time but it can still do considerable damage by smothering seaweeds and animals. The dispersants spread after an oil spill and their combination with the oil has been known to be more lethal than the oil itself. Oil reaching low-energy coasts stays there for some time and disappears into the substratum where the low

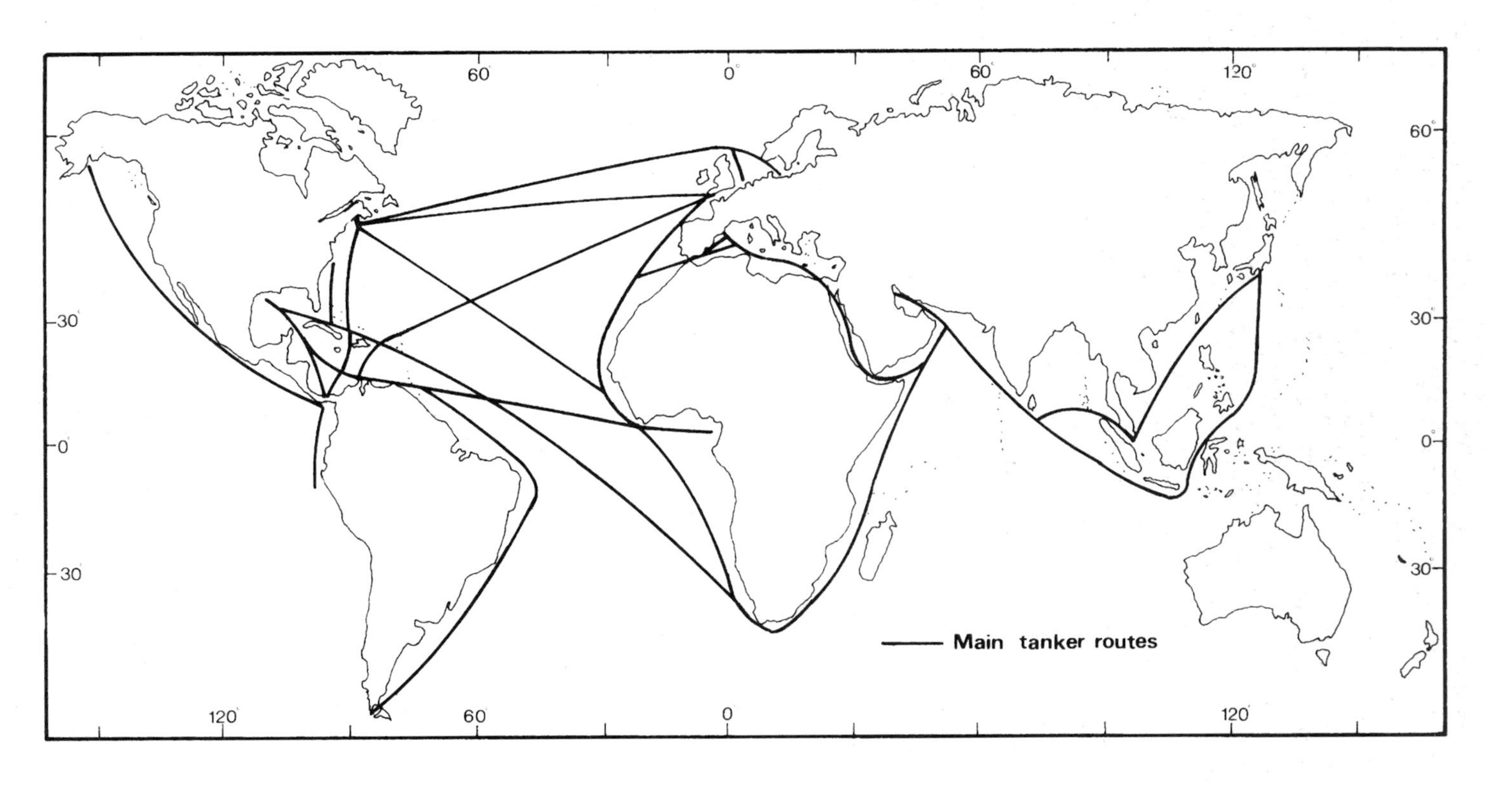

Figure 11.5 *Main tanker routes*

level of oxygen allows little chemical change and the toxicity may persist. Contamination may happen again through this oil emerging later from the subsurface. Fish and types of shellfish (particularly bivalves, crabs, lobsters and shrimps) are killed by oil pollution. Spilled oil may persist in mangroves and salt marshes as these areas are low-energy environments. In the Niger Delta, well known for its oil resources, spills are frequent, the oil persists for several years on land, and may cause destruction of the mangrove community (Moffat and Lindén, 1995). Sea birds and mammals (especially sea otters) are strongly affected, both in marine waters and on coastal lands. Although it is very difficult to determine the number of birds affected, Clark (1992) has estimated that in the North Atlantic alone, 10^4–10^5 sea birds have suffered from oil pollution of different intensity. It is the coating of oil on plumage that becomes disastrous for the birds by destroying its water-repellent property. Coastal fisheries or shellfish farms are extremely vulnerable if penetrated by a nearby oil spill. The effect of oil on fish in the open sea is variable.

Various techniques have been tried for cleaning up after an oil spill. Chemical dispersants sprayed from ships or planes are effective on light varieties of oil. Coastal installations or intakes can be protected by a floating boom which stops the oil from spreading. V-shaped booms have been used to concentrate the oil prior to pumping it out. It is far more difficult to clean beaches as huge quantities of sediment have to be moved in order to physically remove the oil. Oil coming onshore at least temporarily hinders tourist activities, if nothing else.

Conservative wastes

Certain types of wastes do not decompose or disintegrate but remain in an unaltered state for a long time. Organisms that intake such matter often do not excrete the wastes which continue to accumulate in their bodies. The physiological accumulation also increases along a food chain, which may cause toxic reactions, even death, to end-predators including humans. Conservative pollutants therefore cause serious problems.

Two types of conservative pollutants are encountered in the seas: (1) heavy metals (e.g. cadmium, copper, lead, mercury and zinc) and (2) halogenated hydrocarbons (e.g. DDT and other insecticides, and industrial chemicals such as polychlorinated biphenyls, PCBs).

Metals enter sea water in several ways. Small amounts are added naturally by volcanic eruptions, wind, or erosion of ore-bearing rocks. Some metals may reach the sea as atmospheric inputs. A small amount is added by direct outfall from industries and sewage treatment plants or by river discharges. By far the biggest addition comes via the dredging of shipping channels in estuaries. Metals are deposited on the river bed as part of the active sedimentation that goes on in estuaries as described earlier. With the location of industries and ports on an estuary, not only the metal concentration in the bottom sediment increases but such sediment is also frequently dredged up to keep the shipping lanes free. The dredged material is then dumped at a location in the sea, leading to a build-up of metals in the water and sediment. The average global annual dredging volume for the early 1980s was about 900 million m^3 (Frankel, 1995), and it occurs across the globe.

Not all the metals have the same effect on marine organisms and the animals at the top end of the food chain. It is the concentration of mercury in large long-living fish that is the most hazardous. Over years, deep-ocean (pelagic) fish build up a high concentration of mercury in their bodies as these are large carnivores and also swim with their mouth open thereby taking in water which may contain dissolved mercury. Large sea birds, especially those that feed close to the shore, tend to build up an accumulation of mercury in themselves. The effect of mercury on people (progressive brain damage) has been tragically illustrated at the small Japanese town of Minamata in the 1950s, and near the mouth of the Agamo River, Japan, in the 1960s. In both cases mercury poisoning occurred after eating fish taken from water contaminated by the discharge from industrial establishments which used mercury compounds. Following Minamata, standards have been set in several countries for the amount of mercury allowed in food sold for consumption. The WHO standard is a maximum of 0.2 mg of methyl mercury in food, or 0.3 mg of total mercury in a week (Clark, 1992).

Halogenated hydrocarbons are manufactured in large quantities as pesticides, polychlorinated biphenyls (PCBs), and industrial solvents. Such compounds are now found in all parts of the world including in the penguins of Antarctica. These reach oceanic waters in three ways:

- They can cover long distances as aerosols after being sprayed as pesticides from the air.
- River discharges from both agricultural and industrial areas carry halogenated hydrocarbons, especially if laden with silt eroded from agricultural fields that carry adsorbed pesticides.
- In dry conditions fine soil grains carrying pesticides can be transported in strong winds.

As these tend to accumulate higher up the food chain, pesticides can have drastic impacts. Halogenated hydrocarbons have killed fish and large predatory birds, and possibly sea mammals.

Plastics and nets

Plastics, generally being non-biodegradable, accumulate in the coastal and oceanic waters as perhaps the most visible form of marine litter. Plastic may last for nearly half a century and during this time it is carried by ocean currents everywhere. Plastics come from three main sources:

- plastic packing material including container rings for drink cans, sheeting and pellets are carried out from land to the sea by rivers;
- the fishing industry loses more than 150 000 t of fishing gear, mostly in synthetic material, each year including nets, lines, buoys, etc.;
- material used for banding and strapping cargo in shipping.

Although bits of plastic and polystyrene balls are seen floating even on the surface of remote parts of oceans, it is the fishing lines and nets which do most harm. They tend to entangle fish, mammals and birds in the oceans, leading to their death. Marine mammals have died after eating plastic sheeting (World Resources Institute, 1990).

Miscellaneous wastes

Large amounts of solid wastes, often contaminated with heavy metals, clays, coal dust and fly ash, are dumped into the sea either as a result of dredging near harbours or from the washing of industrial wastes. In general, the effect of such solid waste is to smother the bottom-dwelling fauna. Oceans have been used to dispose of munitions and radioactive material. Localized heating of sea water occurs when cooling water from thermal or nuclear power plants and industrial establishments is discharged. The effect is heightened in the tropical seas or in the subtropics during summer when the temperature of the ambient sea water is already high. Such releases in the subtropics have destroyed sea grass and benthic species locally. Often the heated water arrives with a dose of chlorine to keep it clean. Both affect the organisms present. Progressively, sediment fluxes are affecting the coastal waters and the coastal geomorphic systems. This can happen from both reduction and addition of sediment to the ambient supply (Box 11.1). Sand mining for building and excavating artificial pools in hard rock shore platforms for aquaculture also lead to coastal degradation. Accidents happen. For example, in 1985, the *Ariadne* was wrecked and sank in Mogadishu Harbour. She was carrying several hundred tonnes of toxic chemicals and pesticides (Ngoile and Horrill, 1993).

Box 11.1

Sediment Fluxes in the Sea

Due to the large number of dams and reservoirs currently in operation (see Chapter 7), sediment from the upper parts of drainage basins is trapped behind dams and does not travel to the sea. Rivers are a prime source of sediment supply to the coastal waters. The sediment is then deposited to extend and maintain beaches, spits and other coastal depositional features and to form deltas. If there is a loss in sediment supply, such depositional features are not built back after coastal erosion and a progressive loss of land takes place. The erosion which the delta of the Nile River is undergoing after the closure of the Aswan Dam (Chapter 7) is a striking example. It is feared that with the increase in the number of dams and the global warming driven rise in sea level, the world's beaches, islands and deltas could be at risk from coastal erosion in the near future.

On the other hand, the addition of extra sediment is not beneficial. Deforestation, agricultural expansion, extractive industries, and urbanization are drastically eroding the surficial material on slopes, especially in the developing countries where such activities are common due to economic pressures and inequalities (Chapters 5 and 6). This, especially when occurring near coasts and on islands with some relief, results in large sediment plumes flowing out of river mouths, particularly where the coast is shallow. This has been mapped for Southeast Asia from satellite imagery (Gupta, 1996) where sediment plumes of neighbouring streams may join to cover several hundreds of kilometres of coast (Figure 11.6). Figure 11.6 also illustrates that when basin sediment yield (the volume of annual sediment divided by basin area, a standardization technique) data collected from a number of sources are plotted against basin area, well-vegetated basins invariably fall well below the regional norm (Gupta and Krishnan, 1994).

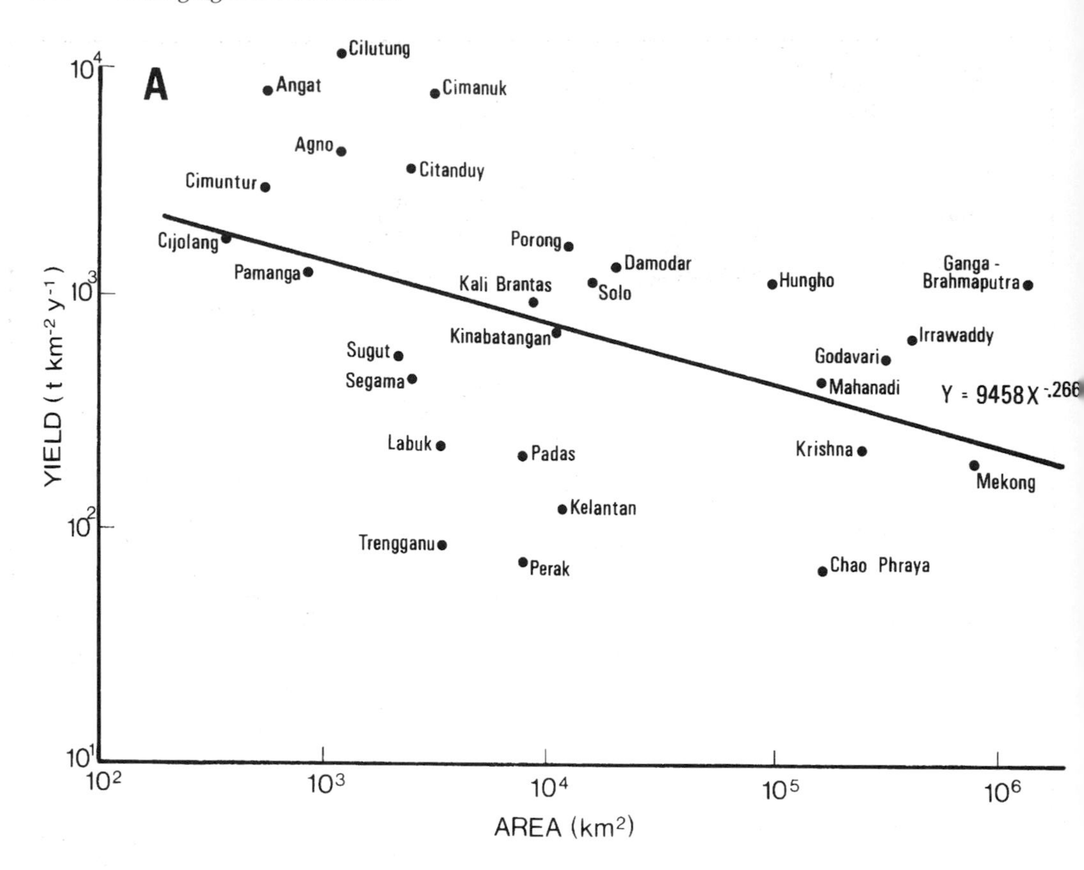

Figure 11.6 *Coastal sediment plumes, South and Southeast Asia. (A) Deviations in sediment yield from area-based regional expectations. (B) Map of coastal areas with high sedimentation. From Gupta, A. and Krishnan, P. (1994) Spatial distribution of sediment discharge to the coastal waters of South and Southeast Asia, in (L.J. Olive, R.J. Loughran and J.A. Kesby, Eds)* Variability in Stream Erosion and Sediment Transport, *IAHS Publication 224, IAHS Press, Wallingford, pp. 457–63.*

11.4 Marine Resources

Overuse of marine resources is the other crucial problem of the seas and oceans. Mostly, this implies overfishing, but destruction of mangroves and coral reefs in the tropical coastal areas also falls into this category.

Fishing in the seas has increased to meet the rising demand of the growing global population. This increase, however, involves greater use of the less valuable species, as a number of the more valuable ones have already been overfished (World Resources Institute, 1994). The annual global catch has been estimated to be between 62 and 87 million tonnes, which is already near the potential sustainable yield

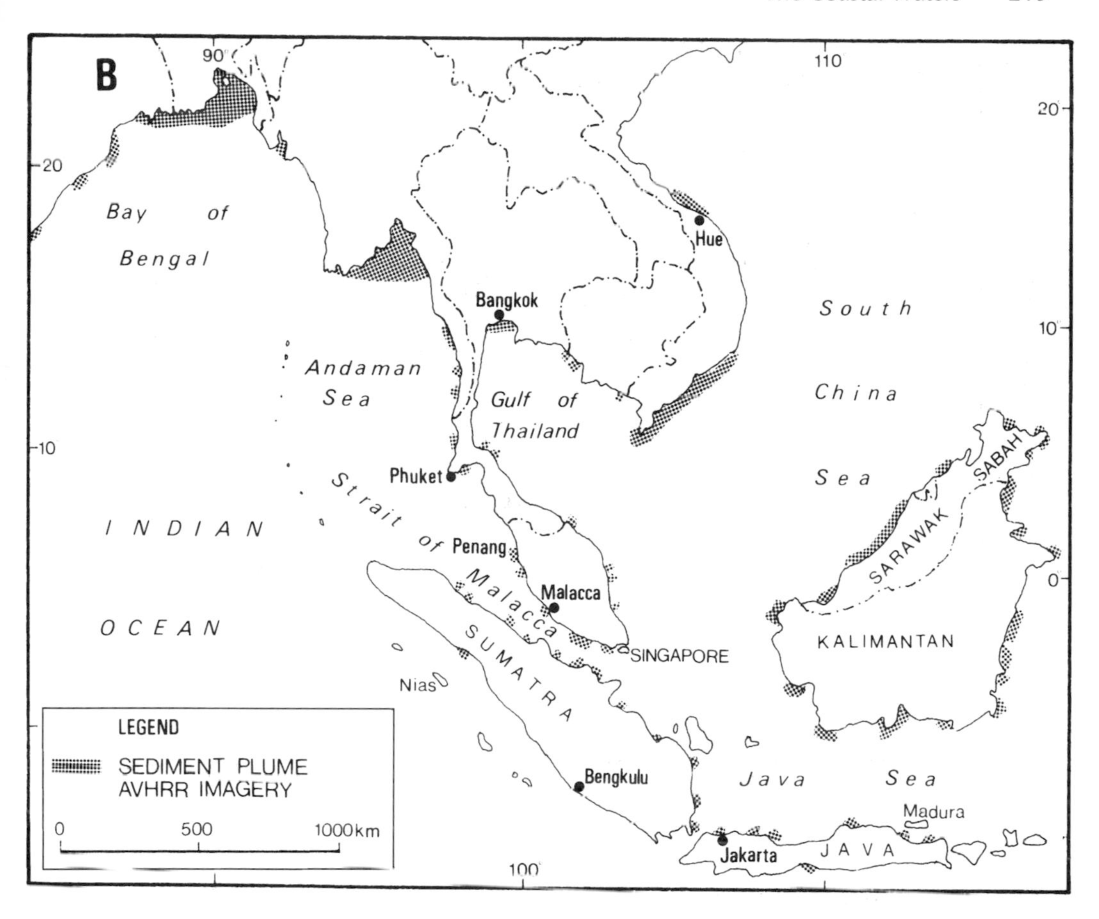

limit. It is therefore necessary to control marine fishing globally in order to avoid serious overexploitation. Such exploitation will disrupt ecosystems in the future and ultimately will negatively affect about a million people who depend on fishing. Excessive exploitation of fisheries has been attributed to a combination of overfishing during the periods of high fish catch and being supported by governmental subsidies during periods when fish populations are in decline (Ludwig, Hilborn and Walters, 1993).

Overfishing has become a serious concern in international waters beyond the territorial limits. The availability of good boats, financial backing, and modern technology results in fishing by boats of various countries operating thousands of kilometres away from home ports and often with little concern about sustainable catch. This has necessarily given rise to friction, especially between the coastal government and the fishermen from other countries. The North Atlantic provides a well-known example where the conflict is between Canada and the United States on one hand, and the long-distance fishing fleets from certain European countries on the other. Attempts at establishing some control such as a quota system on catch and restrictions on fishing gear, especially nets, have not been very successful. The better quality catch continues to deplete. For example, between the early 1960s and 1990, the fish catch over the Georges Bank in the North Atlantic showed a marked change in its composition, from 54 percent cod-like fish and 2 percent dogfish to 15 percent cod-like fish and 45 percent dogfish (World Resources Institute, 1994).

The problem of marine resource depletion is accentuated at times by natural changes. Possibly the best known example of this is the effect of the El Niño on the Peruvian anchovy fishing. El Niños were discussed in Chapter 2.

Upwelling brings in cold nutrient-rich water from the deep along the Pacific coast of Peru. The water is rich in phytoplankton and zooplankton on which fish thrive. In the 1950s, Peru started to build fish-meal processing plants which used anchovies from the Pacific coast for oil and fish-meal which are exported. This was a profitable venture and fishing increased rapidly in volume beyond the sustainable limit. By the early 1970s, the total anchovy catch was above ten million tonnes per year. The arrival of the 1972 El Niño was the trigger action that led to the collapse of the Peruvian fishing industry. El Niños persist for months and not only do they reduce the food available for the small fish, but the warm water that El Niño brings to the Peruvian coast also tends to bring in fish which are anchovy predators. This combination of overfishing, lower anchovy biomass, and predation drastically affected the fishing industry of Peru.

Overfishing has seriously depleted fish stocks in many other instances but in certain parts of the world the physical environment periodically makes fishing a fragile occupation. A very large number of people are dependent on marine fishing. A proper management strategy, as in the agriculture of marginal lands, needs to take into account such environmental variations.

Fishing may also destroy coastal ecosystems. For example, coral reefs which as ecosystems are extremely rich in biodiversity are being dynamited in order to stun fish. Coral reefs are also destroyed by siltation from accelerated land erosion, increased coastal pollution, and direct coral mining. This in turn increases coastal erosion as well as habitat and resource depletion. Mangroves are being destroyed extensively for urbanization, land reclamation, construction of prawn and fish ponds, fuelwood, and construction lumber. An unusual type of destruction was the US Army's use of defoliants and bombing of the Mekong Delta in the 1960s: it has been estimated by local studies that about half of all the mangrove forests of Vietnam were destroyed during the war, and recovery took decades and consistent effort (Viles and Spencer, 1995). Coastal estuaries are currently undergoing progressive pollution by sewage, toxic metals and oils. The coastal habitats are under tremendous threat, and as most of the fish catch comes from coastal waters a serious depletion of marine diversity is extremely plausible. In retrospect it is easier to control and manage coastal environments and the fish that live there, but it is much more difficult to prevent excessive fishing from the commons, the international ocean beyond the territorial waters of the coast. At present, only a limited number of international agreements and marine conventions are available for this difficult task.

11.5 The State of the Inland Seas

The state of large lakes and completely or partially enclosed seas is controlled by the amount of water, sediment and pollutants. In the 1960s, Lake Erie was heavily polluted by industrial discharge. At present, the large inland bodies of water which are in an extremely degraded state are the inland seas of west-central Asia and Eastern Europe: the Aral, Azov, Black and Caspian Seas.

The Aral Sea used to be the fourth largest inland waterbody. Two large rivers of Central Asia, the Amu Daria and the Syr Daria, used to flow into the Aral. Extensive diversion of water by huge canals for irrigating the dry basins of these rivers has expanded the area of irrigated cropland in Central Asia from nearly 300 000 km^2 to over 700 000 km^2. However, the rivers currently discharge only about 10 percent of their former flow into the Aral. In dry years, as in 1986, hardly any flow reaches the Aral Sea. Between 1960 and 1989, the level of the Aral Sea fell by 14 m, shrinking it to a little more than half of its former areal extent and reducing its volume by about two-thirds. Consequently, its salinity increased to 28 g l^{-1}.

The Aral is now divided into two separate parts. Its once considerable fishing and muskrat pelt collection have both disappeared. The regional climate has become harsher; the growing season has shortened; and on the exposed lake bed, winds have moved the sand into dunes. The state of the Aral is of regional and international concern but it will be extremely difficult to restore the Aral to its former condition.

The other seas are degraded but not to the same extent. Both the Black Sea and the Azov Sea have suffered catastrophic ecological damage due to various reasons. The enormous increase in the nutrient load has led to at least the northwestern part of the Black Sea becoming critically eutrophic (Mee, 1992). This is most likely due to the widespread use of phosphoric detergent and agricultural fertilizers in the neighbouring regions. To this are added urban sewage and industrial effluents discharged directly into the waters of the Black Sea. Mee has referred to the intensification of phytoplankton blooms. The falling of the decaying organic matter to the sea bottom has resulted in the disappearance of macrobenthic organisms and rich fisheries. This has led to socio-economic problems due to the loss of both the fishing industry and tourism. Irrigation projects on major rivers and associated construction of dams have reduced the runoff, causing an increase in salinity and modification of the migratory pattern of certain types of fish including the sturgeon. The growing use of high technology in fishing, especially on the Turkish coast, has apparently led to overfishing, especially of anchovies. About two million people depend on fishing in this region. Not only has this source of livelihood been seriously damaged, but marine diversity has also been much reduced. Hard data are scarce, but serious chemical and biological pollution is apprehended for the Black Sea. The pollution is primarily land-based. The Danube alone has been estimated to discharge annually 6000 t of zinc, 4500 t of lead, 1000 t of chromium, 900 t of copper, nearly 60 t of mercury, and 50 000 t of oil (Mee, 1992). Chlorinated hydrocarbons are common and there have been occasional discharges of radioactive material. Mee refers to 16 official dump sites for the disposal of dredging spoils on the western continental shelf. It is crucial that a proper assessment of the state of the Black Sea is immediately carried out followed by its restoration. The discussions and agreements in the 1980s and 1990s among the countries which govern parts of the Black Sea and its coastal areas hold out hope towards the application of a feasible regional policy. 'Environmental Management and Protection of the Black Sea' and 'Environmental Management in the Danube River Basin' are both projects supported by the Global Environment Facility (Chapter 16). The Caspian Sea is also suffering from similar problems, particularly regarding the degradation of its fisheries.

These case studies illustrate the problems that affect large land-locked bodies of water when development projects are undertaken in the coastal zone.

11.6 Management Strategies and Agreements

The quality of the world's oceans and coastal waters is reviewed periodically by a group of marine scientists known as the Group of Experts on the Scientific Aspects of Marine Pollution (GESAMP). The group was first formed in 1969 for a number of United Nations agencies and its first report was published in 1982. The second GESAMP group was convened in 1986, and in a report produced in early 1990, they concluded that marine pollution has spread across the world but unevenly (GESAMP, 1990). Some of the coastal waters are heavily polluted whereas the open oceans are still relatively clean. The most serious and extensive pollution sources are sewage from coastal areas and sediments from accelerated land erosion (World Resources Institute, 1990).

The cleaning of the seas, the management of marine resources, and the planned sustainable use of the seas are desirable but difficult to execute because of the lack of data, the generally fragmented and restricted nature of the regulations, and the shortage of financial resources. A large number of the regulations and standards are either ineffective or designed for reacting to polluting events rather than for preventing their occurrence (Frankel, 1995). Frankel, for example, suggests that preventing shipping accidents from happening is a better strategy than concentrating primarily on reducing post-impact oil spills. The question of responsibility has also not been clearly resolved. Legislation and policies may be formulated

- at the national level by individual countries
- among several countries at the regional level
- at the global level with the involvement of the International Maritime Organization (IMO) which is a technical agency of the United Nations, and other relevant United Nations agencies

Examples of such instruments are given in Box 11.2. The lack of power in the enforcement of these regulations is a prime weakness in protecting the seas. The usual procedure after a mishap is to assess fines and taxes from the polluter. Unfortunately the polluter is rarely caught and penalized, especially on the open seas where the economic advantage of cutting corners frequently becomes extremely attractive (Frankel, 1995). Satellite images are increasingly being used to record such acts of pollution, especially of ships discharging oil illegally.

The problem of coastal zone management is complicated by the varying nature of ownership. Exclusive Economic Zones (EEZ), by international consent, extend 200 nautical miles (about 370 km) from national coastlines. This gives the states effective jurisdiction over the continental shelf, and small islands control large oceanic areas in their periphery. In narrow straits such arrangements are modified. Within a specific country, problems are created by varying responsibilities and ownerships involving central government, state governments, and individual owners.

Box 11.2

A Selected List of Maritime Environmental Instruments

The conventions and agreements fall into several categories (World Resources Institute, 1990; Frankel, 1995).

International Agreements

1954: International Convention for the Prevention of Pollution of the Sea by Oil (OILPOL); amended in 1962, 1969 and 1971.
1963: International Convention for the Prevention of Pollution from Ships; amended in 1973 and 1978.
1969: International Convention Relating to Intervention on the High Seas in Cases of Oil Pollution Casualties.
1972: The Convention for the Prevention of Marine Pollution by Dumping from Ships and Aircrafts.
1972: Convention for the Prevention of Marine Pollution by Dumping of Wastes and Other Matter (London Dumping Convention); operational in 1975.
1973: Protocol Relating to Intervention on the High Seas in Cases of Marine Pollution by Substances Other than Oil.
1978: International Convention on Standards of Training, Certification and Watch Keeping for Seafarers (STCW).
1982: United Nations Convention on Law of the Sea (UNCLOS); deals with all aspects of marine pollution and provides for an efficient enforcement system.
1984: International Convention of Civil Liability for Oil Pollution Damage of 1969 and the 1984 Protocol.
1984: International Convention on the Establishment of an International Fund for Oil Pollution Damage of 1971 (Fund Convention) and the 1984 Protocol.

Regional Agreements

1969: The Agreement for Co-operation in Dealing with Pollution of the North Sea by Oil.
1974: The Convention for the Protection of the Marine Environment of the Baltic Sea Area.
1974: The Convention on the Protection of the Environment between Denmark, Finland, Norway and Sweden.
1976: Convention for the Protection of the Mediterranean against Pollution (Barcelona Convention).
1983: The Agreement for Co-operation in Dealing with Pollution of the North Sea by Oil and Other Harmful Substances.

The two major areas of concern are the coastal pollution which also tends to destroy coral reefs, mangroves and marine life, and the loss of marine resources by excessive fishing. Given the predicted rise in sea level in the future due to global warming, coastal problems will be aggravated (Chapter 14). The agreements at various levels need to deal effectively with such issues as soon as possible.

It is interesting to note that the coastal areas are more degraded than the open seas. The causes of their degradation are mostly land-based and anthropogenic, as discussed earlier. Sewage and siltation are the two common problems in the coastal areas of the tropics, affecting coral reefs and beaches. Algal blooms are becoming more common in tropical coastal waters (Lundin and Lindén, 1993). The protection of the coast should therefore start a long way inland, reducing the accelerated erosion and sediment transfer. It is not only the excessive sediment accumulation that is problematic but also the amounts of pesticides, fertilizers and heavy metals that ride the sediment grains. Sewage, which is probably the most polluting of the items listed here, can only be tackled by proper treatment (either in a plant or directed to flow through a succession of natural water bodies) before its discharge. The pressure on the coastal areas, particularly the well-populated coasts of the developing countries, can only grow more serious unless proper steps are taken immediately and certainly before the imminent sharp rise in urbanization (Chapter 10). Coastal pollution is both a health hazard and aesthetically displeasing. Both have considerable relevance as coastal areas are increasingly used for recreation and to attract tourism.

Viles and Spencer (1995) use a study in the Philippines published by the University of Tokyo to illustrate the effect of rising population on the coastal areas. Approximately 65 percent of the total population live in coastal areas and the 10 largest cities are coastal. This has in particular affected the mangroves, coral reefs, and Manila Bay. As well as the usual degradation, the mangroves have been widely replaced by fish ponds or suffer from mine tailings disposal. Coral reefs are damaged by siltation, pollution and fishing practices that physically destroy the reefs. Manila Bay is highly polluted from domestic sewage and airborne particles whose sources are the four thermal power stations in the vicinity. It is not enough to have laws to protect coastal areas; their implementation and involvement of public concern are essential. As Rajasuriya, De Silva and Öhman (1995) point out, the coral reefs of Sri Lanka have been significantly destroyed by fishing practices such as the use of explosives and bottom-set nets, mining coral for lime production, the uncontrolled collection of ornamental reef fish and invertebrate species, excessive sedimentation in coastal waters from accelerated land clearance, industrial pollution, runoffs carrying agricultural pesticides and fertilizers, and discharge of untreated sewage. In spite of the fact that coral reef protection has been a national policy, and regulations do exist, little has been done to prevent large-scale reef destruction. Examples like this are too common in the tropics.

The development of the coastal areas, stretching right to the beach itself, also disrupts the local pattern of erosion and deposition. Coastal resorts and associated developments and attempts at engineering protection fail from time to time because of a lack of information and understanding of the coastal processes. Coastal erosion and deposition occur rapidly, often eroding buildings and beaches in months or years. It is therefore necessary to build up our knowledge of coastal environments and measures that prevent degradation of the coastal environment. Coastal zone management is fast developing into an important component of sustaining the environment at its current quality. The responsibility and execution of coastal zone management lie with national governments which may require background research in earth and environmental sciences and technology transfer. Some information is available from the Regional Seas Programme of the United Nations Environmental Programme (UNEP). UNEP established 10 such programmes, involving 120 countries (World Resources Institute, 1990). The regional seas are as follows:

- East Asian Seas
- Southwest Pacific
- Southeast Pacific
- Wider Caribbean
- Southwest Atlantic
- West and Central African Seas
- Mediterranean
- Red Sea and Gulf of Aden
- East African Seas
- Kuwait Action Plan Region

It is, however, necessary to accelerate the study of the coastal environment and related problems of the developing countries. This is imperative not only due to the degradation associated with growing urbanization, industrialization, pollution and resource depletion, but also because of the expected changes in the coastal zones in the decades following the predicted sea level rise from global warming (Chapter 14). Such information should help to protect the coastal areas, especially those undergoing rapid development. Environmental legislation and its strict enforcement should be in place, along with careful planning such as dividing the coastal region into environmental zones which would prevent land development and environmental stress in areas of fragile ecosystems such as coral reefs (Odum, 1976). Such areas have begun to be designated as marine parks which at least has the potential to provide them with state protection.

Exercise

1. Figure 11.5 shows the main tanker routes. Choose an area within the tropics where the tanker route approaches the coast, such as in the Malacca Strait or parts of the Caribbean Sea. Imagine a significant oil spill from a holed tanker at a location selected by you. What would be the effects? You may need to do some library work to gather the background information for the area concerned before you can do this exercise.

2. The Mekong flows through a number of countries of Southeast Asia. A large delta has formed at its mouth which is heavily populated and acts as a rice granary. The Mekong discharges large quantities of sediment at its mouth (Figurc 11.6). Parts of its basin are still forested and the Mekong is a relatively unused resource. Plans are being discussed for various development projects on the Mekong, such as a series of dams and reservoirs for power generation and irrigation. Such development will certainly destroy significant amounts of natural vegetation and bring in more settlers. Assuming that the basin remains primarily non-urban and the sea level rises 50 cm by the middle of the next century due to global warming, what environmental problems do you envisage around the Mekong River delta-face in another 50 years? What steps would you suggest to ameliorate the impacts of the projects in the Mekong Basin? (In case you do not have enough material for the Mekong to answer this question, choose a comparable river to work out the problem.)

12

Techniques for Environmental Evaluation

12.1 Environmental Evaluation

We have discussed various types of environmental degradation and the procedures for ameliorating such conditions. In almost all cases serious environmental degradation is anthropogenic and associated with spontaneous or planned attempts to improve existing economic conditions by individuals, organizations or the State. Problems associated with development projects executed by the State or a community often tend to be large-scale and serious. Techniques have therefore evolved to identify serious environmental degradation at the planning stage; and also to use such findings to modify these development projects to ones which are, at least, less damaging to the environment. A selection of such techniques, currently used as tools for environmental management, is presented in this chapter. The techniques, however, continue to evolve over time, attempting to optimize existing environmental knowledge, resources, and ability to evaluate project-related environmental degradation.

The better known of these techniques include the following:

- an environmental evaluation of development projects at the early planning stage, known as the environmental impact assessment (EIA), or an extension of it known as the environmental impact statement (EIS);
- various types of environmental assessment, less in-depth than a standard EIA, and often used for a rapid assessment; the term environmental assessment (EA) is used to cover both types of study;
- a rapid assessment of the general environmental conditions of a country, carried out by the relevant government or a donor agency such as the World Bank or the United Nations Development Programme (UNDP);
- a detailed assessment of the present state of the environment of a country known as the environmental inventory or auditing; environmental audits could also be carried out for organizations such as industrial or utility companies;
- risk assessment.

EIA is the earliest technique of all such formal appraisals, being developed in the United States in the late 1960s. Although the basic objective and methodology of each type of appraisal remain the same, all techniques are flexible and subject to improvizations and innovations. They deal with both the ambient conditions (such as a special class of physical environment, or the lack of an important data set), and the development projects. The techniques change over time to optimize the resources available for environmental evaluation. Furthermore, the techniques vary in importance, usage and presentation among countries. EIAs, for example, are required for important projects in certain countries, but not all. Countries also differ significantly in public participation and openness in preparation of EIAs.

Environmental impact assessments and risk assessments will be discussed in this chapter. Two other techniques, rapid assessment of the environmental impact of a development project and the environmental signature of a country are discussed in Chapter 16. These assessments are relatively recent in origin, and a discussion of the current state of environmental governance is necessary before their introduction.

12.2 Environmental Impact Assessment

It is difficult to provide a clear definition of an environmental impact assessment, although its objectives can be easily listed. An EIA is expected to have the following characteristics (Ahmad and Sammy, 1985):

- It is a study of the effect of a proposed development project on the environment, the term 'environment' including all aspects of natural and human environments.
- It considers both environmental and economic costs and benefits.
- It is a predictive mechanism; it estimates the changes in the quality of the current environment as a result of the proposed action.
- It compares all the alternatives to the proposed project and evaluates them according to both environmental and economic costs and benefits.
- It is a decision-making tool which the decision-maker uses to arrive at the optimal selection of all possible alternative forms of the project, including the abandonment of the project.

The growth of EIAs is associated with the legislation passed in the United States on 1 January 1970, arising out of the National Policy Act (NEPA) of 1969, requiring environmental impact assessment for major projects. This led to three developments:

- a number of EIAs were prepared in the United States;
- a host of methods and methodologies were proposed and out of these a mainstream technique for preparing an EIA emerged;
- a rapid growth of legislation took place in many countries requiring an EIA.

Not all countries require an automatic EIA preparation for development projects. But in a number of countries where EIAs are not compulsory, as in Singapore, some form of environmental assessment is carried out before approving a project. This has given rise to a terminological confusion. In the United States an EIA is an introduction to a full-scale environmental report. If the EIA indicates that a detailed study is required, it is carried out and the final report is called an environmental impact statement (EIS). Not all EIAs lead to an EIS. In many countries, the EIA is the complete document including the full technical report. The term EIS is rarely used, and if used, tends to represent a summary of the EIA and the final recommendations forwarded to the decision-making authority. In this book we use EIA to represent the complete assessment document. From the very beginning, as the United States Council on Environmental Quality described it, EIAs were designed to provide a full and fair account of environmental impacts. Furthermore, their function was to indicate reasonable alternatives to a project which would avoid or minimize unfavourable impacts or improve the quality of the environment (United Nations, Department for Development and Management Services, 1994). In general, an EIA is an objective evaluation of the environmental problems that may arise with the implementation of a project, and it should provide the public with this information, clearly and completely.

12.3 The Structure of an EIA

A number of textbooks and research papers have been written on EIAs, and the number of actual EIAs that have been produced runs into tens of thousands worldwide. Not all EIAs have the same structure; this varies according to the project, location and the authorship. An EIA may deal only with project impacts; include in addition an account of the current environmental conditions (known as the baseline study); or in the full version, contain accounts of project impact, baseline conditions, and available alternatives to the proposed project. EIAs differ in both their structure and their emphasis. A general design can be constructed by dividing the EIA into several sections, which are often called steps. The following discussion deals with such a step-wise structure (Table 12.1) which is modelled after Ahmad and Sammy (1985) who summarized this outline with the EIAs for the developing countries in mind.

Table 12.1 *Structure of a mainstream EIA*

1. Preliminary activities and terms of reference (TOR)
2. Scoping
3. Baseline study
4. Environmental impact evaluation
5. Mitigation measures
6. Assessment of alternative measures
7. Preparation of the final document
8. Decision-taking
9. Monitoring of project implementation and impacts

Modelled after Ahmad and Sammy (1985)

Preliminary activities include defining the project (TOR) and determining the personnel involved. A brief summary of the project is extremely helpful at this stage. This summary should be clear and unambiguous, and list exactly what the development project entails, e.g. construction of a dam at a particular location and generation of so much power for such an area; providing technical and financial support for updating a fishing fleet which is expected to fish in a specified area of the coast; constructing a highway through a mountain, etc. The existing laws and regulations that are applicable to the project should also be reviewed along with the regulating authorities. The same should be carried out for available technical, financial and managerial resources.

A very important section of these activities includes identifying the team that will carry out the EIA along with a co-ordinator and the decision-maker who will read the final report. The composition of the EIA team varies from country to country. In the United States, the developer carries out the functions of both EIA preparation and decision-taking with the Environmental Protection Agency (EPA) in a review and monitoring capacity. In other countries a government organization may actually carry out the EIA. Some countries may not have the appropriate machinery, and a team of consultants may be employed. The latter often turns out to be an expensive proposition. It could be a team partly of local government personnel and partly of consultants. Similarly, the decision-maker may be a person, a committee, a number of organizations, etc. Members of the EIA team should have different expertise. For example, a team may include an engineer, an economist, a physical geographer and a sociologist, with a senior government official in the role of co-ordinator. All these need to be identified and declared before the actual EIA may begin.

Scoping is also known as impact identification. At a very early stage in the preparation of an EIA, the impacts that the project will have on the environment are identified. In cases where the list is very large, a selection of significant impacts may be chosen. This determines the limits or scope of environmental degradation. By delineating the extent of the EIA, scoping controls the cost and time of the assessment. It is therefore a very important step, both identifying the impacts and controlling the size of the EIA.

Scoping is usually a two-step technique. First, an exhaustive list is prepared of all possible impacts for a project of this type. The list may come out of the experience of the team members, or from previous EIAs done for this kind of project, or from a reference system such as INFOTERA of the United Nations Environmental Programme (UNEP). The second step consists of examining this list and selecting only the significant impacts. In general, the significance of an impact is determined by four factors (Ahmad and Sammy, 1985):

- the magnitude of change;
- the significance of change; some changes may be small in magnitude but disastrous for certain environmental components such as a fragile ecosystem or an unstable mountain slope;
- the geographical area over which the impact will be effective;
- the special sensitivity of certain areas to particular types of change; for example, coral reefs may be affected by excessive sediment production on land.

Scoping therefore requires some amount of knowledge or experience of the country concerned, and is usually carried out by a team of people with different expertise. It is best done immediately after engineering and economic studies for the project have been completed. The scoping team then can identify important environmental impacts with greater certainty.

Various techniques are used for identifying environmental impacts. These include

- checklists
- matrix
- networks
- overlay techniques

Checklists used for identifying project impacts are lists which are expected to encompass all possible impacts from a project of this kind. Table 12.2 illustrates a section of a checklist designed for a small water power project. Simple checklists which only included a list of the environmental indices that could be affected were used extensively in the earlier days of EIAs. Later, detailed descriptions of the impact of each environmental factor (often arranged as a questionnaire which elicits the necessary information) were added to the selection of environmental indices. This type is called a descriptive checklist. It is also possible to use checklists which not only include an exhaustive list of environmental factors but also rank the impact on all its significant environmental factors by the designed project and also the alternatives. In more sophisticated checklists, the relative importance of individual factors can be shown by varying weights being given to different factors, along with the total ranking calculated for all the versions of the project.

Matrices are very suitable for EIAs as they link a particular environmental factor to a specific action of the development project, thereby explaining the nature of the impact. Leopold and his associates, who in the late 1960s designed techniques for evaluating landscape aesthetics (Leopold, 1969; Leopold and Marchand, 1968), provided probably the most used matrix in EIAs (Leopold *et al.*, 1971). The Leopold matrix is a large arrangement with 88 environmental factors along the vertical axis and 100 development characteristics along the horizontal one, giving rise to 8800 cells (Figure 12.1). Each cell is divided by a diagonal line, and magnitude and importance of the impact are written (one in each half of the cell) on a scale of 1 to 10, 10 being the maximum. This type of matrix is called an interaction matrix.

The Leopold matrix has certain properties which are extremely useful:

- It is excellent as a basic scanning tool.
- Looking at the entire matrix, the assessor gets a visual impression of the total impact of the project on the environment, and also the part of the project causing most impact.
- It allows the application of only a section of the matrix, as required.
- It can be used to indicate both beneficial as well as adverse impacts by writing a plus or minus sign in front of the entries in the cells.

Table 12.2 *Parts of a checklist which could be used for evaluating a water power project*

A. Physical impacts
- change in sediment transport downstream
- change in water quality
- change in river hydrology
- river erosion
- slope instability
- deforestation
- increased erosion on slopes
- increased seismic disturbance in the area
- inundation of significant amounts of natural vegetation
- disturbance of the local fauna
- alteration of aquatic life
- alteration of soil fertility
- loss of rare or unique features or life-forms

B. Impact on local population
- inundation of settlements
- resettlement of people
- problems of sanitation
- availability of river water downstream
- spread of endemic diseases
- introduction of new diseases
- psychological stresses due to changes in lifestyle and economy

C. Impact on local economy
- disturbances in local economy
- inundation of existing transport network
- disruptions in river transport
- submergence of objects of cultural heritage
- inundation of historical sites
- availability of electricity
- improvement of agriculture
- industrial expansion

Note: This is an incomplete list, produced here to illustrate a checklist.

- It can be used repeatedly throughout a project, to estimate the impact at different stages and also for different parts of the project area.

Modifications of the Leopold matrix have been used by many agencies. Such modifications involve redesigning it, condensing it, or giving the impact rating in codes which summarize the nature of the impact and also indicate whether the negative impacts can be mitigated (Canter, 1996).

There are other types of matrices. At times, a particular development step may affect an environmental factor which in turn alters other factors. Such changes are examined by using a stepped interaction matrix. An assessor of the project impact may find it necessary to either modify one of the standard matrices (e.g. the Leopold matrix) or write a new one to better suit the local conditions.

II. PROPOSED ACTIONS WHICH MAY CAUSE ENVIRONMENTAL IMPACT

INSTRUCTIONS

1. Identify all actions (located across the top of the matrix) that are part of the proposed project.
2. Under each of the proposed actions, place a slash at the intersection with each item on the side of the matrix if an impact is possible.
3. Having completed the matrix, in the upper left-hand corner of each box with a slash, place a number from 1 to 10 which indicates the MAGNITUDE of the possible impact; 10 represents the greatest magnitude of impact and 1, the least, (no zeroes). Before each number place + if the impact would be beneficial. In the lower right-hand corner of the box place a number from 1 to 10 which indicates the IMPORTANCE of the possible impact (e.g. regional vs. local); 10 represents the greatest importance and 1, the least (no zeroes).
4. The text which accompanies the matrix should be a discussion of the significant impacts, those columnns and rows with large numbers of boxes marked and individual boxes with the larger numbers.

SAMPLE MATRIX

	a	b	c	d	e
a		2/1			8/5
b		7/2	8/8	3/1	9/7

II. Proposed actions

A. MODIFICATION OF REGIME: a. Exotic flora or fauna introduction; b. Biological controls; c. Modification of habitat; d. Alteration of ground cover; e. Alteration of ground water hydrology; f. Alteration of drainage; g. River control and flow modification; h. Canalization; i. Irrigation; j. Weather modification; k. Burning; l. Surface or paving; m. Noise and vibration

B. LAND TRANSFORMATION AND CONSTRUCTION: a. Urbanization; b. Industrial sites and buildings; c. Airports; d. Highways and bridges; e. Roads and trails; f. Railroads; g. Cables and lifts; h. Transmission lines, pipelines and corridors; i. Barriers including fencing; j. Channel dredging and straightening; k. Channel revetments; l. Canals; m. Dams and impoundments; n. Piers, seawalls, marinas and sea terminals; o. Offshore structures

I. EXISTING CHARACTERISTICS AND CONDITIONS OF THE ENVIRONMENT — A. PHYSICAL AND CHEMICAL CHARACTERISTICS	PROPOSED ACTIONS	A.a	A.b	A.c	A.d	A.e	A.f	A.g	A.h	A.i	A.j	A.k	A.l	A.m	B.a	B.b	B.c	B.d	B.e	B.f	B.g	B.h	B.i	B.j	B.k	B.l	B.m	B.n	B.o
1. EARTH	a. Mineral resources																												
	b. Construction material																												
	c. Soils																												
	d. Land form																												
	e. Force fields and background radiation																												
	f. Unique physical features																												
2. WATER	a. Surface																												
	b. Ocean																												
	c. Underground																												
	d. Quality																												
	e. Temperature																												
	f. Recharge																												
	g. Snow, ice and permafrost																												
3. ATMOSPHERE	a. Quality (gases, particulates)																												
	b. Climate (micro, macro)																												
	c. Temperture																												
4. PROCESSES	a. Floods																												
	b. Erosion																												
	c. Deposition (sedimentation, precipitation)																												
	d. Solution																												
	e. Sorption (ion exchange, complexing)																												
	f. Compaction and settling																												
	g. Stability (slides, slumps)																												
	h. Stress-strain (earthquake)																												

Figure 12.1 *Part of the Leopold matrix. From Leopold, L.B. et al. (1971) A procedure for evaluating environmental impact,* US Geological Survey Circular 645

Networks are used to show interrelationships among the different components of the environment of the area and also to indicate the flow of energy or impact throughout the environment such as an upland ecosystem or a drainage basin. They resemble networks used in ecological studies. Different types of networks are known as sequence diagrams, directed diagrams or impact trees. Again, networks can be used to show both temporal and spatial flows of impacts.

Overlay techniques were being used in planning before formal EIAs were designed. Individual impacts such as affecting soil, water, settlements and noise conditions are individually summarized and mapped over the area using chloropleths (shaded zones) to indicate the intensity of the impact. These maps are transferred on to transparencies which are then laid over one another to produce a composite effect, summing up the total impact (Figure 12.2). Necessarily, only a limited number of impacts can be shown, although it is possible to summarize a large amount of information within each transparency. With the advent of the Geographic Information System (GIS) and computer technology, the physical constraint on this method has been removed. It is also possible now to revise the raw data in files and do the overlays very easily and repeatedly, showing either temporal changes or projected environmental modifications.

Details of techniques used in scoping are in standard books on EIA such as Canter (1996) or Wathern (1988). Checking past EIAs, especially those dealing with similar types of environment or project, is always helpful. In the early days of the EIA, impact identification was often done on an *ad hoc* basis, using the

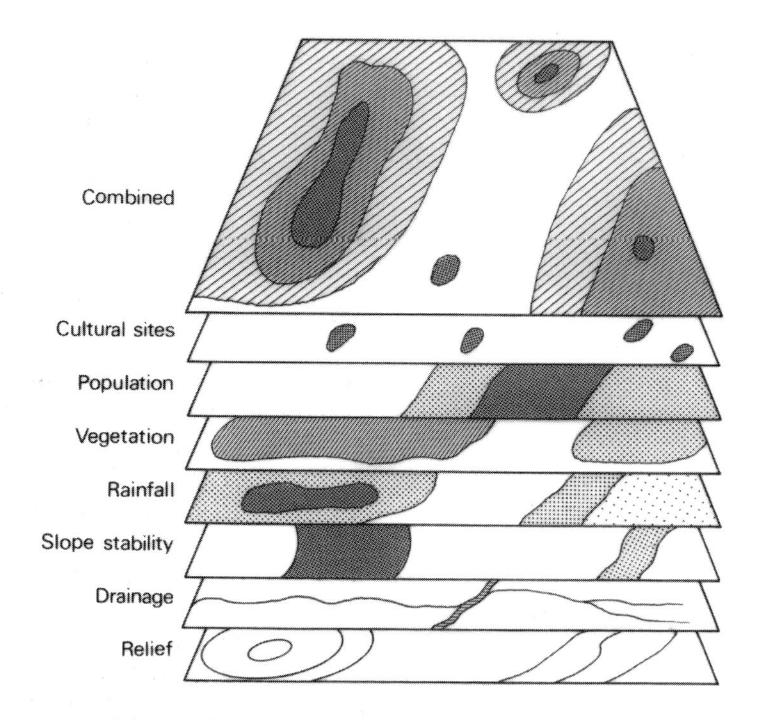

Figure 12.2 *Use of the overlay technique in an EIA*

experience of the EIA team. A lack of data may still force the assessors to do so in certain parts of the world. Impacts are also currently identified and evaluated by various kinds of computer programs developed by a number of agencies which specialize in EIA work. Scoping also requires active participation by the public in order to determine the impact of the project on the people of the area. This contact may come about in many forms: direct dialogue in public hearings, questionnaires, field surveys, etc. Direct contact usually works best, and it is often possible to involve the community in the scoping operation. EIAs were originally designed to keep the public informed.

A *baseline study* is the realization of the state of environment prior to development. This serves as the datum against which projected changes are measured. Baseline studies require experience as they should not include all environmental items, but should cover everything important with reference to the project impact. On the other hand, a proper scoping necessarily highlights significant factors of the local or regional environment. Hence, scoping and baseline studies often flow into each other. It is also obvious that baseline studies are relatively easier to carry out in countries which enjoy technical expertise, long-term data for environmental measures (e.g. riverflow), and existing reports and research papers on the area concerned. They are therefore more easily prepared in developed countries. In developing countries, local expertise can be utilized instead, such as consulting a forester or a soil scientist. This will at least fill in some of the gaps, provided the team responsible for the EIA has the ability or good fortune to reach the correct personnel for information. Failing that, consultants can be employed, but consultants, especially overseas consultants, often are expensive and drive up the cost of the EIA. It is the price a country pays for not knowing itself.

Environmental impact evaluation grows out of scoping and baseline study. Primarily, this is a quantification of the different impacts of the project on the environment. This is the most technical, difficult and controversial part of the EIA. Not every impact (especially natural and social impacts) can be quantified. For example, it is very difficult to agree on a number which sums up all the negative impacts of deforestation. Occasionally, it is possible to use surrogate measures, such as the amount of money required to mitigate the damage or the amount of money local inhabitants are willing to pay to clean up the river. But the accuracy and appropriateness of such techniques can be questioned. Again, an existing data set is extremely useful for impact evaluation, but it is also costly. Impact evaluation therefore calls for very careful consideration of the significant impacts, their numerical evaluation, and an understanding of the level of accuracy of the quantification of impacts. This has to be done not only for the proposed project but for all the possible alternatives also, so that a balanced decision can be taken regarding the final shape of the project. The quality of scoping done earlier, is obviously extremely important for the impact evaluation stage.

Mitigation measures are considered as a post-impact evaluation step. These are proposed measures to reduce the impacts. For example, one may suggest that a thermal power plant is constructed with scrubbers and uses washed coal. This of course incurs some costs, but it is expected that such measures will, in the long run, reduce the impacts so much as to make the project both economically and environmentally viable. Mitigation measures, strictly speaking, are part of environmental impact evaluation. The EIA team, however, may have to decide between two alternatives regarding this thermal power plant; for example, having a high cost and low pollution programme or a low cost but not a low pollution situation.

An *assessment of alternative measures* becomes possible at this stage. The proposed project and all relevant versions thereof by now have been examined for environmental impacts, corrected for mitigation measures, and some kind of standardization such as impact quantification has been carried out to make them comparable. The next step is to combine environmental degradation and improvement with economic losses and gains. In standard EIAs a summary for each version of the project is presented, and the comparison between them is often carried out using benefit–cost analysis. Benefit–cost analysis was introduced in Chapter 4. The advantage of doing a benefit–cost analysis is the familiarity of this technique across a wide range of people: engineers, economists, bureaucrats, etc. On the other hand, not everything is quantifiable. There is no numerical category for a beautiful view, although even that has been attempted. Not all assessments therefore use net benefit criteria. Some have used a subjective ranking of the alternative projects or a graphical or weight system for the final selection. The benefit–cost analysis, if used, has to be done for all the options. For example, is it more desirable to put up with limited pollution at a lower cost of mitigation or to remove the pollution completely at a much higher cost?

Preparation of the final document should meet two objectives. First, a complete and detailed account of the EIA has to be prepared. Second, a brief summarized account should be drafted for the decision-maker who may not be a technical person. The detailed account has been described as the *reference document*. It is used by the technical personnel associated with the project, and also for preparing future EIAs in the same geographical area, or for the same type of project in a different area. This part of the documentation is normally supported by technical calculations, graphs, and field and laboratory measurements. The summarized non-technical account is described as the *working document*, the objective of which is to communicate to the decision-maker the findings of the EIA team clearly and without using technical language. This document is extremely important as informed, prompt and correct decisions can only be taken if the findings and recommendations of the EIA are properly understood by non-technical decision-makers.

Decision-taking is a post-EIA mechanism. Decisions could be taken by a manager or a committee, or personnel from a ministry who had not been associated with the EIA during its preparation. Preparation of EIAs involves consideration of technical and economic aspects of project alternatives. These of course are still considered at this stage but, at times, political expediency and project feasibility control the final choice. In general, a decision-maker has three choices:

- accepting one of the project alternatives;
- returning the EIA with a request for further study in certain specific areas;
- totally rejecting the proposed project along with its alternative versions.

As Ahmad and Sammy (1985) have described it, EIAs are expected to aid decision-making. The preparation and drafting of an EIA should be carried out with this objective clearly and always in mind.

Monitoring of project implementation and impacts is carried out while the selected project is on stream. This is basically an inspection to satisfy that the project is following the guidelines and recommendations stated in the EIA. Such inspections may be carried out also after a time lapse following the completion of the project to determine the accuracy of impact prediction by the EIA. This could be a valuable lesson for environmental impact assessors. The decision-making process related to an EIA is shown in Figure 12.3.

EIAs should be considered as an evolving concept. As such, they vary between two extreme viewpoints. EIAs are seen by some practitioners as having a highly focused purpose: the production of information to enable appropriate personnel to identify optimal project design. This is a project-oriented and highly technical approach. On the other hand, EIAs are seen by others as techniques for consensus building and integrating environmental with economic and social planning and management. To them, EIA is an open-ended, process-focused technique for identifying and exploring issues and also resolving them (Smith and van der Wansem, 1995). Most EIAs probably do a bit of both.

12.4 EIAs: Misunderstandings, Shortcomings and Problems

Three misunderstandings exist about EIAs. First, EIAs have been interpreted as holding up development, or at least delaying it. Second, they are seen as expensive exercises. Third, EIAs have been viewed as a mere bureaucratic process which does not really influence decision-making.

None of these is necessarily correct. EIAs are designed not to delay development but to make sure that a large-scale development project does not degrade the environment severely and unnecessarily. In this book so far we have provided examples of projects that would have been far more environment-friendly if EIAs had been in vogue at the time of their inception. The establishment of unlined canal irrigation in sub-humid and arid areas is a striking example. EIAs are not designed to prevent the establishment of a project unless the project is extraordinarily hazardous to the environment. Furthermore, EIAs are supposed to start very early, at the prefeasibility stage of the project, so that project design and EIA preparation can go on simultaneously.

More information exists regarding the costs of EIAs in developed countries than in developing ones. Ahmad and Sammy (1985) estimate that a median cost for EIA would range from 0.5 to 1 percent of the total project cost. The percentage comes down for bigger projects, due to economies of scale. The cost of an EIA is therefore reasonable, considering what it achieves in return. Whether an EIA is a mere formality or whether its recommendations and findings are used in project design depends not on the EIA but on the decision-making authorities and the overall policy environment. EIAs are more effective when they are not reactive EIAs, i.e. commissioned after the project has been fully designed and a large amount of money and effort has already been spent in anticipation. It is easier to incorporate the recommendations of an EIA if it is carried out simultaneously with the

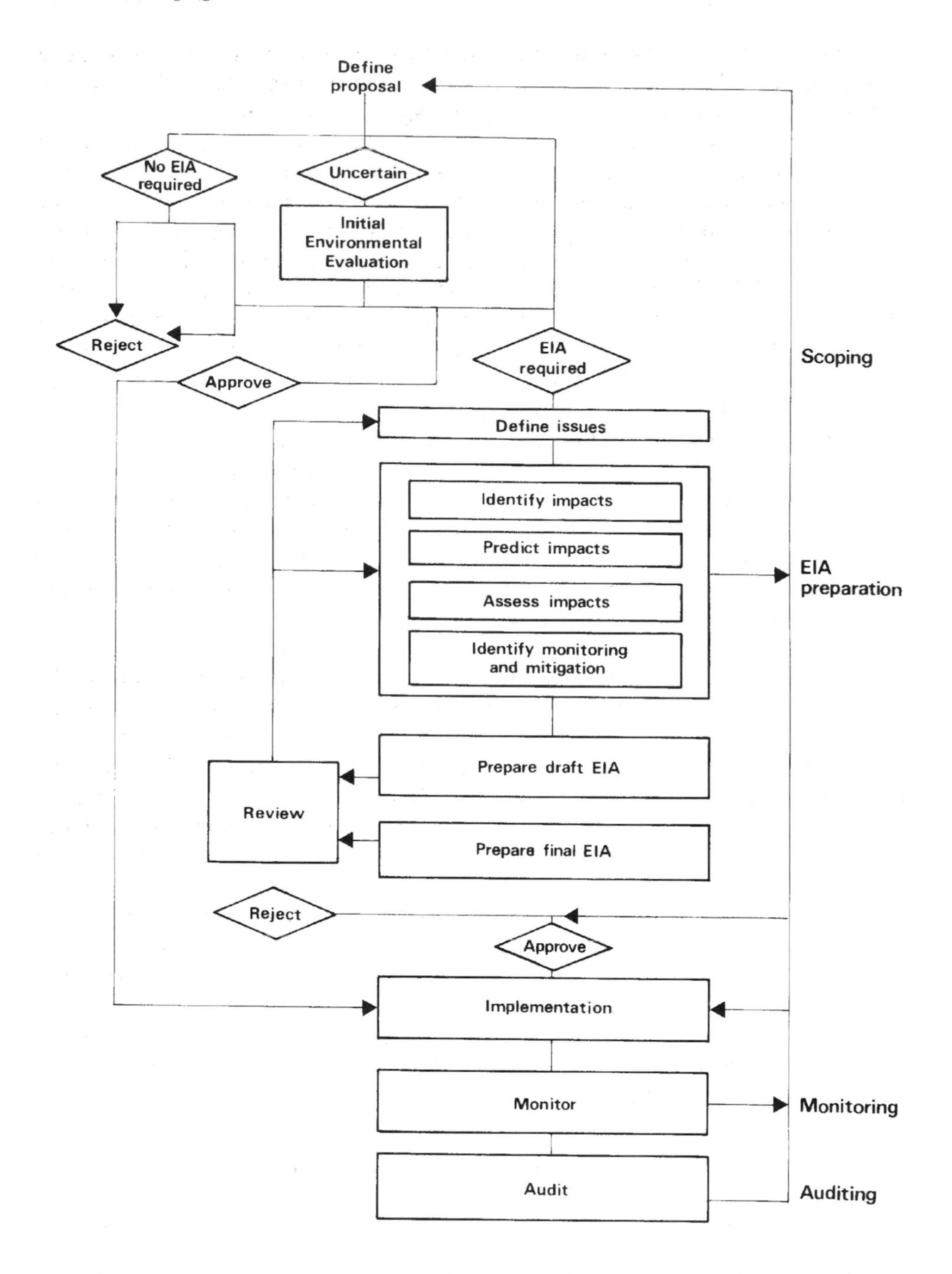

Figure 12.3 *Flow diagram showing the decision-making process related to an EIA. From Wathern, P. (Ed.) (1988)* Environmental Impact Assessment: Theory and Practice, *Routledge, London*

technical design. An EIA is not supposed to be a justification of the proposed project or of decisions already made regarding the project (United Nations, Department for Development Support and Management Services, 1994).

The major problem with EIAs is their size. Over time they have evolved into large reports which require the expenditure of considerable financial, technical and human resources. Such resources are not always available in a number of developing countries. Again, not all projects require a proper EIA for a decision on the approval, alteration or rejection of the proposed design. Donor agencies, such as the World Bank or the UNDP, have designed in-house concise environmental assessments which are used to determine whether the project requires any environmental investigation at all, whether the environmental evaluation can be done with the in-house system, or whether the project is so large and/or complicated that a full-scale EIA is essential. A full-scale EIA is therefore put into operation only after a preliminary screening. These in-house techniques are discussed in Chapter 16.

Other practical problems that may come up include too many possible alternatives, a wide range of impacts, lack of necessary data, lack of technical and managerial expertise, difficulties with quantification of impacts, and the appropriateness of the benefit–cost analysis in a particular situation. Some of these problems have been discussed in the previous section. Ahmad and Sammy (1985) provide several examples of dealing with these problems. Some of these problems have well-tried solutions such as scoping to reduce the number of types of impact, although one may question whether it is worthwhile to put forward a project that triggers off a wide range of negative environmental impacts. Scoping may, however, be used for weeding out impacts that are of little importance. The techniques for measurement of physical or biological impacts are more sophisticated at present than those for measuring socio-economic impacts. This also has to be considered. All these problems are perhaps to be expected as environmental decision-making deals primarily with uncertainties and intangibles, and its correctness can only be checked at least several years into the future.

Environmental impact assessment also requires the support of a national policy, proper legislation (which makes them mandatory at least for certain projects), available background information, personnel with technical and managerial capabilities, and public participation. EIAs are incomplete if people whose lives are touched, either beneficially or adversely, by the project are not given a chance to transmit their reactions. The optimal mechanism for such involvement, however, varies from country to country.

12.5 EIAs in the Developing Countries

The developing countries have a wide range of usage of EIAs. In certain countries, the use of EIAs started very early, only a few years after their formal beginning in the United States. For example, EIAs became a requirement in the Philippines and Thailand in 1977 and 1984 respectively (Gilpin, 1995). EIAs have certainly been conducted in the developing countries since the early 1970s. A number of countries have made EIAs mandatory for certain types of projects (Table 12.3). Other countries have not done so but a range of environmental assessments, ranging from a formal EIA to an in-house variation, are used routinely depending on the development project. At the other end of the scale, some countries only do an assessment under unusual circumstances. The use of EIAs, however, is increasing in the developing countries, especially if the project is a large one with considerable potential impact or if the project receives assistance from an international donor agency. Currently, the donor agencies require environmental assessments for many kinds of project, some of which may even need a full EIA. In 1980, a Declaration of Environmental Policies and Procedures related to Economic Development was accepted by UNDP, UNEP, the World Bank, the Inter-American Development Bank, the Asian Development Bank, the African Development Bank, the Arab Bank for African Development, the Caribbean Development Bank, the Organization of American States, and the European Economic Community (Ahmad and Sammy, 1985). This agreement and subsequent ones together ensure that environmental impacts are studied prior to funding a development project. Some of these organizations also prepare a country-wide evaluation of environmental strength and vulnerability. This is used for reference and is periodically updated. This will be discussed later, along with the policies of the major donor agencies, in Chapter 16. In any case, for certain types of projects (e.g. power generation) environmental impact studies are required as part of the overall project planning (United Nations, Department for Development Support and Management Services, 1994). It is now expected that an EIA or some kind of environmental assessment will be done for every large-scale or environmentally sensitive project in the developing countries. Biswas, in a publication dated 1992, estimated that by then over 11 500 EIA studies had been completed in the developing countries of Asia, of which several thousand environmental impact studies came from two countries, the Philippines and Thailand (Biswas and Agarwala, 1992). The problem lies not so much in the acceptance of the EIA but more regarding the existence of appropriate laws and regulations and basic environmental information. For example, proper measures of the possible impact of air pollution from a power station may require long-term wind direction and wind speed data. Similarly, water quality data may be essential for estimating the impact of a new industrial or urban site on a river. This kind of long-term baseline information is not available for most of the countries.

Smith and van der Wansem (1995) reviewed the EIA capacity in Asia, including detailed studies of three countries: the Philippines, Indonesia and Sri Lanka. Although all three countries require EIAs, the procedural arrangements are partially different. In the Philippines, the Environmental Management Board (EMB) and the regional offices of Environmental and Natural Resources (DENR) are primarily responsible. Projects which are environmentally critical or located in environmentally sensitive areas are evaluated. Projects cannot be implemented unless an Environmental Compliance Certificate has been issued by the EMB. Two types of documents are produced as part of the EIA process: project descriptions and environmental impact statements. Projects are usually reviewed only after the major site and design decisions have been taken. Alternative project designs are not considered important. A public hearing is conducted if the project is large or

Table 12.3 *Thailand: categories of development projects that require EIA*

- Dam or reservoir with a storage surface greater than 15 km^2 or a storage volume over 10^7 m^3
- Irrigation project serving an area greater than 128 km^2
- Airport
- Ports and harbours for vessels with a capacity greater than 500 gross tons.
- Communication systems of the Expressways and Rapid Transit Authority of Thailand
- Hotel and resort facilities with more than 80 rooms in environmentally sensitive areas (near to rivers, coasts, lakes, beaches and national parks)
- Mining
- Thermal power plants
- Industrial estates
- Industrial plants for
  - petrochemical industry including oil refinery
  - natural gas separation or refining
  - types of chemical industry
  - iron and steel
  - smelting of other metals
  - cement
  - wood pulp

All industries on this list have a size limit beyond which EIAs are required.

Source: summarized from Htun (1988).

located in a particularly sensitive area. If the project has been approved, the compliance to the EIA recommendations is monitored usually by the DENR regional offices, although a mixed team, which also includes the project organizers, the local community and Non-Government Organizations, is sometimes used.

Indonesia simplified the EIA procedure in 1993. The EIA system is known as AMDAL in Indonesia and its overall co-ordination rests with BAPEDAL, the Environment Impact Management Agency, a division of the Ministry of Environment. Under the new regulation, the AMDAL committee now includes greater NGO representation, has extended the review concept to regional planning activities, and has speeded up the procedure. BAPEDAL has also been given greater responsibilities for supervising the EIA procedures. All projects that have been implemented since 1987 need to be reviewed if they have not been done already. Public participation is allowed at any stage of the EIA process, at the discretion of the committee. Authorized government agencies are also required to keep the public informed of the decisions in most of the cases.

In Sri Lanka, the National Environmental Act was amended in 1988 to provide for regulations pertaining to environmental impact assessment. The Central Environmental Authority is responsible for guidelines, project selection, and co-ordination of various project-approving authorities related to EIA. Many of the procedures and requirements are new and yet to be worked out properly. The procedures include provisions for public hearings, public announcement of approvals, and appeals against unsuccessful projects.

Smith and van der Wansem (1995) have come to certain conclusions which are probably valid for many other Asian countries. The conclusions are as follows:

- EIAs are gaining acceptance but not universally, and success stories alternate with setbacks and other problems.
- Existing laws and regulations are basically adequate.
- Compliance and enforcement of EIA requirements should be more strict.
- There is a need for reporting on EIA legal decisions and interpretations.
- Although a central EIA agency may exist, its role is poorly defined and the organization is poorly developed.
- EIA authority is too centralized.
- The inadequacy of a career in environmental operations makes it difficult to recruit staff which may lead to over-reliancy on temporary consultants.
- EIA funding is generally inadequate.
- Non-Governmental Organizations are weak in many countries.
- More baseline information should be available.
- Adequate technical and procedural guidance materials are in short supply.
- EIA practitioners should have more access to existing guidance materials.
- There is inadequate interagency co-ordination and co-operation.
- EIAs should be better linked with planning and policy development and also with enforcement activities.
- More emphasis should be given to public participation.
- A great need exists for capacity-building which would develop skilled human resources, technical laboratories, libraries, data centres and computer facilities.
- Data collection and mapping programmes should be developed in conjunction with national and regional land-use planning agencies, resource management agencies, and private industries.

Some of these are also applicable to the developed countries. Environmental assessment, as a procedure, is clearly established in the developing countries. The next stage lies in improving the technique and using the EIAs meaningfully for environmental management.

12.6 Risk Assessment

Risk assessment is a technique that has evolved and continues to evolve regarding the assessment of environmental risks such as the breakout of a disease or an explosion in a power station. Unlike EIA, this is a very precise and numerical technique which determines the probability of the occurrence of a hazard at a particular scale. It is a highly numerical appraisal and even non-quantitative parameters need to be converted into numerical measures, which may or may not be possible. Risk assessment may be considered along with EIA in certain instances, such as the siting of hazardous technologies.

Risk analysis is used by both private and public sectors. For example, it is common in the chemical industry and insurance (Andrews, 1988). It is used by the state to determine health hazards and energy production. Like EIA, risk assessment is a tool for prediction, although with risk assessment, the prediction has a probabilistic measurement. Risk assessment is a technique for the uncertain world we live in. It started in the mid-1970s as an administrative requirement to build extensive documentation to justify proposed risk regulations but also in order to balance risks against economic costs and benefits. A small number of recent EIAs have incorporated risk assessment into the report. This is essentially for projects dealing with hydrocarbons, nuclear power, and forestry. It is possible that in future a greater number of EIAs will incorporate risk assessment within their structure. Andrews (1988) provides a detailed discussion of risk assessment.

12.7 Conclusions

Environmental impact assessment and risk assessment have been discussed in this chapter. Other techniques have evolved for quick assessment of the environment, both country-wise and project-wise. These became necessary primarily because of the large number of development projects currently in operation. As the major donor agencies are involved simultaneously with a very large number of projects, they have designed their own in-house techniques. The techniques first carry out a check on a particular project to determine which of the following is applicable:

- no environmental assessment is required;
- the in-house assessment method will be sufficient;
- the scale and sensitivity of the project requires a full and proper EIA.

These in-house techniques of donor agencies, however, are recent developments and their evolution is related to current developments in global environmental management. They are therefore discussed in Chapter 16, following an account of the history of environmental management and global environmental governance. However, in all cases, local participation remains extremely important. Standard EIAs are currently widely used for the evaluation of projects, and certainly in the developing countries.

Exercise

1. Select a large-scale development project such as a multipurpose river development project, or a transmigration project, or an urban development project. Study it in detail. Prepare the following steps of an EIA for the project: scoping, baseline study, environmental impact evaluation.
2. Construct an imaginary development project in an area you are familiar with. For example, if you live in a city you may think of a new housing project planned to replace an open space. Write a project proposal. Scope the proposal.

Part III

The Global Issues

13

The History of Current Environmental Awareness

13.1 Introduction

It is disturbing to realize that our attitude towards our environment has not changed significantly throughout our existence. It has remained the same mixture of concern and reckless exploitation. The arrival, in the Pleistocene, of *Homo sapiens* in what is now the American Southwest resulted in one of the most impressive losses of global biodiversity. This act of destruction, carried out with primitive weaponry, is well documented (Martin and Klein, 1984). As Harvey (1993) has pointed out, the ecologically sensitive traditions of Taoism, Buddhism and Confucianism in China co-existed with deforestation, land degradation and increased flooding (Lin, 1956; Richardson, 1956; Smil, 1984). The extensive knowledge of natural science in ancient Greece is evidenced by Theophrastus' *Enquiry into Plants*, which arrived several years after Plato's lament on the consequences of land exploitation in Attica (Darby, 1956; Theophrastus, 1916 trans.). Both religious and poetic texts in Sanskrit attribute unusual empathy and importance to trees but, on the other hand, aided by divine powers, the heroes in the *Mahabharata* bravely burn down an entire forest with all the animals inside.

The difference between us and our ancestors lies in our increased ability both to degrade and to manage the environment. We have extensively modified the land cover of the earth's surface, altered the quality of its air and water, and eventually created situations such as stratospheric ozone depletion which have put the entire planet in a critical state. All this necessarily has led to an increased awareness of the deterioration of the environment. No single date can be attributed to the growth of such awareness in the modern times. We know, however, that a concern for the protection of nature was present in the middle of the nineteenth century, a concern which arose from two widely different types of background: the growth of natural sciences following Humboldt, Darwin and Wallace; and the romanticism and educational value of nature as espoused in the writings of Wordsworth, Emerson and Thoreau. Environmental awareness grew out of both types of fascination with nature (McCormick, 1989).

The Origin of Species was first published in 1859. Possibly the earliest environmental group – the Commons, Open Spaces and Footpath Preservation Society – was established in Britain in 1865. The first anti-pollution law, the Alkali Act, was passed by the British Parliament in 1863. The East Riding Association for the Protection of Sea Birds, formed in 1867, campaigned for the preservation of wildlife. In the United States, the efficient clearing of forests by the settlers and the beauty of the natural landscape, in combination, had a similar effect: the development of environmental awareness. Such awareness again grew out of both scientific and romantic needs, the romanticism at times bordered on treating nature as an object of worship. The beautifully illustrated volumes of *The Birds of America* by John James Audubon and the writings of Ralph Waldo Emerson and Henry David Thoreau are creations of this period (McCormick, 1989).

The wonderful landscape of the American West created a demand for preserving areas of scenic beauty and a 'primitive' style of life (McCormick, 1989). John Muir, who later in 1892 was involved in the founding of the Sierra Club, was the leader of the movement to preserve the wilderness. An act of the US Congress in 1864 set aside the Yosemite Valley as a recreational area, the first step towards preserving a unique landscape. This was followed by the establishment of the first national park at Yellowstone in 1872.

13.2 George Perkins Marsh and Dietrich Brandis

1864 was also the year of publication of George Perkins Marsh's seminal work *Man and Nature*, the title of which was later changed to *The Earth as Modified by Human Nature*. Marsh was a geographer who was familiar with rural New England, knew a number of languages, had travelled extensively, and had held American consular positions in Europe. The book was a distillation of his knowledge from his extensive readings and travels. In the preface to the first edition, Marsh wrote:

> The object of the present volume is: to indicate the character and, approximately, the extent of the changes produced by human action in the physical conditions of the globe we inhabit; to point out the dangers of imprudence and the necessity of caution in all operations which, on a large scale, interfere with the spontaneous arrangements of the organic or the inorganic world; to suggest the possibility and the importance of the restoration of disturbed harmonies . . .
>
> (Marsh, 1898: VII)

A hundred years later such words read very true. Although *Man and Nature* was widely read, the concern and wisdom expressed in the book unfortunately did not have as much impact as it should have had for the good of our planet. It was among the foresters of various countries that Marsh made the most impact. A quote from Marsh turns up often in books dealing with environmental issues:

> Man has too long forgotten that the earth was given to him for usufruct alone, not for consumption, still less for profligate waste.
>
> (Marsh, 1898: 33)

His evidence of anthropogenic modification of the environment came from a wide range of locations. Here is Marsh describing sedimentation from the Po and the building of the plains of Lombardy:

> We know little of the history of the Po, or of the geography of the coast near the point where it enters the Adriatic, at any period more than twenty centuries before our own. Still less can we say how much of the plains of Lombardy had been formed by its action, combined with other causes, before man accelerated its levelling operations by felling the first woods on the mountains whence its waters are derived. But we know that since the Roman conquest of Northern Italy, its deposits have amounted to a quantity which, if cemented into rock, recombined into gravel, common earth, and vegetation mould, and restored to the situations where eruption or upheaval originally placed or vegetation deposited it, would fill up hundreds of deep ravines in the Alps and the Apennines, change the plan and profile of their chains, and give their southern and northern faces respectively a geographical aspect very different from that they now present.
>
> (Marsh, 1898: 270)

Like Marsh, the leaders of the environmental movement in the United States were mostly professional people: foresters, hydrologists, geologists. To them, protection of nature was not mere preservation but planned conservation and efficient use of natural resources. Forestry and water management were the two principal issues. The beginnings of environmental awareness in the United States were associated with two schools of thought: conservation of nature with a view to careful management, and preservation for its own sake. The movement was divided as to the strategy to be adopted; whether to preserve the wilderness as proposed by John Muir or to manage the environment scientifically, as planned by Gifford Pinochet who was trained as a forester and Charles Sprague Sargent, a botanist. This divergence in attitude continued for quite a long time.

The second half of the nineteenth century saw attempts to preserve and plan the use of forest resources in many countries. Scientific forestry was well established in Germany by this time and the influence of German foresters spread throughout the world. This happened both directly, e.g. the appointment of the German forester Dietrich Brandis as the Inspector General of Forests for India in 1864; or indirectly, as the British foresters trained in Germany or after a spell in India moved to other countries of the British Empire. The coming of the European imperialism devastated the flora, fauna and human societies of the New World and altered the traditional relationship between the environment and the people in the Old World. In the New World, new extensive and productive agricultural systems made a firm beginning, which Crosby (1986) has identified as ecological imperialism. In the Old World, as in India, the effect was less drastic but did radically alter existing ecological arrangements and food production systems (Gadgil and Guha, 1992). Gadgil and Guha traced the British influence on the management and utilization of forest resources in India. The justification of a forest service supervising the forests came from a commercial demand for timber, first for ships and then for railways. In Madras Presidency alone, 35 000 trees were felled to produce 250 000 railway sleepers every year. Not only were forests depleted, but the species were replaced. As the oak–pine forests on the Himalayan slopes were harvested, pines were planted in their place. The old practice of allowing the local villagers to use the forest for grazing or shifting cultivation was severely curtailed, leading to unrest and periodic violence. The arguments of Dietrich Brandis to allow the village communities to manage their own grove of trees or those of J. A. Voelcker for building up local fuel and fodder reserves, a practice that would now be described as social forestry, were superseded in favour of recognizing forests as state reserves controlled by the forest department and not, as the then Agricultural Secretary Allan Octavian Hume pointed out in rebuke of a rather vociferous official, for the 'semi-savage denizens of the Kanara forest' (Gadgil and Guha, 1992). Undoubtedly this policy led to financial success and better knowledge of Indian forests, but it also sowed the seeds of local popular unrest arising from economic deprivation.

Forestry, land management and the preservation of wildlife became important issues over time, following widespread destruction of woods, soil and animals. The movement involved people such as President Theodore Roosevelt and the German naturalist Carl G. Schilling who commented on the propensity of young men from Britain or continental Europe who had failed in love, to go to Africa and shoot big game. Aldo Leopold's essays, which were later published together in *A Sand County Almanac*, indicate the harmony that was sought between people and nature (Leopold, 1966). The sanitation movement of the late nineteenth and early twentieth centuries was also a significant development as attempts were made to have some control over the quality of water and air concerning both industries and settlements.

13.3 Ecological Disasters and International Organizations: The First Half of the Twentieth Century

The ecological destruction continued in the twentieth century. Detailed accounts of environmental degradation and its consequences in the first half of the twentieth century are given by Eckholm (1976) and McCormick (1989). Of these, the blowing away of the topsoil from the Great Plains of the United States is perhaps the best known. This caused great misery on a regional scale, especially as the drought arrived concomitant with the depression economy of the United States. The poorer sections of the farming community were displaced from their homes and travelled westward, which was the backdrop of John Steinbeck's *The Grapes of Wrath*.

This ecological disaster followed the common pattern of extending agriculture into hazardous areas in the quest for monetary gains. A combination of bad farming practices, a strong moisture deficit, and a depressed national economy left the topsoil vulnerable against wind erosion in the years of drought that occurred repeatedly in the 1930s. The total area affected was more than a million square kilometres. The fine windblown particles and humus from the topsoil crossed the east coast into the Atlantic. This required immediate action, and in 1935 the Soil

Conservation Service was established, with Hugh H. Bennett as its chief, in order to promote proper soil management and to prevent accelerated erosion. Eckholm (1976) perceives the Dust Bowl experience as a catalyst in the ecological history of the United States. The importance of soil conservation and proper agricultural practices was also subsequently realized in other countries where agriculture was carried out under similar hazardous conditions. This type of agricultural disaster, however, had happened repeatedly on the Great Plains but had not necessarily damaged the productivity of the fields (Malin, 1948). A litany of common problems has occurred in the drier parts of the world. Major examples include salinization of agricultural fields following canal irrigation in dry areas, accelerated slope failures due to cultivation of steeplands, and the spread of deserts in the wake of risky land use practices on their fringes. Similar problems happened repeatedly and in different locations so long as the physical geography remained similar. The spread of salinization, for example, is as old as the Sumerian Civilization of 3000 BC. It also happened in the twentieth century in West Asia, the Indian Subcontinent, and Australia. It happened in Egypt after the Aswan Dam and Lake Nasser behind it became functional in 1971 (Kishk, 1986).

A number of international organizations came into existence after the Second World War with the establishment of the United Nations in 1945. Several of these organizations were concerned with themes such as preservation of nature, solving the food problem in a rapidly populated world, and better land management – themes from earlier times. A list of these organizations and the year of their establishment includes the Food and Agricultural Organization (1945); the United Nations Educational Scientific and Cultural Organization (1946); the World Health Organization (1948); the International Union for Conservation of Nature and Natural Resources (1956), the precursor of which was the International Union for Protection of Nature; the World Wildlife Fund (1960), currently known as the World Wide Fund for Nature; and the United Nations Development Programme (1966). Several conferences and discussions, such as the United Nations Scientific Conference on the Conservation and Utilization of Resources (UNSCCUR) in 1949, or the Arusha Declaration of 1961 on wildlife protection, from time to time highlighted the themes mentioned earlier. The continuation of scientific interest and research was often due to the leadership of individuals such as Julian Huxley who was the director-general of UNESCO for some time. The International Monetary Fund, The International Bank for Reconstruction and Development (IBRD), popularly known as the World Bank, and other multilateral development banks were also established about this time. These global institutions later played important roles in both environmental and developmental spheres, ultimately linking the two concerns.

13.4 *Silent Spring* and Vocal Ecologists

The thrust of the environmental movement changed in the early 1960s. The publication of Rachel Carson's *Silent Spring* was a benchmark in the history of the environmental movement. She spoke above all of the disastrous consequences of using dangerous pesticides, including DDT, indiscriminately; a timely warning that reached an extremely wide readership. Her writings, which were serialized in the *New Yorker*, became a best seller in book form, subsequently published in a number of countries outside the United States. It made a powerful impact on a very important sector of society, the affluent educated white Americans, who could recognize in their everyday world Carson's description of the effects of uncontrolled use of dangerous pesticides. This is reflected in the writing of the present Vice-President of the United States:

> Our farm taught me a lot about how nature works, but lessons learned at the dinner table were equally important. I particularly remember my mother's troubled response to Rachel Carson's classic book about DDT and pesticide abuse, *Silent Spring*, first published in 1962. My mother was one of many who read Carson's warnings and shared them with others. She emphasized to my sister and me that this book was different – and important. Those conversations made an impression, in part because they made me think about threats to the environment that are much more serious than washed-out gullies – but much harder to see.
>
> (Excerpt from Gore, 1992, *Earth in Balance*. Copyright © 1992 by Senator Al Gore. Reprinted by permission of Houghton Mifflin Co. All rights reserved.)

The 1950s and 1960s were also the time when unhappiness and worry about the state of the environment remained in the forefront following a number of atomic tests carried out by the USA, the USSR, Britain and France, with accompanying instances of radioactive contamination of land, water and people. Furthermore, a series of environmental disasters in the 1960s indicated repeatedly the increasingly mismanaged state of the world's environment. These included repeated pollution of extensive coastal waters from large-scale oil spills such as from the tanker *Torrey Canyon* off the southwest coast of England or the oil platform off Santa Barbara, California; the mining spoil heap landslide at Aberfan, Wales; or the discharge of effluents containing mercury in the waters of Minamata Bay, Japan, with disastrous consequences. Many of these instances are still coming to light. This was also the period when extensive protest movements took shape: the Campaign for Nuclear Disarmament, the Civil Rights movement in the United States, and the anti-war movement following the escalation of American involvement in Indo-China. The need and the mechanism for a large-scale movement to prevent environmental degradation were thus present (McCormick, 1989). People started to realize that the degradation was affecting their everyday lives and they were not in control.

The existing conservation organizations such as the Sierra Club enjoyed an impressive increase in their membership over this period and new environment-oriented groups were established. Such organizations subsequently became known as non-governmental organizations (NGOs) in the environmental arena. NGOs since then have been at the forefront of environmental movements. Certain characteristics of NGOs were evident from the beginning. Among their members, the groups include (a) people who are directly at the receiving end of an environmental disaster; (b) people who have scientific and technical expertise; and (c) people who are capable of mounting serious political lobbying in support of their cause. These three properties – grassroots involvement, scientific backing and political activism –

have usually been associated with NGO movements regarding environmental issues, and also account for their success in a number of such movements.

The strength of the United States environmental movement which grew in the 1960s was evident on Earth Day, 22 April 1970, when meetings were held across the country with leadership from Senator Gaylord Nelson and federal funding. Teach-ins and gatherings took place in thousands of schools and colleges. New York banned cars from Fifth Avenue for part of the day. Earth Day is celebrated every year although such celebrations no longer have the impact of the very first one.

David Harvey has described his experience on that day:

> Around 'Earthday' 1970, I recall reading a special issue of the business journal *Fortune* on the environment. It celebrated the rise of the environmental issue as a 'non-class issue' and President Nixon, in an invited editorial, opined that future generations would judge us entirely by the quality of environment they inherited. On 'Earthday' itself, I attended a campus rally in Baltimore and heard several rousing speeches, mostly by middle class white radicals, attacking the lack of concern for the qualities of the air we breathed, the water we drank, the food we consumed and lamenting the materialist and consumerist approach to the world which was producing all manner of resource depletion and environmental degradation. The following day I went to the Left Bank Jazz Club, a popular spot frequented by African-American families in Baltimore. The musicians interspersed their music with interactive commentary over the deteriorating state of the environment. They talked about lack of jobs, poor housing, racial discrimination, crumbling cities, culminating in the claim, which sent the whole place into paroxysms of cheering, that their main environmental problem was President Richard Nixon.
>
> (Harvey, 1993: 1)

This conflicting nature remains a characteristic of the environmental movement. The common concern of environmental degradation brings together individuals, communities and even national governments. On the other hand, strong disagreements over the priorities evolve between the prosperous and poorer sections of a population, and also between the developed and developing countries.

A number of books based on ecology, pollution and the rising population appeared in the 1960s and early 1970s (Ehrlich, 1968; Ehrlich and Ehrlich, 1970; Commoner, 1971), a few of which sold prodigiously. Stuart Udall wrote *The Quiet Crisis* (Udall, 1963), an account of the environmental concern in the United States. A few books of this time, however, were of an alarmist nature and predicted various types of doom and gloom for the world, predictions which fortunately for us have appeared to have been highly exaggerated. Such books generally do not discuss in any detail the environment of the developing countries and at times have been known to offer impractical and offensive solutions to their doomsday scenarios. The books, however, kept alive the academic and popular focus on environmental issues. Even computer models were used to determine the future of the earth (Meadows *et al.*, 1972), leading to more controversies. Certain terms such as Spaceship Earth and the Commons became popular. The idea of a spaceship indicated both the integral and isolated nature of our environment, the term acquiring special significance because of the pictures of the earth which started to arrive from satellites, which in 1970 were something of a novelty. The concept of shared finite resources began to make headway. The tragedy of the commons was examined by Hardin (1968). E. F. Schumacher in 1973 wrote *Small is Beautiful* which called for a re-examination of our values and lifestyles. In the middle of all this, an international conference in Stockholm in 1972 changed the environmental scene, which became more global in nature and less concerned with pollution and the loss of lifestyles in developed countries.

13.5 The Stockholm Conference

The Stockholm Conference of 5–16 June 1972 is a benchmark event. It helped the environmental movement to develop into a global movement; it altered the thrust of the environmental concern; it correlated good environment with development; and it admitted that the concern and responsibility should be shared by both officials and citizens. To a large extent its success was due to a number of meetings which served as precursors to the Stockholm Conference, clearing the air and working towards a general pre-Stockholm consensus. The first of these meetings, the Biosphere Conference, had been organized by UNESCO in Paris in 1968. Apart from other outcomes, this conference had stressed the interrelatedness of the environment, and the needs for global movements, research and social approaches for proper management of the ecology.

The Stockholm Conference was originally planned as an international conference organized by the United Nations on pollution. The developing countries stressed from the beginning that environmental internationalism could only be successfully achieved if the objective of the conference were widened and the topics of development and transfer of knowledge and technology were also considered. This probably arose because of the emphasis placed by the developed countries on pollution as the most important environmental factor and the continuing debate on doomsday hypotheses and growth limitation. A resolution was passed in the General Assembly ensuring that no environmental policy should adversely affect the future development policies of the developing countries. The discussions at two meetings (Founex, Switzerland and Canberra, Australia) in 1971 helped to resolve the difference between the nations. A consensus started to emerge which accepted the compatibility of development with environmental improvement, and also the global nature of environmental degradation. Another preparatory meeting, with the presence of Maurice Strong, the designated secretary-general for Stockholm, expanded the agenda of the conference to include issues which recognized the relationship between development and environment. Such meetings, seminars and reports preparatory to Stockholm provided the conference with much clearer objectives and kept controversy to a low level so that it did not jeopardize the discussions. The conceptual framework of the conference was further clarified with the publication of *Only One Earth* by Barbara Ward and René Dubos, who were commissioned by Maurice Strong to provide a conceptual framework for the conference (Ward and Dubos, 1972). The sequel to this volume came out seven years later, *Progress for a Small Planet* (Ward, 1979).

Only One Earth stressed our collective global responsibility and the need for a co ordinated international policy for tackling issues of global proportions such as climate or oceans. It also pointed out the difference in environmental problems that plague the developed and the developing world.

The Stockholm Conference, thanks to all these efforts, brought in representatives from 113 countries. It was attended by a number of international organizations and, for the first time, NGOs were officially involved. Stockholm is a landmark meeting for various reasons: the globalization of the environmental concern; the linkage of environment with development; the recognition given to non-governmental organizations; and the building of hope that finally environmental issues are being tackled. It set in motion activities that culminated in the meeting at Rio de Janeiro twenty years later, but unlike Rio, Stockholm did not see a great procession of political leaders.

Indira Gandhi, the sole important political leader to attend Stockholm, emerged as an articulate spokesperson for the developing countries. She identified the profit motive, individual or collective, as the base cause of the crisis in ecology. Speaking of India, she said, 'we do not wish to impoverish the environment further, and yet we cannot for a moment forget the grim poverty of large numbers of people. Are not poverty and need the greatest polluters?' (D'Monte, 1985)

A declaration, a number of principles and an action plan came out of Stockholm. In summary, the principles were as follows:

- protection and conservation of natural resources
- an interrelationship of development and environmental concerns
- the establishment of environmental standards and resource management at a country scale without endangering other nations
- limited pollution
- use of science, technology, education and research in the field of environmental concern

The hopes and aspirations arising out of Stockholm were only partially fulfilled. In retrospect, the conference did succeed in strengthening the environmental movement to a considerable extent. The global nature of the environment was recognized by the participants, the linkage between environment and development was established, the important role of the NGOs was admitted, the practice of discussing and resolving environmental issues in meetings became acceptable, and lastly, Stockholm led to a new mind-set, priorities and institutions within the United Nations. The best known of these new institutions is the United Nations Environmental Programme (UNEP), set up in 1972 to monitor the state of the world's environment. All this was achieved without intense controversy. Stockholm modernized the environmental movement.

14

Current Global Events and Projected Effects

14.1 Introduction

The Stockholm Conference was originally planned to discuss the spread of pollution across international boundaries. It later developed into a meeting with a much wider agenda, linking environment with development and also building mechanisms for international collaboration. The years that followed indicated that many of the environmental problems have indeed become global in nature and their solutions require international efforts simultaneously in various fields: technical, economic, social and political.

An environmental problem becomes global in scale due to two factors: the areal linkages between the physical processes which operate at or near the surface of the earth (such as wind circulation, sediment transfer by streams, movement of subsurface water, etc.); and the sheer size of the problem. The effect of large-scale degradation of the environment is carried to all parts of the world. Destruction of the tropical rainforest (Chapter 5) with its attendant increase of atmospheric CO_2 and loss of global biodiversity is a good example of such linkage. From time to time the world has experienced environmental problems that did not stay confined within national boundaries. Acid rain following the burning of sulphur-bearing coal, and the downwind drifting of radioactive material after a nuclear disaster are examples of such international problems. Two events of environmental degradation, however, have become much more widespread than others, and are currently threatening to affect the entire earth. Even sections of the world that had little part in generating the problems have become vulnerable.

The two events are *stratospheric ozone depletion* and *global warming*. Our knowledge regarding the background and causes of stratospheric ozone depletion is clearer than that concerning global warming, but in both cases, it is manifestly evident that precautionary measures need to be taken, and taken immediately, in order to prevent catastrophes of a global dimension. The solutions involve understanding, agreements and sacrifices among nations. These are global ecological problems but the developed countries are the chief contributors to their build-up. They also have much easier access to the future solutions. Due to this disparity between the developed and the developing countries, global agreements have not been easy to ratify. In this chapter we examine the causes, effects and possible solutions of these two potential global catastrophes.

14.2 Stratospheric Ozone Depletion

Atmospheric temperature profile and ozone distribution

The temperature of the atmosphere decreases upwards from the surface of the earth as the air absorbs heat from long-wave terrestrial radiation (Chapter 2). At about 20 km above the earth's surface, the temperature change reverses direction and starts to increase. This level is known as the *tropopause*. The lowest part of the atmosphere below the tropopause is known as the *troposphere* and the zone immediately above it is known as the *stratosphere* (Figure 14.1). Beyond the stratosphere the temperature reverses its trend two more times, giving rise to two other zones: the *mesosphere* and the *thermosphere*. It is the troposphere which mostly affects our climate, and we enter the stratosphere when we travel in jet aeroplanes.

From Table 9.1 we know that ozone is a rather minor constituent of the atmosphere. About 90 percent of it is concentrated in a stratospheric belt around the earth. This is caused by the breakdown of oxygen molecules by UV-C radiation into O atoms. Most of this conversion happens between altitudes of 35 and 50 km in the upper stratosphere. The O atoms then meet up with other O_2 molecules to form O_3 molecules, creating a belt of ozone in the lower stratosphere. The ozone belt varies in height, the peak concentration (10 ppm) being 25 km above the earth's surface at the equator and 15 km above the poles (Rowland, 1990). This zone, often called the ozone shield, absorbs much of the ultraviolet rays of the solar radiation and thereby reduces their arrival at the surface of the earth, with attendant possible harmful effects. The absorption of UV radiation by ozone generates heat, which explains the rise in temperature in the stratosphere. In the 1970s, high-flying supersonic transport aeroplanes were planned in large numbers. It was feared that nitrogen oxide from their exhausts would react with stratospheric ozone and destroy the shield. Fortunately, only a limited number of such aircrafts were built and the problem did not become critical.

Chlorofluorocarbons: background and use

Chlorofluorocarbons (CFCs) were developed during the 1920s in the United States to replace the use of ammonia and SO_2 as coolant gases in refrigerators. They were produced to meet the demand for a cooler that was safe by being chemically inert, odourless, non-corrosive and non-toxic (Rosemarin, 1990).

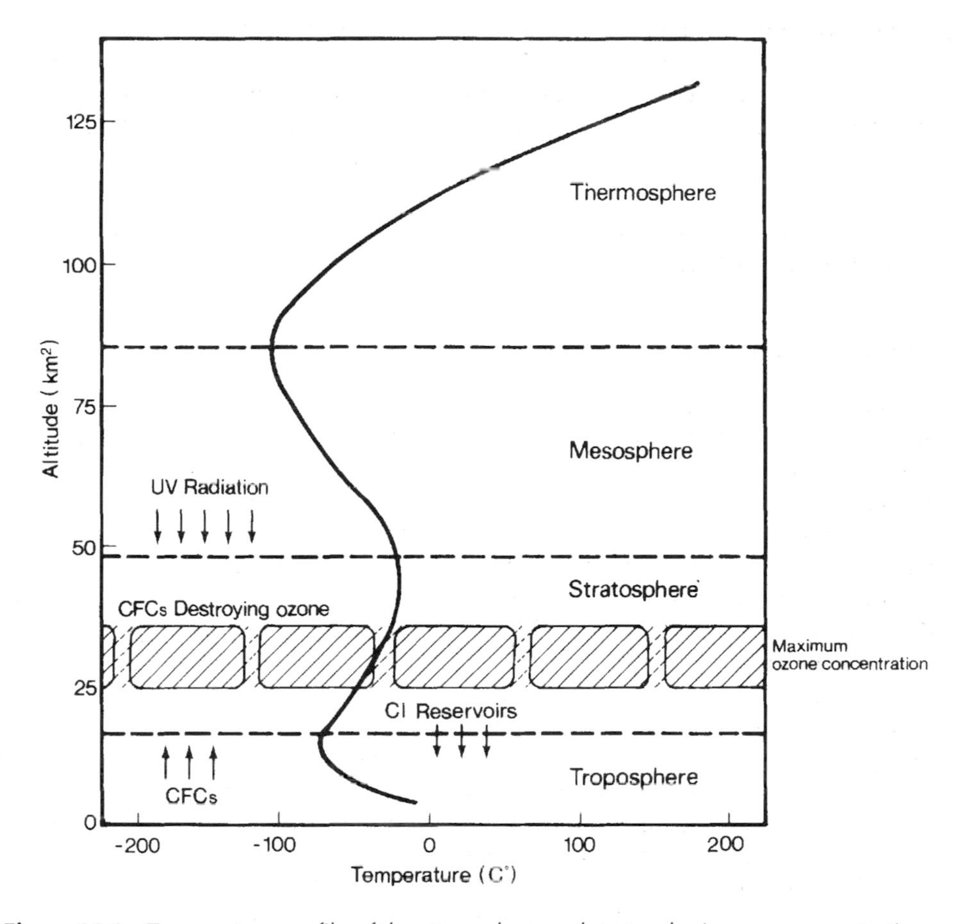

Figure 14.1 *Temperature profile of the atmosphere and stratospheric zone concentration*

CFCs are numbered according to their chemical composition. The acronyms, formulae and the full chemical names for selected CFCs would be as follows:

CFC-11	CCl_3F	(trichlorofluoromethane)
CFC-12	CCl_2F_2	(dichlorodifluoromethane)
CFC-22	$CHClF_2$	(chlorodifluoromethane)
CFC-113	$C_2Cl_3F_3$	(trichlorotrifluoromethane)

The numbering system originated from the chemical company Du Pont, and has subsequently been adopted world-wide. The digit to the extreme right is the number of fluorine atoms in a particular CFC. The second digit from the right comes from the number of hydrogen atoms plus one. The third one from the right is the number of carbon atoms minus one. Zeros are not listed. The final number is then added to the acronym CFC (Miller and Mintzer, 1986).

Over time, the demand for CFCs increased tremendously. CFCs were used not only in refrigeration but also in air-conditioners, automobile car cushions, open-celled plastics, cleaning of electronic equipment including silicon chips and aerosol sprays (Table 14.1). The use and production of CFCs spread quickly from the 1930s onwards, first in the USA and then in Europe, and continued to increase because of their usefulness, up to the last rapid increase in the mid-1980s. The global production jumped in the mid-1980s to a figure of more than one million tonnes. A very large percentage of the CFCs tends to escape into the atmosphere from the containers and drift randomly upwards. At least 85 percent of all the CFC-11 and CFC-12 manufactured so far has escaped into the atmosphere. In absolute terms, the volume of CFC-11, CFC-12 and CFC-113 annually released reaches several hundred thousand tons for each type (Rowland, 1990). Although CFCs escape more to the skies of the industrial countries, CFCs are now found in the air over extremely remote parts of the world due to atmospheric diffusion. The atmospheric concentration of the different types of CFCs over time is shown in Figure 14.2. As the CFCs are chemically inert, they drift randomly upwards until they reach the ozone shield in the lower stratosphere.

Chlorine chemistry and ozone depletion

The ozone-destructive property of the CFCs was proposed by Molina and Rowland in 1974 in *Nature* (Molina and Rowland, 1974) for which they, with Paul J. Crutzen, were awarded the Nobel prize in chemistry in 1995. The normally inert CFC

Table 14.1 *Selected CFCs and their properties*

Type	Use	ODP	GWP	Lifetime (years)
CFC-11 (CCl_3F)	Foam blower, aerosol, refrigerant, solvents	1.0	1.0	60 (75?)
CFC-12 (CCl_2F_2)	Refrigerant, foam blowing, aerosols, air conditioning	0.9–1.0	2.8–3.4	120
CFC-113 ($C_2Cl_3F_3$)	Solvent and cleaner for microelectronics	0.8–0.9	1.3–1.4	90
HCFC-22 ($CHClF_2$)	Refrigerant, air conditioning	0.04–0.06	0.32–0.37	15.3
HFC-134a (CH_2FCF_3)	Replacement for CFC 12	0	0.24–0.29	15.5
HCFC-141b (CH_3CCl_2F)	Replacement for CFC 11	0.07–0.11	0.084–0.09	7.8
Methyl chloroform (CH_3CCl_3)	Solvent	0.10–0.16	0.022–0.026	6.3
Halon 1301 (CB_2F_3)	Fire extinguisher	7.8–13.2		
Halon 1211	Fire extinguisher	2.2–3.0		

Note: ODP: ozone depletion potential compared to that of CFC-11; GWP: global warming potential compared to that of CFC-11, discussed in Section 14.3; Lifetime: survival period in the atmosphere.
Sources: Rosemarin (1990), Rowland (1990).

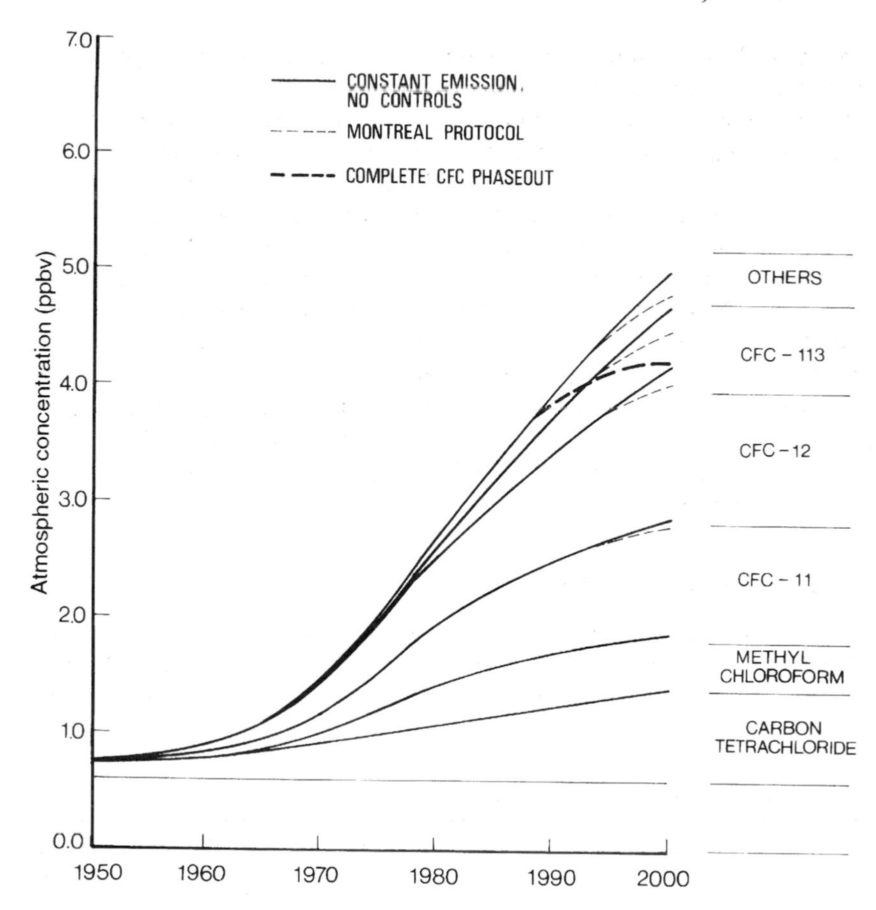

Figure 14.2 *Calculated atmospheric concentration of different CFCs over time prior to significant changes following agreements at Montreal and subsequent international meetings. After such agreements the concentration of CFCs has started to diminish and the peak concentration is already behind us (see text). The line marked 'complete CFC phaseout' indicates the projected change if CFCs were totally banned at Montreal. The lines of 'constant emission, no control' indicate (a) the rapid rise of CFCs in the atmosphere, and (b) the extent of the threat which has been averted since Montreal. The basal section below carbon tetrachloride indicates the presence of methyl chloride whose major sources are in the natural environment. From Rowland, F.S. (1989)* The role of halocarbons in stratospheric ozone depletion, *in National Research Council,* Ozone Depletion, Greenhouse Gases, and Climate Change, *National Academy Press, Washington, D.C., pp. 33–47. Reprinted with permission. Copyright (1989) The National Academy of Sciences, Courtesy of the Joseph Henry Press, Washington, D.C.*

molecules break down above the stratospheric ozone belt, where they come into contact with the extremely energetic UV radiation from the sun.

In the mid-stratosphere, in contact with the solar high-energy ultraviolet radiation, CFCs decompose to release atomic chlorine. This atomic chlorine is highly reactive and almost immediately combines with an ozone molecule:

$$Cl + O_3 = ClO + O_2 \tag{14.1}$$

The reactive ClO then reacts with another atomic O:

$$ClO + O = Cl + O_2 \quad (14.2)$$

The atomic chlorine is thereby available to convert another ozone molecule to oxygen, and if there is more than one Cl atom (as in CFC-12, CCl_2F_2), the second one goes through the same reaction. The process takes only a few minutes in sunlight. The two-stage chemical reaction (known as the ClO_x chain) keeps on repeating until the chlorine is bound into a reservoir molecule such as HCl or $ClONO_2$. This previously known basic chemistry of the ClO chain was shown to occur in the atmosphere by R. J. Cicerone and R. S. Stolarski in 1973 (Rowland, 1990). The abundance of atomic O in the upper stratosphere determines that most of the reaction happens at this level. Due to the repeated nature of the reaction, a single Cl atom may destroy about 100 000 ozone molecules before the chlorine enters a reservoir (usually HCl); diffuses down into the troposphere; and is returned to solid earth by rainout (Rowland, 1990).

Brominated fluorocarbons (Halon-1301, $CBrF_3$ and Halon-2402, $CBrF_2CBrF_2$) destroy stratospheric ozone even more effectively, but their limited industrial and commercial use restricts the damage. The widespread use of CFCs, the time taken for their release to the atmosphere, and their slow random rise to the stratospheric levels which takes several years, indicate that even if the manufacturing of all CFCs stops immediately, the process of ozone destruction will continue for some time. For example, CFC-11 and CFC-12 stay in the atmosphere for about 75 and 120 years respectively. In 1990, Rowland estimated that the highest rate of stratospheric ozone destruction would occur early in the next century even if all production of CFCs is halted. About 6 percent of the CFC-12 molecules in the atmosphere are expected to persist to the beginning of the twenty-fourth century (Rowland, 1990).

Ozone in the atmosphere has been systematically measured since the 1920s. Extensive data were collected in a network during the International Geophysical Year (1957–58). The measurements were done with Dobson ultraviolet spectrometers, which measure the ratio between UV-A and UV-B radiation reaching the earth's surface. UV-B is partially restricted by the ozone layer but UV-A is not seriously impeded. The network was expanded in the 1960s, and continuous measurement of ozone has been carried out for several decades in some areas, although most of the data come from a belt between 30° N and 65° N parallels.

Measurements from satellites started in 1970 with Nimbus-4 and continued with Nimbus-7 in 1978 which carried two instruments: a Total Ozone Mapping Spectrometer (TOMS) and a Solar Backscatter Ultraviolet (SBUV) satellite instrument. The instruments use reflected UV-wavelengths from the earth, and therefore, unlike ground-based instruments, can continue to take measurements in darkness. Ozone measurements have also been carried out from balloons and aeroplanes. It is now accepted that since 1970, a loss of 2–3 percent has occurred over the northern temperate latitudes. This loss happens mostly in winter. Satellite data indicate ozone loss at all latitudes but again the data are best between 30° N and 64° N.

The loss of ozone shield leads to stronger UV-radiation reaching the surface of the earth, affecting humans, plants and animals (Table 14.2). It is, however, the discovery in 1985 of drastic ozone loss over Antarctica, popularly known as 'the

Box 14.1

The Antarctic Ozone Hole

In 1985, observers at a British Antarctic Survey coastal station located at Halley Bay found that there had been a pronounced drop in atmospheric ozone over Antarctica in the previous October, which is spring in Antarctica. The ozone, measured with a Dobson spectrometer, dropped from the mean value of 320 Dobson units (DU) in October 1960 to 180 DU in October 1984 (Levi, 1988). One Dobson unit is defined as the amount of ozone in a vertical atmospheric column equivalent to a 0.01-mm-thick layer of ozone at standard pressure and temperature. It can be taken as approximately 1 ppb by volume. The loss was confirmed in 1986 by NASA from the TOMS measurements aboard Nimbus-7. Ozone had virtually disappeared between 15 and 20 km elevations, with only about 5 percent of its late-1970 concentration remaining. Further investigations by British, Japanese and US scientists used instruments aboard satellites, balloons, and aircrafts flying from Punta Arenas, Chile. The satellite observations indicated ozone loss over an area as big as Western Europe, with a sharp boundary demarcated by a very steep change in ozone concentration. This has been described as the ozone hole. Balloon flights indicated no ozone loss during the long dark Antarctic winter. The loss was found to start in late August with the first appearance of the sun and to build up extremely rapidly in September, i.e. early spring (Figure 14.3). The loss stops by mid-October but the hole remains. By November, with a weakening of the air vortex over the South Pole, the ozone-poor air is displaced northwards by air with a normal concentration. In 1987, this ozone-poor air drifted over to Melbourne causing some consternation. The hole re-emerges in the following spring, and the ozone concentration progressively drops over Antarctica (Figure 14.4).

The explanation of this dramatic ozone loss over Antarctica lies in a combination of chlorine chemistry and polar meteorology. During winter the winds over polar regions circulate tightly, and especially so over Antarctica (Hamill and Toon, 1991). This creates an isolated but substantial mass of stratospheric air which remains in total darkness for several months but becomes steadily colder. When the temperature drops below –78 °C, clouds begin to form in the stratosphere. The temperature may drop to –90 °C when clouds become abundant.

Two types of polar stratospheric clouds (PSCs) form over Antarctica in winter. Type I PSCs are made of crystalline nitric acid trihydrates, $HNO_3(H_2O)_3$. When the temperature drops below –83 °C, Type II PSCs form by accumulation of water ice over the cloud surface to form larger particles. Almost the entire volume of nitrogen oxides remains in the clouds in solid phase, which prevents reactions with chlorine to form temporary reservoirs, and the Cl stays in the gas phase (as Cl_2 and HOCl) on the cloud surface. The clouds remain in darkness all through the winter.

With the coming of sunlight in early spring (late August), a very large fraction of the chlorine changes into atomic chlorine, which reacts at once with ozone to form ClO. However, as there is very little atomic oxygen present, the concentration of ClOs rises to such high levels that the molecule ClOOCl is formed, which disintegrates in sunlight to form chlorine atoms and an oxygen molecule. This reaction over Antarctica is different from the one described by Molina and Rowland (1974) but is highly effective in removing the ozone. The rapid loss of stratospheric ozone continues until the sun is high enough to warm up the stratosphere (in mid-October) and dissipate the polar stratospheric clouds. The released nitrogen oxides begin to form reservoirs with chlorine, and the process reverts to that taking place over the lower latitudes.

ozone hole', which introduced a sense of global crisis and prompted international efforts at halting the entire production of CFCs.

Temperature over the Arctic normally will not fall as low as that over Antarctica, nor will the polar vortex work with the same efficiency. Occasionally, however, the temperature is low enough to form PSCs, and chemical reactions like those over Antarctica occur for example, an Arctic ozone hole appeared in 1997. However, the PSCs tend to disintegrate rapidly, and ozone depletion at a significant rate is only temporary. The loss over the Northern Hemisphere indicates a seasonal cycle with a high winter loss. Data from other parts of the world are too limited to form comparable conclusions. A conspicuous drop in ozone, as happens in Antarctica, in any of these areas will cause even greater consternation due to their proximity to populated areas.

CFC Replacement: substitution, agreements, and technology transfer

Stratospheric ozone depletion by CFCs and the inherent dangers from it (Table 14.2) were well understood by the mid-1970s. This led to attempts to restrict the production of CFCs. The early attempts were piecemeal in nature but gradually the search for a global arrangement for banning CFCs developed. Although CFCs are primarily put out into the atmosphere by the industrial countries, their spread into the atmosphere and the effects are global in nature. This factor, together with the growing need for refrigeration, insulation, and cleaning of electronic components as countries develop, ensured the need for a global agreement for the successful control of chlorofluorocarbons. The agreement, as to be expected, did not materialize without prolonged negotiation between the governments.

The use of CFCs as propellants in aerosols was the earliest form to be banned in a number of countries where substitutes such as manual pumps came into use. The

Table 14.2 *Effect of ultraviolet radiation on people, plants and aquatic ecosystems*

On people

- Sunburn and thickening of the skin
- Nonmelanoma skin cancer
- Malignant melanoma
- Cataracts and retinal damage
- Suppression of the immune system
- Aggravated incidence of infectious diseases

On terrestrial plants
(not all species are affected or affected at the same level)

- Decreased photosynthesis
- Decreased water efficiency
- Decreased leaf area
- Variable intraspecific effects

On Aquatic ecosystems

- Affects both orientation and motility in phytoplankton, resulting in reduced survival rates
- Reduction in phytoplankton production
- Damage to fish, shrimps, crabs, and amphibians: reduced reproductive capacity and impaired larval development

Note: This is a generalized summary. The effects of ultraviolet radiation are still being assessed.

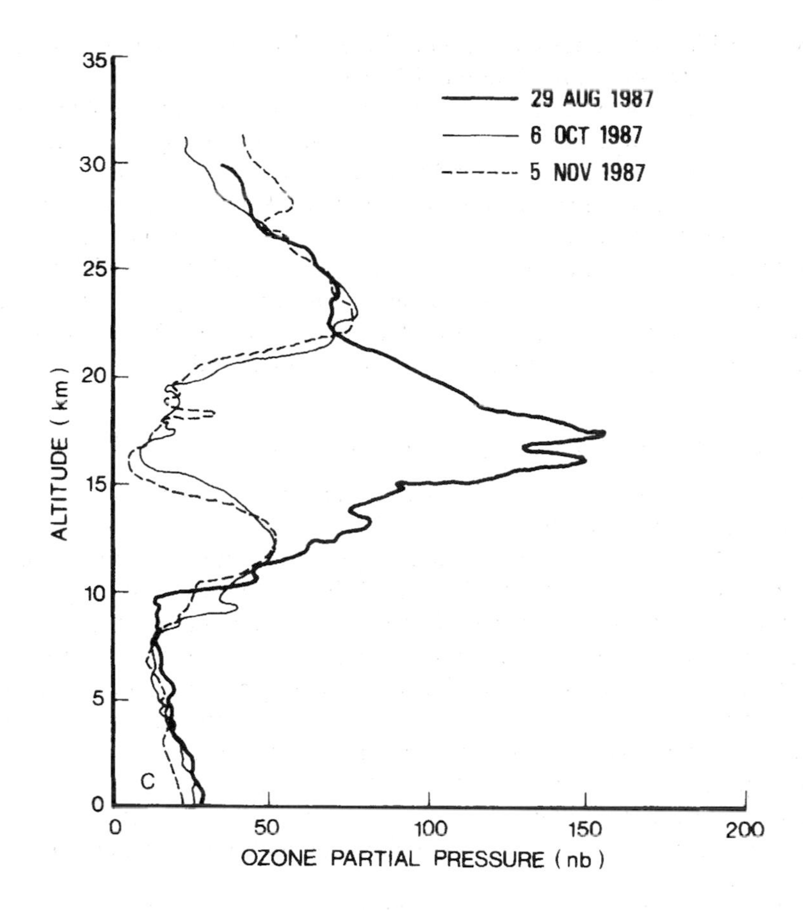

Figure 14.3 *Ozone loss over Antarctica in spring time. Modified from Watson, R.T. (1989) Stratospheric ozone depletion: Antarctic processes, in National Research Council,* Ozone Depletion, Greenhouse Gases, and Climate Change, *National Academy Press, Washington, DC, pp. 19–32. Copyright (1989) the National Academy of Sciences. Courtesy of the Joseph Henry Press, Washington, D.C.*

use of ozone-depleting CFCs can be controlled in four ways. First, by replacement, such as the use of a manual pump or a roll-on in the case of aerosol sprays, or cardboard instead of styrofoam as a packing material. Second, a new brand of CFC gas which is less threatening to the ozone layer can be used in substitution; for example, one of the possible replacements of ozone-depleting CFCs is HCFC-22 (currently one of the main refrigerant gases used in home air conditioning). Containing hydrogen, it tends to break down in the troposphere, unlike the main CFC gases. Another alternative is to use a CFC which carries fluorine instead of chlorine, and is about 1000 times less efficient in ozone depletion (Rowland, 1990). Substitution is therefore possible but it could be costly, especially at the beginning. It also raises the question of who is to develop the substitute and who is going to pay for the substitution. Third, it is possible to capture and recycle CFC gases used in industrial production or refrigeration. Fourth, their losses to the atmosphere could be minimized.

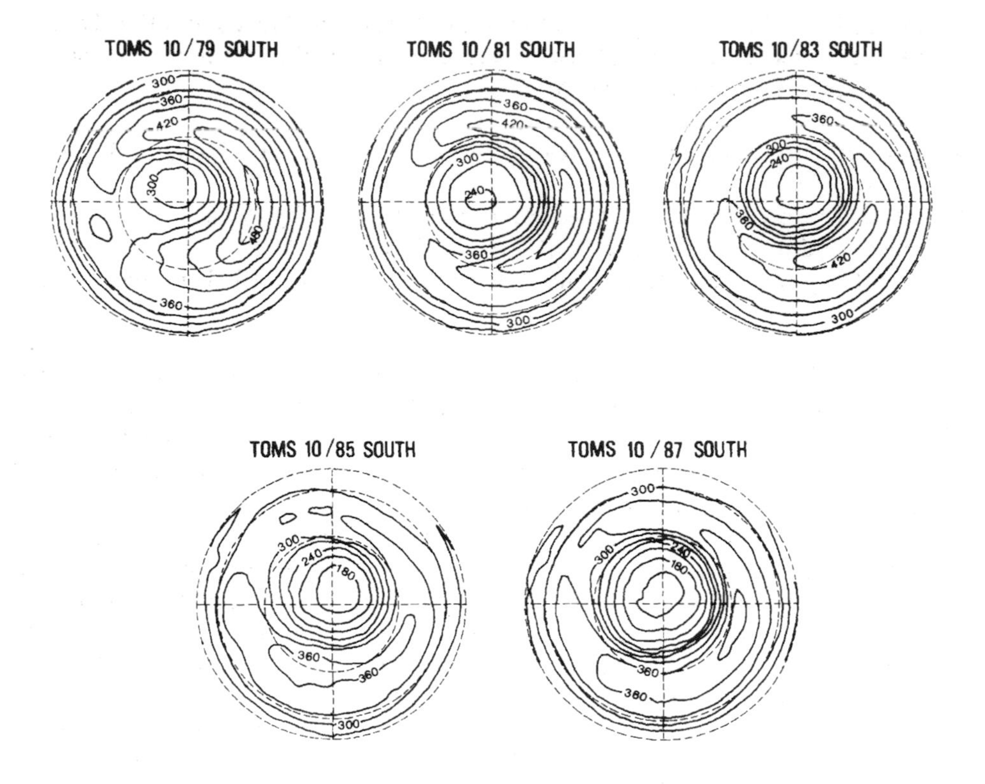

Figure 14.4 *Southern hemisphere: monthly mean total ozone for October, 1979–87 (odd years only). From Watson, R.T. (1989) Stratospheric ozone depletion: Antarctic processes, in National Research Council,* Ozone Depletion, Greenhouse Gases, and Climate Change, *National Academy Press, Washington, DC, pp. 19–32. Copyright (1989) the National Academy of Sciences. Courtesy of the Joseph Henry Press, Washington, D.C.*

The paper which attributed stratospheric ozone depletion to CFCs was published in 1974 (Molina and Rowland, 1974). Representatives of major CFC-producing countries met several times in the next few years and national restrictions were placed on CFC use, mainly concerning aerosols. CFC use in industrial countries dropped by about a quarter from this ban (Miller and Mintzer, 1986). However, the change of government in the United States in 1981 reduced for the time being the possibility of further bans on a national scale. International efforts, however, continued with UNEP encouragement. This resulted in the 1985 Vienna Convention for the Protection of the Ozone Layer which was signed by 20 countries, and also involved industrial and environmental organizations. The Convention urged the countries to control activities adverse to the existence of the ozone layer, and also initiated international co-operation in research and the exchange of information.

The late 1980s saw various types of piecemeal CFC-controlling legislation coming into force in many industrial countries: Australia, Japan, West Germany, Norway, Sweden, France, Netherlands, UK, USA and Canada. However, the critical nature of the depletion of the ozone shield required an international protocol which was attempted in 1987 in Montreal. The Montreal Protocol on Substances that Deplete the Ozone Layer set out a number of specific deadlines. For example, the signatories to the protocol were required to

- freeze production and consumption of five CFCs (11, 12, 113, 114 and 115) at the 1986 level by 1989;
- reduce their production to 80 percent of the 1986 level by 1993;
- reduce production to 50 percent of the 1986 level by 1998;
- freeze the production and consumption of three halons (1301, 1211 and 2402) at the 1986 level.

However, in light of the ongoing scientific investigation, this agreement was found to be inadequate. In fact, even with the Protocol in operation, the chlorine level was predicted to double over the next 50 years. The list of signatories was limited and the Protocol did not take into consideration the other harmful effect of the CFCs: their global warming potential.

Two other important controls on CFCs were imposed in the United States immediately afterwards. In 1989, US Congress imposed an excise tax on the sale of ozone-depleting chemicals. In 1990, the US Senate passed legislation to control the sale and use of all ozone-depleting compounds, the Stratospheric Ozone and Climate Protection Act. Apart from controlling the total production of CFCs, these acts made the use of substitutes financially more feasible.

The next international deliberation was at the London Meeting of June 1990. This conference proposed a much tighter restriction on the CFCs, and also attempted to involve the developed countries. In brief, the London meeting required that by 2000

- five CFCs (11, 12, 113, 114 and 115) and three halons (1301, 1211 and 2402) should be totally phased out;
- all other fully hydrogenated CFCs should also be phased out;
- the use of carbon tetrachloride should be eliminated;
- the use of methyl chloroform should be reduced by 70 percent.

Furthermore, it was decided to monitor all CFC substitutes with a view to their annual production and consumption, to issue strict guidelines for their use, and to fulfil a commitment to phase out the CFC substitutes within a stipulated time period. About 60 countries, whose total CFC production and consumption was 90 percent of the world figure, signed the agreement. Nearly one hundred countries, including a large number from the developing world and including India and China, did not sign the Protocol at this stage. The total CFC consumption by the developing countries with their huge population was only 16 percent of the world's total at the time of the London Meeting (Rosencranz and Milligan, 1990). The huge population and the ongoing economic development implied that even a modest increase in the use of refrigeration units and consumer goods would mean a very large volume of CFC gases in use. This is particularly true for China and India. The reason for the developing countries not signing the Montreal Protocol at that time

Table 14.3 *Ranked global use of CFCs by categories, 1985*

Category
Aerosols
Rigid foam insulation
Solvents
Air conditioning
Refrigeration
Flexible foam
Others

Modified from: Worldwatch Institute (1989) State of the World, *W.W. Norton & Co, New York.*

Note: 1985 was a year of high CFC use.

lay in their belief that as the developed countries are responsible for the damage to the ozone layer it should be their responsibility to rectify the damage. Added to this was the apprehension of the cost for switching over to the ozone-friendly technologies (Rosencranz and Milligan, 1990).

The financial burden was recognized at the London Meeting where it was proposed to establish a CFC fund for providing financial support to the developing nations. This proposal finally evolved into a financial arrangement called the Global Environment Facility (GEF) where the funding was supplied by a group of both developed and developing countries and the administration was carried out by the World Bank, UNDP and UNEP. Some controversy still exists regarding the GEF, which is discussed in detail in Chapter 16. More developing countries have since signed the Montreal Protocol, usually contingent on financial assistance and technology transfer. Currently the total number of signatories is about 150. It is obvious that the prevention of stratospheric ozone depletion requires global arrangements but such arrangements can only rise from trust, openness and a sense of urgency.

Phasing out CFCs will require striking adjustment efforts from the chemical industry. The world production of CFCs carries an estimated value of more than two billion dollars. However, the stratospheric ozone depletion has caused the chemical industry to organize research and to come up with CFC substitutes. Du Pont, the world's largest producer of CFCs, alternated at the beginning between searching for CFC substitutes and opposing CFC regulations on grounds of uncertainty. However, after the Montreal Protocol of 1987, Du Pont decided to phase out the production of CFCs, and subsequently developed the first reliable two-dimensional model of stratospheric ozone which indicated significant depletion. About this time, a NASA report showed ozone depletion even over temperate areas. In 1988, Du Pont decided to stop manufacturing CFCs entirely by 1999.

Du Pont has planned to build four world-scale plants to manufacture CFC substitutes which contain hydrogen for tropospheric breakdown or fluorine but no chlorine. Carrier Corporation in 1990 announced a recovery and recycling system for refrigerants, thereby reducing CFC losses to the atmosphere. In 1991, Du Pont announced a set of refrigerator and car air-conditioning coolants which are to be used by General Motors (Buckholz, 1993). A number of industrial establishments in various countries now manufacture non-CFC home refrigerators and automobiles. It is now possible to hope that with international agreements and funding, technology transfer, industrial efforts and governmental supervision, the stratospheric ozone depletion will slow down and ultimately the ozone shield will start to recover.

Stratospheric ozone depletion by CFCs has clearly been shown to happen. It has led to consternation and international agreements, and a search for substitutes and recycling technology. The Montreal Protocol, as reorganized in London (1990) and subsequently in Copenhagen (1992), has accepted the goal of phasing out CFCs in several years. Du Pont further announced that they would phase out CFC sales to the industrialized countries by 1996, halon sales by 1994 (both pre-schedule), and speed up the elimination of HCFC-22 (World Resources Institute, 1992). The United States banned the manufacture of CFCs from January 1996. A number of other countries have taken similar action. However, the existing CFCs in the atmosphere are still depleting the stratospheric ozone and expanding the hole over Antarctica. Such degradation is expected to continue until 1998. Currently a gradual recovery is predicted, starting from about AD 2000, but it will be slow and the ozone layer is not expected to recover for about half a century (van der Leun, Tang and Tevini, 1995). This prediction is more positive than those made about five years ago but encouraging signs have started to emerge. Recently, Ronald Prinn of the Massachusetts Institute of Technology has shown an annual decrease of approximately 2 percent of the level of methyl chloroform – the first decrease noted of a substance restricted by the Montreal Protocol (Anon., 1995).

14.3 Global Warming

Global warming is a planetary phenomenon

Like stratospheric ozone depletion, excessive global warming is another example of anthropogenic environmental degradation that affects the entire planet. Global warming occurs when part of the long-wave terrestrial radiation escaping into the atmosphere is trapped by various gases present in the air. The heat energy is then returned towards the surface of the earth. Such gases include water vapour, CO_2, O_3, CH_4, N_2O and CFCs. This phenomenon is popularly known as the greenhouse effect because of the analogue with a cold climate greenhouse where the glass roof and walls trap the heat which comes in as solar radiation, thereby raising the temperature inside the greenhouse high enough for growing plants. The analogue is not quite perfect. In an ordinary greenhouse the heat transfer from the roof and the walls takes place via conduction, convection and radiation, whereas in the atmosphere heat transfer by conduction is negligible.

Such gases have been described as greenhouse gases. Under natural conditions water vapour is the most efficient greenhouse gas. The natural global warming raises the average temperature of our planet to a comfortable 15 °C instead of –18 °C. In comparison, on Mars, where the atmosphere is thin and water vapour is absent (the planetary water lies frozen in the Martian regolith), the average surface temperature is approximately –50 °C, whereas Venus, with a heavy atmosphere with high concentrations of CO_2 in the atmosphere, has an average surface temperature of around 430 °C. Living on the Earth, we are much better off.

The problem of global warming lies in the anthropogenic acceleration of this phenomenon. By producing and releasing to the atmosphere quantities of all the major greenhouse gases except water vapour, we have increased the greenhouse effect of the atmosphere, resulting in a slow rise of the global temperature. This in turn alters certain environmental components (e.g. the level of the sea surface) thereby threatening to degrade the quality of life across the world in the near future. Unlike stratospheric ozone depletion, however, the signal for global warming was not strong at the outset, resulting in acrimonious controversy and confrontation. A general consensus has emerged among scientists regarding the occurrence of global warming but not about all its effects.

Measurement of global warming

Ideally, to determine global warming, we should have accurate measurements of temperature from many weather stations spread around the world and with the record going back at least beyond the Industrial Revolution. We should also have similar data for the concentration of greenhouse gases in the atmosphere to determine their rate of change. This of course is not available. Excellent records of atmospheric CO_2 exist from 1957–58 from the Mauna Loa Climate Observatory in Hawaii because of the foresight of Roger Revelle who was instrumental in starting the project during the International Geophysical Year (1957–58). Direct measurements of past concentrations of other greenhouse gases only go back several years, and the data need to be supplemented by indirect means such as the examination of air bubbles trapped in Antarctic ice.

At present, land temperature records are collected and maintained at a number of weather stations, and the sea surface temperature is collected by ships sailing across oceans. Global temperature measurements are compiled through the World Weather Watch (WWW) system. The surface temperature of our planet is also measured from satellites such as the Earth Radiation Budget Experiment (ERBE) satellites. The past records, however, did not provide a representative coverage of the world. In the middle of the nineteenth century, the existing temperature records were first compiled and sorted by the German meteorologist Heinrich Wilhelm Dove. The data were land based and mostly covered the northern temperate areas. Most of the stations were located in towns and thereby possibly modified by the urban heat island effect. The American naval captain Matthew Fontaine Maury, who standardized the weather maps, also organized the measurement of the temperature of the surface waters of the ocean by ships. In the beginning it was done by raising a canvas bucket of water to the deck and dipping a thermometer in the bucket, which did not provide very accurate measurements. These days, such temperature is measured near the entrance to the water intake pipe of ships and is therefore more accurate.

Jones and Wrigley (1990) investigated and processed the existing temperature record. The standardized data showed that the global climate has generally become warmer since the 1880s with a few variations such as a cooling period between 1940 and 1970. Global temperature in a particular year may deviate from the rising trend due to three factors: sunspot cycles, volcanic eruptions, and the El Niño Southern Oscillation (ENSO) effect. The variation in solar energy during sunspots is assumed to change the amount of radiation normally received by the earth. Large eruptions of pyroclastic volcanoes release very high amounts of aerosols into the atmosphere, as happened during the eruptions of El Chichon in 1982, and Pinatubo in 1991 and 1992. The aerosols envelop the earth and tend to reflect and scatter the incoming solar radiation thereby lowering the global temperature. The release of sulphate aerosols in the atmosphere from the El Chichon eruption was about 12 megatons which has been estimated to have lowered the global temperature by 0.2 °C for several years. In the case of Pinatubo, the amount of aerosols released was more than double the El Chichon figure, cooling the surface temperature by about 0.5 °C. An empirical relationship exists between the occurrence of El Niño events and cooler years. Correcting for these anomalies, Jones and Wrigley estimated a 0.5 °C warming since the 1880s.

This is a modest increase, but subsequent evidence from other sources indicates the occurrence of accelerated global warming. Eight of the hottest years on record happened between 1980 and 1992, when records are corrected for the ENSO events. This particular sequence caught the popular imagination regarding global warming. Again, both 1995 and 1997 were particularly warm years. Glacial melting has increased in recent years, especially on the mountains (Oerlemans, 1994). This is particularly true for the tropical mountains, and measurements exist for the Andes (Schubert, 1992), Mount Kenya (Hastenrath and Kruss, 1992), and Central Asia (Thompson *et al.*, 1993). Measurement of the temperature of the top 100 m of water off the coast of California has indicated an increase of 0.8 °C over the last 40 plus years. The use of long-range low-frequency acoustics is planned, in order to determine accurately the warming of the sea temperature change in Arctic Ocean temperatures and sea ice thickness and roughness (Mikhalevsky *et al.*, 1995). Investigation of ice cores from Antarctica also shows a warming trend.

The Antarctic data usually come from three well-known cores. These are the 2160 m Byrd core from West Antarctica; the 900 m deep Dome core from East Antarctica; and the Vostok core which took 20 years to complete. Vostok is a series of cores up to 2500 m deep, and at its base the core is in ice formed more than 200 000 years BP.

The snowfall of a year turns into glacial ice by partial remelting and refreezing at depth and by the pressure of subsequent accumulation on top. This ice includes trapped tiny air bubbles. Examination of this air reveals three pieces of information:

- the composition of the atmosphere at the time when the bubble was trapped;
- the date of the ice layer which contains the bubble;
- the temperature at that time.

The last of these is revealed by calculating the ratio between two isotopes, ^{16}O and ^{18}O, of the oxygen in the bubble. An increase in the proportion of the ^{18}O isotope indicates a cooling period. This isotope technique is commonly used to study the sequence of Quaternary glaciations but it has also been used to determine the relationship between the CO_2 content of the atmosphere and the ambient temperature at a particular time.

All investigations tend to show an increase in global temperature. The amount of increase, however, is debatable, especially the projected future change.

The agents of global warming

An increase of carbon dioxide in the atmosphere is the principal cause for accelerated global warming. The other significant greenhouse gases are nitrous oxide, methane, ozone and CFCs-11 and -12. Atmospheric growth of these gases is summarized in Table 14.4.

In 1895, Svante Arrhenius linked CO_2 accumulation with an increase in atmospheric temperature in a paper to the Royal Swedish Academy of Sciences. Carbon dioxide in the earth's atmosphere shows an increasing trend as determined indirectly from the Antarctic ice cores and later directly from the Mauna Loa measurements (Figure 14.5). The rate steepened considerably from the 1950s. Long-term examination of the Vostok ice core shows a marked relationship between CO_2 concentration in the atmosphere and global temperature. Greenhouse heating due to CO_2 enrichment of the atmosphere is estimated to be about 50 Wm^{-2}, a far greater contribution than that of the other gases which range between 0.06 and 1.70 Wm^{-2}.

Methane is produced from wet paddy-fields and also by ruminating animals. Its greenhouse effect is enhanced by the location of its absorption and emission bands at wavelengths which complement those of CO_2 and water vapour. Its contribution to global warming has recently been scaled down (Prinn, 1994). Ramanathan (1988) showed that CFCs operate as greenhouse gases by absorbing long-wave emissions around 12 μm wavelength. The large amounts of CFC-11 and CFC-12 in the atmosphere has made them significant greenhouse gases but global agreements are currently in place on phasing out CFCs. In sum, carbon dioxide is very much the chief greenhouse gas and apparently will remain so at least in the near future.

Figure 14.6 is a pictorial presentation of human activities that may cause global warming. The production of energy, agricultural practices, modification of land use

Table 14.4 *Growth of greenhouse gases in the atmosphere*

Gas	Atmospheric concentration		
	Pre-industrial	1986	1989
Carbon dioxide (CO_2)	275 ppm	346 ppm	354 ppm
Methane (CH_4)	0.70 ppm	1.65 ppm	1.70 ppm
Nitrous oxide (N_2O)	280 ppb	305 ppb	306 ppb
Tropospheric ozone (O_3)	not known	35 ppb	
CFC-11 (CCl_3F)	0	0.23 ppb	0.28 ppb
CFC-12 (CCl_2F_2)	0	0.40 ppb	0.47 ppb

Source: World Resources Institute (1990, 1992); compiled from various sources.

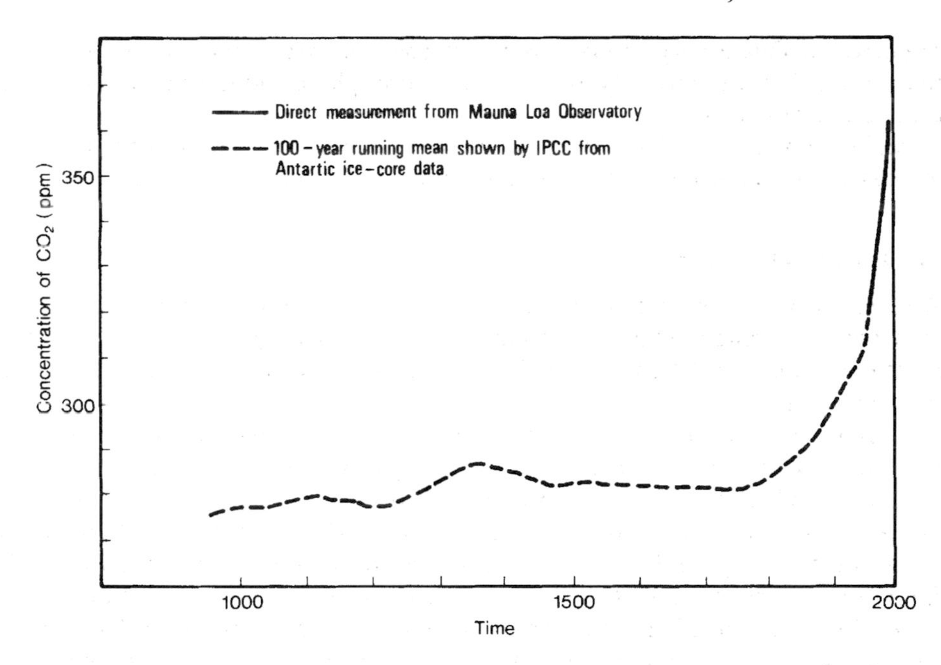

Figure 14.5 *Atmospheric CO_2 concentration for the past 1000 years. The earlier part is a 100-year running mean determined from ice core records. The end portion is from Mauna Loa observations. Note the steep rise since industrialization. Based on IPCC (1995)* Climate Change 1994, *Cambridge University Press, Cambridge. Copyright by Arthur N. Strahler*

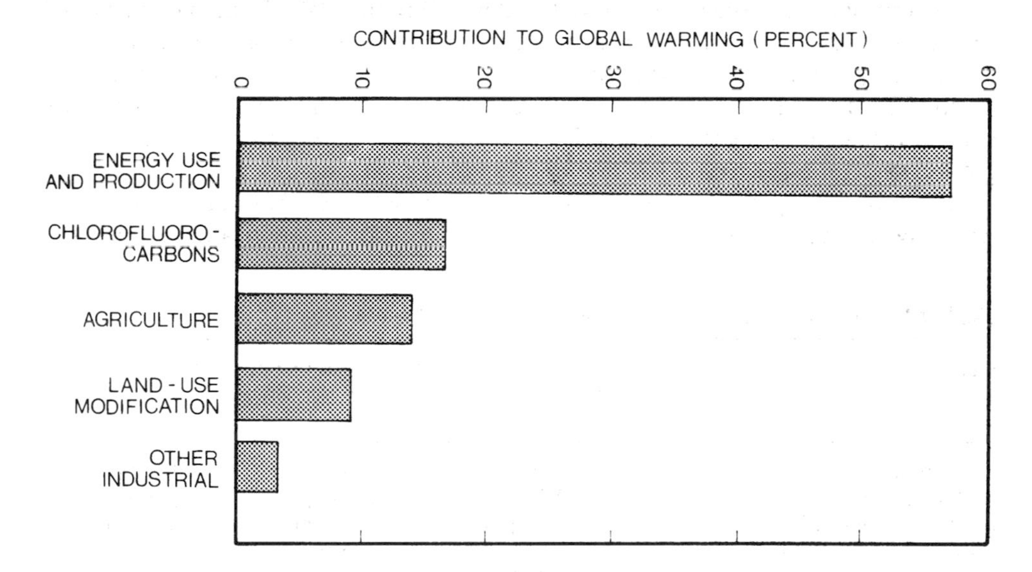

Figure 14.6 *Anthropogenic activities that lead to global warming. From White, R.M. (1990) The great climate debate,* Scientific American, ***263****(1), 18–25. Copyright © 1990 by Scientific American, Inc. All rights reserved*

such as destruction of rain forests, and industrial activities all produce CO_2. It has been estimated that the increase in long-wave heating of the atmosphere has been about 1.9 Wm^{-2} since the Industrial Revolution, and the increase continues. Henderson-Sellers (1994) summarized the phenomenon of global warming and the use of numerical models.

The climate models and global warming

The effects of global warming in the future are estimated in three ways. First, a set of numerical computer-based models, known as the *General Circulation Models* (GCMs), have been designed to reproduce climatic conditions under various concentrations of greenhouse gases in the atmosphere. Second, the concept of geographic analogue has been used. For example, if an area is expected to have less rainfall in the future, we can look at places which are very similar but currently have precipitation amounts close to the predicted value and extrapolate the environmental conditions. Thirdly, we can use the time machine technique: we can look through the geological records to discover the environmental conditions for the place with a climate comparable to the predicted one. This is helped by our growing knowledge regarding the climate changes in the Pleistocene period. So far all techniques provide trend estimates on a regional basis. It is not yet possible to forecast the future conditions at a local level.

General circulation models are very useful numerical three-dimensional models which have been developed from the pioneering work of J. von Neumann and J. G. Charney at the Institute of Advanced Studies, Princeton. The first climate model was designed in the Geophysical Fluid Dynamics Laboratory (GFDL), a NOAA laboratory, by S. Manabe and R. T. Wetherwald in 1975. This is known as the GFDL model, the current version of which was developed from the 1975 prototype. Since then several other models have appeared and been used for predicting the effect of global warming. The list includes the GISS model from the Goddard Institute for Space Studies; the NCAR model used by the National Center for Atmospheric Research at Boulder, USA; the UKMO model designed by the Meteorological Office, UK; and the OSU model proposed by Oregon State University at Corvallis. The GISS model could identify regions where the effects of global warming will appear early (Hanson *et al.*, 1988). The NCAR model suggested the importance of clouds in atmospheric modelling.

All these three-dimensional models tend to use big grids. A spatial resolution size of 5° of latitude and 7° of longitude, or approximately 500 km by 500 km, is typical. The coarse resolution is due to the very large number of calculations required for each step in time. Between two and eleven vertical levels can be considered, although only a limited relationship between climatic phenomena can be used. The models probably work better for land-based situations and for elucidating the general pattern rather than synoptic situations. Details of these models are given by Mitchell (1989) and Henderson-Sellers (1994).

The models currently predict climatic conditions under situations such as the effect of instantaneous doubling of the CO_2 in the atmosphere on a large-scale basis. They are still not precise enough to answer specific questions regarding the conditions in a particular town in the year 2050. The models tend to agree, although not always. This is illustrated in Figure 14.7. Table 14.5 summarizes a common set of predictions from the general circulation models.

Table 14.5 *GCMs on regional effects of global warming*

Generalized conclusions
- The largest temperature changes are expected in the high latitudes.
- Mid-latitude changes will be greater in the Northern Hemisphere.
- Winters will show a greater temperature change than summers.
- Latitudinal frostlines will move between 250 and 400 km polewards.
- Major grain-producing areas in the temperate zone will move about 325 km polewards.

Temperature
Seasonal temperature changes are estimated to be
- 60°–90° latitude: 0.5x–0.7x (summer); 2.0x–2.4x (winter)
- 30°–60° latitude: 0.8x–1.0x (summer); 1.2x–1.4x (winter)
- 0°–30° latitude: 0.7x–0.9x (summer); 0.7x–0.9x (winter)

Precipitation and storms
- 60°–90° latitude: winter increase.
- 30°–60° latitude: summer decrease.
- 0°–30° latitude: increased precipitation in current areas of high rainfall.
- More rainfall expected over the low latitudes and Western Europe.
- Less rainfall expected over the Mediterranean Sea area, the Mid-West and part of the west coast of the USA.
- Snow will melt early in the Rockies and the Himalaya.
- The interiors of the northern continents will experience summer desiccation.
- Increased frequency of extreme events (droughts, floods, storms and tropical cyclones).

Source: Generalized from Schmandt and Clarkson (1992).

The effects of global warming

The controversy which raged in the 1980s regarding the occurrence of global warming has nearly died down. There is now a general acceptance of climate change due to the emission of the greenhouse gases, and governments have signed an agreement to control atmospheric emission of the greenhouse gases that cause climatic change, in Rio de Janeiro in 1992 (Chapter 15), although the agreement has not always been maintained. Our knowledge about the effects of global warming, however, is imperfect and this imperfection has given rise to numerous debates. The Intergovernmental Panel on Climate Change (IPCC) is a panel of hundreds of atmospheric scientists brought together by the World Meteorological Organization and UNEP. Recently, a consensus regarding the viewpoints of the scientists on global warming has been attempted. Their review opinions have been classified as *virtually certain* (near unanimous agreement), *very probable* (approximately about 0.9 probability), *probable* (about 0.67 probability), and *uncertain* (lacking in evidence) (World Resources Institute, 1994). The consensus about our state of knowledge regarding the physics of global warming and the associated climate change is summarized in Table 14.6.

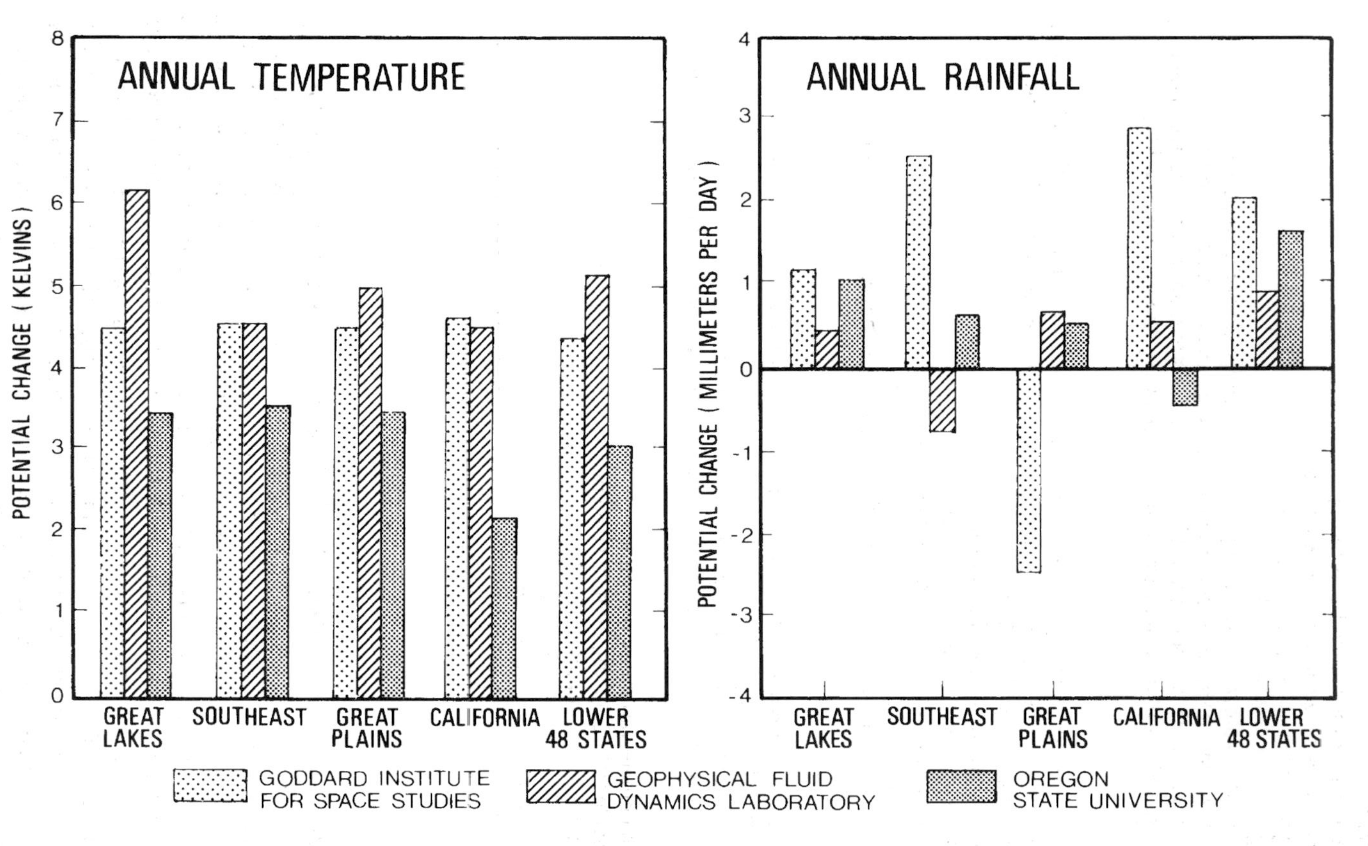

Figure 14.7 *Comparison of predictions from three computer models for regions of the United States assuming a doubling of global CO2. From White, R.M. (1990) The great climate debate,* Scientific American, ***263****(1), 18–25. Copyright © 1990 by Scientific American, Inc.*

Table 14.6 *Current consensus on climate change*

Virtually certain
- The fundamental physics of global warming.
- A number of gases may cause global warming.
- Anthropogenic increase of greenhouse gases.
- Anthropogenic increase of aerosols.
- Sulphate aerosols (volcanic and anthropogenic) contribute a cooling influence.
- Significant lowering of the level of most greenhouse gases may take centuries.
- The average surface air temperature is about 0.5 °C warmer than in the nineteenth century.
- Large stratospheric cooling will result from increased CO_2 concentration and O_3 depletion.

Very probable
- Increase in the rate of warming of the global mean surface temperature by the mid-twenty-first century; unless greenhouse gases are reduced, surface temperature will increase between 1990 and 2050 by 0.5–2.0 °C.
- Globally mean precipitation will increase.
- Reduction of Northern Hemisphere sea ice.
- Arctic land area winter warming and reduction of snow cover.
- Increased rise in global sea level.
- Over the next 50 years, solar variability will not produce stronger forcing than the greenhouse gases.

Probable
- Increase in summer dryness of the mid-latitudes of the Northern Hemisphere.
- Increase in high latitude precipitation.
- Slower warming of the Antarctic and North Atlantic Ocean regions.
- Short-term (several years of) cooling due to the transient eruptions of explosive volcanoes.

Uncertain
- Changes in climate variability over time.
- Regional-scale (100–200 km) climate changes.
- Increase in tropical storm intensity.
- Climate change over the next 25 years.
- Nature of biosphere-climate feedback.

Source: World Resources Institute (1994), Barron (1995).

Climate changes

A temperature increase of about 0.5 °C since the 1880s has been calculated by Jones and Wrigley (1990). James Hanson and Sergej Lebedeff of the NASA Goddard Institute of Space Studies have put the figure at 0.6 ± 0.2 °C (Hansen and Lebedeff, 1987). A further increase of between 0.5 and 2 °C by 2050 is now expected to occur due to increasing greenhouse gases in the atmosphere. As this will give rise to enhanced evaporation, global mean precipitation is also expected to rise. However, the distribution of the new precipitation pattern is yet uncertain. One of the main problems of working out the future climatic scenarios is determining the role of the clouds. Precipitation is expected to rise in the high latitudes, whereas the middle of the northern temperate land masses will probably experience drier summers. An increase in tropical storm intensity is likely but uncertainties remain until the following are worked out: the effects of future Hadley cell circulation (Figure 2.4) in a warmer world; the changes in poleward heat flux; and tropical sea surface temperature responses. Global oceanic circulation, including the mixing of surface waters downwards and upwelling of deeper cold waters, makes modelling the latitudinal heat flux in a warmer world a difficult proposition. All this may, in turn, usher in the following changes.

Sea level changes

Sea level is expected to rise although earlier estimates have been scaled down. Currently, the most reasonable estimates are between 5 and 40 cm by 2050. This rise is initiated by thermal expansion of the sea water but uncertainties remain as to the melting of the land and sea ice, which may cause changes at a bigger scale. With global warming, the mountain glaciers are expected to melt and the rate of their retreat can be reasonably estimated. The uncertainty lies more with the huge volumes of ice sheets over Greenland and Antarctica. The problem not only lies in determining the mass balance of the ice sheets but also regarding their melting rates. The rates originally calculated for the Antarctic ice sheet (30.1 million km^3) indicated a slow rate of melting, but later observations suggested much faster rates for the ice over Greenland (2.7 million km^3). Recent work by Kurt Cuffey and his colleagues indicates that the temperature in Greenland rose by 20 °C (starting from a low of –52 °C and rising to –32 °C) between late Pleistocene and mid-Holocene. Within the same time span, the tropical and temperate areas went through warming which was about a third of this value (i.e. the temperature rose by around 6–7 °C) (Carlowicz, 1995). The Greenland ice cap is considered climatologically to be in a precarious position, and the possibility of its rapid melting leading to a large-scale sea level rise, with the associated changes, is one of the major worries at present. A recent study by the US Environmental Protection Agency projects a most probable global temperature rise of 1 °C by 2050 and 2 °C by 2100. However, a rise of 4 °C carries a 0.1 probability. Such changes may have an enhanced effect near the poles, leading to the melting of both Greenland and Antarctic ice sheets. This is the disaster scenario. If it happens, the sea level has an even chance of rising 80 cm by 2200, and a very low probability (0.025) of rising as much as 3 m by 2200 (Carlowicz, 1995, referring to a report by J. Titus and V. Narayanan). Controversy also exists over the potential collapse in a warmer earth of the west Antarctic ice sheet which rests on a number of islands and frozen seas, unlike the east Antarctic ice sheet which rests on land (Figure 14.8). If the ice sheets melt at a fast rate the sea will quickly rise to a much higher level than the current estimates suggested above. In 1998, a part of the ice shelf in Western Antarctica (Larsen B) did indeed become detached from the main body of ice.

The low-lying areas of the world are at a considerable risk from the rising sea level. Such areas include low coral islands, river deltas and flat coastal plains. Island groups such as the Maldives, Kiribati and Vanuatu have a very large portion of their area below 1 m, and a rising sea level will reduce their land area, inundate valuable coastal land and tourist infrastructure, and also create saline incursion of the groundwater. All these could make these islands almost uninhabitable, and concern has been expressed as to their ability to support the existing population. Such a scenario may lead to emigration of the islanders to safer places and the term

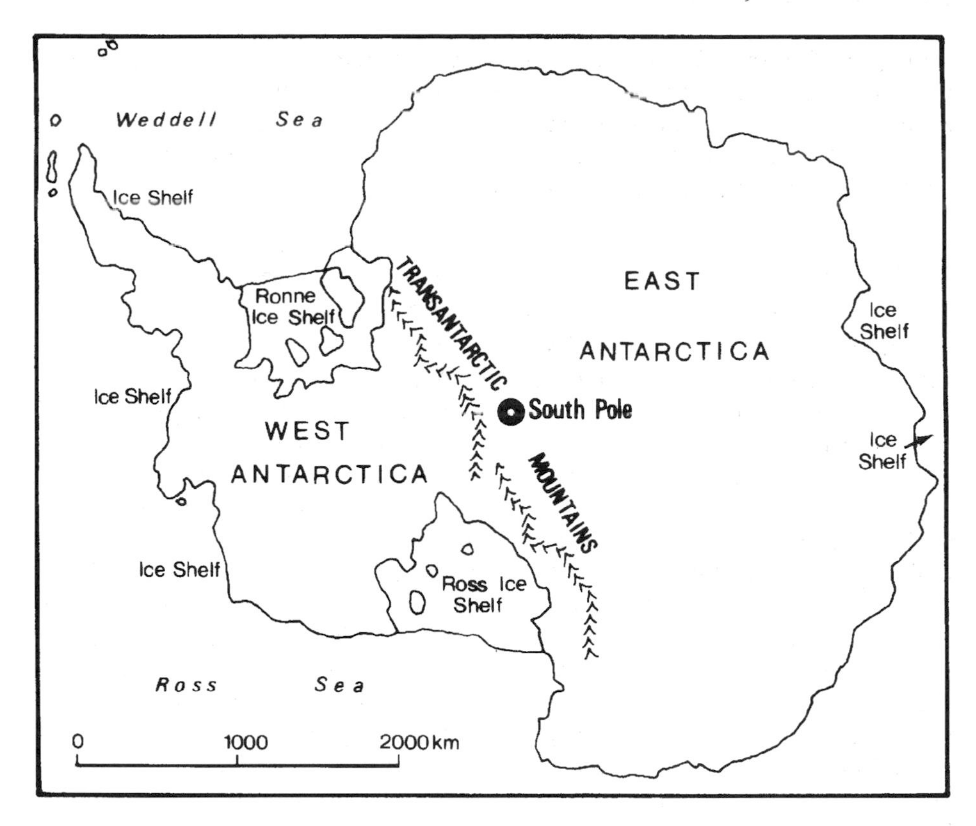

Figure 14.8 Antarctica

'environmental refugees' has been coined to describe people in the future fleeing from a much reduced island state. Coastal deltas, such as most of Bangladesh (Figure 14.9) or the Mississippi delta, and low coastal plains, as in the Netherlands or Guyana, are similarly vulnerable to a rising sea level. The problem is more acute in these areas as they are often densely populated and carry considerable economic investments on which the rest of the country depends. The increased tropical storminess and the extended sea surface area may also make the low coastal plains and deltas between 10° and 30° latitudes more at risk from tropical cyclone damages. IPCC (Tegart, Sheldon and Griffiths, 1990) estimated the costs of protecting the coastlines against a 1 m rise in sea level. The figures are high, the extreme case being that of the Maldives where the estimated costs are about one-third of the GNP. This IPCC estimate shows the protection costs to be at least 1 percent of the GNP for 34 countries. A 1 m rise in sea level is probably an extreme case but the figures do indicate that anthropogenic global warming is a costly misadventure.

Both climate and sea level changes have been studied in some detail. A series of other types of changes are suspected but require more research before conclusions can be drawn with acceptable certainty. Such investigations are, however, extremely necessary, at least for policy formulation. The state-of-the-art situation is briefly reviewed in the following paragraphs.

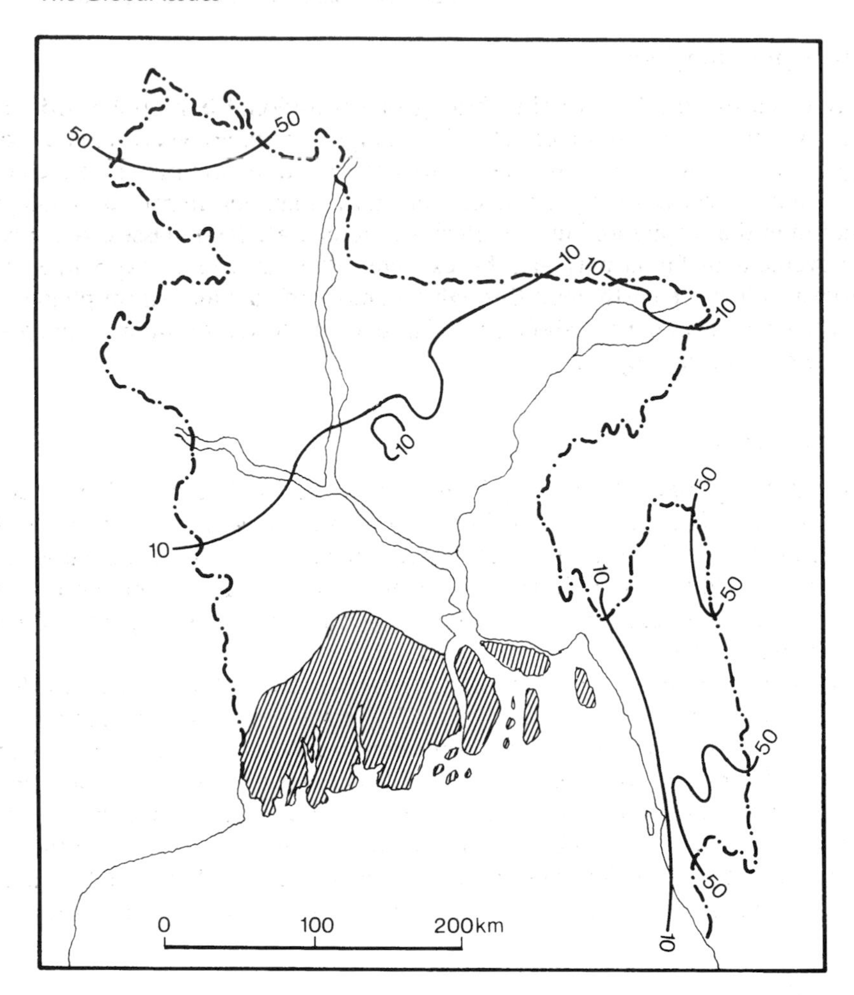

Figure 14.9 *Estimated inundation of Bangladesh from a projected sea level rise of 1 m. (shaded area). Based on Mahtab F.U. (1992). The delta regions and global warming: impacts and response strategies for Bangladesh in (J. Schmandt and J. Clarkson, Eds)* The Regions and Global Warming: Impacts and Response Strategies. *Copyright © 1992 Oxford University Press, Inc. Used by permission of Oxford University Press, Inc.*

Geomorphological changes

The major rivers of the world may undergo adjustments due to both climate and sea level changes. These will involve hydrological changes such as reduced streamflows, increased flooding, and salt water intrusions in their lower reaches. Coastal erosion may increase and also slope failures following changes in soil moisture patterns, vegetation shifts, and ice and snow melting on the upper slopes of mountains. The rapid retreat of mountain glaciers and ice caps will result in changed runoff patterns, with possible socio-economic implications. These may include shortages of water supply, shortfalls in hydroelectricity generation, and the alteration of aquatic ecosystems.

Effects on flora and fauna

A warmer earth may lead to the shifting of vegetational belts polewards. Such changes will probably be more effective in the higher latitudes where small absolute changes in temperature and a higher number of frost-free days may be expected to have greater effects. Similarly, shifts in precipitation amount may cause changes in vegetation in the tropics and in the interior of temperate land masses. A concomitant movement of fauna may also be expected. Parallel events happened in the Pleistocene when the earth went through a number of cold and warm phases. The rise in sea level beyond a threshold should also affect the distribution of mangroves and coral reefs in the tropics.

Agricultural changes

At present, the global pattern of agriculture is driven by climate, soil, market and technology. Certain parts of the world have low farming potential because of their locations in either cold or arid environments. In contrast, other areas such as the American Mid-West or the plains of Ukraine are popularly known as the breadbaskets of the world. Shifting climatic belts associated with future global warming may alter this pattern.

Vigorous agricultural expansion is expected for the northern latitudes due to longer and warmer growing seasons. If the hypothesis of increased rainfall in the semi-arid subtropics turns out to be valid, such areas may also become far more prosperous. In contrast, the present breadbaskets of the world, situated as they are in the middle of temperate land masses, may experience decreased precipitation, reduced soil moisture, and falling crop production due to water stress. One naturally wonders whether technological innovations would be able to ameliorate such distress, or whether the map of world food production would begin to alter.

Political changes

Possible political changes are the most uncertain of all probable outcomes of global warming. In the past, however, shifting climates leading to economic distress and large-scale migration have led to political rearrangements. The movement of people from low-lying areas and the shifting of areas of high agricultural production may require special global arrangements in order to cope with a changed physical world. We are also uncertain at this stage of the effect on cities of shortages of water, increased natural hazards, and industrial pollution. These may give rise to local dissatisfaction.

14.4 Conclusions

As stated earlier, a general consensus seems to have been reached regarding the occurrence of global warming. Our knowledge about its effects, however, is not uniformly certain. It is also generally agreed that preventive steps should be taken regarding the control of global warming by at least reducing the release of anthropogenic greenhouse gases. This requires international agreements and, like the prevention of stratospheric ozone depletion, has given rise to arguments between states. Among the agents of global warming, CFCs are being phased out. The major agent of warming, the excess CO_2 in the atmosphere, originates mainly from industrial effusions, the generation of electrical power from thermal stations, and biomass burning. The choice of energy-efficient policies and of saving the world's vegetation could substantially mitigate problems of global warming in the future. Obviously such a control requires agreements between nations and also arrangements that allow the developing countries to improve their quality of life at the same time. This adoption of a wise course in the face of uncertainty was carried out in 1992 at the United Nations Conference on Environment and Development at Rio de Janeiro, which is discussed in the next chapter. Given the scale of the possible environmental disaster, the shortcomings to our knowledge, and the nature of the challenge to earth and atmospheric scientists, global warming will continue to dominate environmental research and discussion at least for the near future. In these circumstances a risk-averse strategy appears to be a sensible one to adopt, although the current trend (Chapter 17) regarding global warming prevention is less encouraging.

15

Environmental Arrangements: Present and Future

15.1 Introduction: The Post-Stockholm Developments

Environmental arrangements at the global, national and local scales have come a long way since Stockholm. The current environmental awareness and activities involve local organizations, scientific communities and industrial establishments, as well as governments at various levels. Environmental management of the future will continue to extend beyond governance and require popular commitments, industrial innovativeness and international co-operation. This chapter is an account of the historical development of environmental management between Stockholm and the current time, and a description of the major environmental actors and institutions.

The post-Stockholm development of environmental awareness and institutional arrangements began in 1972 with the establishment of the *United Nations Environment Programme* (UNEP) by resolution 2997 (XXVII) of the United Nations General Assembly. UNEP consists of a governing council of 58 nations, a secretariat, an environment co-ordination board, and a voluntary environmental fund. It is based in Nairobi, breaking the precedence of locating major UN agencies operating at a global level in either the United States or Europe. The Programme, in spite of its name, is essentially restricted to collecting, processing and disseminating environmental information on a limited budget. The actual developmental project-related work is generally carried out in the field by various other UN agencies such as UNDP.

UNEP carries out its task of global environmental assessment via a UN-sponsored network (the Earthwatch). Earthwatch includes

- the International Referral System (INFOTERRA) for exchange of environmental information;

- the Global Environmental Monitoring System (GEMS) which is designed to collect information from various governments and compile the overall environmental scenario at both global and national levels;
- the International Register of Potentially Toxic Chemicals (IRPTC), a Geneva-based data bank.

UNEP also collaborates with various other international organizations in coordinating important conventions such as the Convention on International Trade in Endangered Species of Wild Fauna and Flora (CITES). Its Regional Seas Programme brought together 120 countries and 14 UN agencies to discuss shared problems of marine pollution and coastal degradation. UNEP has been involved in administering the secretariats for the 1985 Vienna Convention for the Protection of the Ozone Layer and the 1987 Montreal Protocol on Substances that Deplete the Ozone Layer (Anon., 1994). The Programme also disseminates environmental education. For example, a joint UNEP–UNESCO Environmental Educational Programme was created in 1975, and UNEP has also been prominent in organizing a number of environmental conferences and workshops around the world, usually in collaboration with other international or national bodies. UNEP maintains a register of the Nongovernmental Organizations of the world. The impact of UNEP on the world environmental scene, however, is still limited (McCormick, 1989).

In the two decades following Stockholm, both the number and scope of work of NGOs increased rapidly. By 1988, about 7000 NGOs were on record in the UNEP Environmental Liaison Centre. Membership of NGOs also showed substantial growth. For example, membership of the Sierra Club increased from 68 000 in 1968 to 348 000 by 1984. The corresponding figures for the National Wildlife Federation were 364 000 and 820 000 (McCormick, 1989). The environmental movement became a political issue, and in many countries, notably in Europe, new parties (collectively described as the 'Greens') contested national elections on environmental platforms. The role of the national government in environmental protection was formally organized in the form of a specific ministry of environment or equivalent. This happened early in the United States, principally with the establishment of the Environmental Protection Agency (EPA) in 1969. By the mid-1980s, most countries already had a Ministry of Environment or an equivalent body in place. The power and contribution of such national organizations, however, varied.

Over time, reservations started to build up regarding the effect of international aid to developing countries and the expectation of donor countries and institutions. This resulted in structural alterations of these organizations and also in the establishment of in-house environmental management procedures. Concern towards environmental degradation and institutional structural changes speeded up after the release of a 1987 report produced by a UN commission often known as the Brundtland Commission. The commission had been established to review the environmental and developmental problems of the world and also to recommend future management strategies.

15.2 The Brundtland Report

In September 1983, a United Nations General Assembly resolution called for a special independent commission to direct and manage a global inquiry into the state of the world. Three objectives were listed in the commission's mandate:

- the critical environmental and development issues were to be examined and realistic proposals for dealing with them formulated;
- new forms of international co-operation on these issues were to be proposed in order to direct policies and events so that the necessary changes could take place;
- the level of understanding and commitment was to be raised regarding individuals, NGOs, businesses, institutions and governments (WCED, 1988).

The Commission was known as the World Commission on Environment and Development, and consisted of 23 commissioners from 22 countries, with Gro Harlem Brundtland (the Prime Minister of Norway) as the Chair, Monsour Khalid of Sudan as the Vice-Chair, and Jim MacNeill of Canada as Secretary. The commissioners came from widely different backgrounds, and their countries ranged across the developed and developing worlds and the communist block.

Eight key issues were selected by the Commission for analysis:

- perspectives on population, environment and sustainable development
- energy: environment and development
- industry: environment and development
- food security: agriculture, forestry, environment and development
- human settlements: environment and development
- international economic relations: environment and development
- decision support systems for environmental management
- international co-operation

These issues were examined from the perspective of the year 2000 and beyond. The Commission also agreed that its deliberations would be open, visible and participatory. Furthermore, the broadest possible range of views and advice regarding the key issues would be sought.

In order to achieve these objectives, deliberative meetings were held in various parts of the world, which in turn enabled the commissioners to familiarize themselves with different regional attitudes to environment and development. Public hearings were held during these deliberations where government officials, NGO representatives, scientists, technicians, industrialists and the general public could all express their views. Such meetings were held in Indonesia, Norway, Brazil, Canada, Zimbabwe, Kenya, the USSR and Japan (WCED, 1988). These public hearings demonstrated clearly that by the 1980s, the concern regarding environmental issues had indeed spread right across the world.

The final report of the Commission (popularly known as the Brundtland Report) was published in 1987. Besides its account of the state of the environment of the world, the commission should be remembered for at least four of its observations:

- First, it recommended the concept of sustainable development, defining it as meeting 'the needs of the present without compromising the ability of the future generations to meet their own needs'. It described sustainable development as 'a process of change in which the exploitation of resources, the direction of investments, the orientation of technological development, and institutional changes are made consistent with future as well as present needs'. In the opinion of the Commission such reconciliation is possible, provided fundamental changes take place in managing the global economy. Environment and development are inseparable.
- Second, it emphasized that international co-operation is essential but difficult to achieve. The Brundtland Report succinctly stated this problem as: 'The Earth is one but the world is not'. The earth behaves holistically so far as the physical processes are concerned, and hence the effect of large-scale environmental degradation affects the entire planet. In contrast, the policies of different countries are often governed by local and regional interests which may not coincide with environmental expectations at the global level.
- Third, the commission suggested strengthening of the national environmental agencies.
- Fourth, the Brundtland Commission recommended establishing a UN programme on sustainable development and holding a world conference on environment and development (WCED, 1988).

These recommendations have influenced environmental policies and management at the international level since the production of the report. Sustainable development has become a key concept. Development agencies such as the UNDP and the multilateral lending institutions such as the World Bank became far more environmentally conscious and decided to include possible environmental impacts as part of standard evaluation procedures for approving development projects. The world conference took place a few years later, in 1992. The concept of sustainable development, however, has generated a surprising amount of discussion and critique, although as a working concept it has formed the basis of environmental management policies and techniques used by UN agencies.

15.3 The Concept of Sustainable Development: Meaning and Policy Relevance

The term 'sustainable development' addresses the conflict between economic growth and protection of the environment. The term was popularized by the Report of the Brundtland Commission which promoted the politically brilliant idea of sustainable development. Even though the message in the Report was to a large extent similar to that of the 1972 report of the Club of Rome, *The Limits to Growth*

(Meadows *et al.*, 1972), it received almost universal endorsement. In contrast, *The Limits to Growth* had been sharply criticized, particularly by the economists, and had had no discernible effect on government policies. It was regarded as emphasizing redistribution, whereas the Brundtland Report was interpreted as favouring continued economic growth. While growth is regarded almost universally as at least necessary, if not desirable, recent decades have witnessed an apparent consensus to move away from redistribution both within countries and among countries. The term sustainable development has been subjected to a wide variety of interpretations, particularly concerning its meaning and policy relevance. The thrust towards the new era of economic growth (in countries at all income levels) has even been interpreted as an implicit consent for 'business as usual' with only minor adjustments. This certainly was not the intention of the Brundtland Commission.

The Brundtland Commission argued that,

> humanity has the ability to make development sustainable – to ensure that it meets the needs of the present without compromising the ability of future generations to meet their own needs
> (WCED, 1988: 8).

The terms 'needs' and 'ability' have proved to be difficult to operationalise in practice. For example, it is easy to conceive of unmet material needs of individuals who are poor with extremely restricted consumption choices, but it is harder to do so for people with high incomes. Similarly, we do not know enough about the carrying capacity of ecosystems and the extent to which technology can substitute for resources to formulate a reasonable estimate of the ability of future generations. Even if such knowledge were available, it would only address issues of the physical environment. The word 'ability' should also include economic, political and social aspects.

Sustainability has been conceptualized in different ways, which complicates the issue. Common, for example, views sustainability as 'how to address problems of inequality and poverty in ways that do not affect the environment so as to reduce humanity's future prospects' (Common, 1995: 31). Similarly, Solow states that 'it would be unfair or unsound to use limited resources for current benefit, in a way that will impoverish future generations' (Solow, 1996: 16). This can be viewed as a concern for inequality over a time period. But there are also inequalities among regions and among parts of a region, and also among countries and within parts of a country. It is not immediately evident why the intertemporal inequality should receive particular consideration among all the various types of inequalities that are present. Solow (1996) therefore cautions against regarding sustainable development as an absolute goal.

In a more positive fashion, sustainability has been defined as leaving 'future generations as many opportunities as, if not more than, we have had ourselves' (Serageldin, 1966). Following the traditional practice in economics, opportunity may be measured using the concept of capital, and identifying four types of capital:

- human-made capital (such as machines);
- human capital (comprising skills and knowledge of people);
- natural capital (broadly defined as the environment, including natural resources);
- social capital (the institutions and attitudes necessary for a functioning society).

With any capital, both stock and flow are relevant. Sensible sustainability would require that 'in addition to maintaining the total level of capital intact, some concern should be given to the composition of that capital between natural, man-made, human and social' (Serageldin, 1996: 190). This allows limited substitution of one type of capital for another. Some complimentarity between the different types of capital is also likely. It is therefore necessary to (1) measure and value each type of capital, (2) work out the way in which each type of capital interacts with others, and (3) determine the rate of exchange between each type of capital.

The linkage between economic activities and the natural capital is of crucial importance in analysing sustainability. It is difficult to synchronize linear technological systems with cyclical natural systems. This requires technically promoting the cyclical flow of resources which treats residue materials from various processes as a resource and not as waste (Wijkman, 1996). Some analysts go further and argue the need to fit human activities into nature's patterns (Peet, 1992).

Pearce, Barbier and Markandya (1990) include the following elements in a vector defining *development:*

- increases in real income per capita
- improvements in health and nutritional status
- educational achievements
- access to resources
- a fairer distribution of income
- increases in basic freedom

Sustainable development, then, requires a situation in which the above development vector does not decrease over time. But this definition of development creates several difficulties such as the determination of the weight assigned to each element and its variability across countries and also over time. Furthermore, decision-making takes place over a finite time period whereas the definition of development by Pearce *et al.* implies that time is infinite.

Sustainable development is defined by Pearce *et al.* as

> the general requirement that [*the above-mentioned*] vector of development characteristics be non-decreasing over time where the elements to be included in the vector are open to ethical debate and where the relevant time horizon for *practical* decision-making is similarly indeterminate outside of agreement on intergenerational objectives.
>
> (Pearce, Barbier and Markandya, 1990: 3)

The emphasis on ethical and practical aspects clearly recognizes that although the underlying logic and methodology of economics remain indispensible in analysing sustainable development, the concept extends beyond the limits of such analysis. Common (1995) has regarded the ahistorical, amaterial and apsychological nature of economics as a major limitation in dealing with the sustainability issues. Clearly, no rigorous and universally accepted definition of sustainable development exists.

Regarding the policy relevance of sustainable development, the ambiguities continue:

> Nearly five years after UNCED and ten years after the Brundtland Commission report, the concept of sustainable development is beginning to be reassessed and questioned globally. The high expectations coming out of Rio have, for the most part, not been met . . . Sustainable development is in need of change.
>
> (Hsu, 1996: 168)

Explanations regarding this state of affairs and suggested improvements both differ widely. In general, the lack of international co-operation, commitments, and weak political arrangements have been blamed for this failure (Brundtland, 1996; Haas, 1996). Brundtland (1996) herself has pointed out that a new world order involving burden-sharing, common perceptions and common responsibilities has not made much progress. According to Wijkman (1996), sustainable development should focus on people, especially on poverty, and not on environment *per se* because of the strong link between poverty and environmentally unsustainable behaviour, particularly at the local and regional levels.

At a national level, empirical studies involving a sample of a large number of countries at different stages of economic development exhibit an inverted U-shaped relationship between nominal per capita GDP and various water and urban quality indicators of environmental degradation (Grossman and Kruegar, 1995). Their main finding is that while the turning points in these inverted U-shaped relationships vary for the different air and water pollutants, in almost every case they occur at an income of less than $8000 expressed in 1985 prices. More widespread adoption of cleaner technologies, structural transformation of the economy, and more effective demand on the part of the people for better environmental quality may help to explain the observed empirical relationship.

Grossman and Kruegar, however, also note that the studies by the World Bank and others do not find evidence of an inverted U-shaped relationship for certain other indicators of environmental degradation such as municipal waste per capita and carbon dioxide emission. For these indicators, environmental degradation continues to worsen even in the cases when per capita income reaches a very high level.

Thus, at the national level, the relationship between per capita income and environmental degradation is neither straightforward nor uniform across the different indicators of environmental degradation. This is even more likely to be the case at local and regional levels.

There is nothing automatic about the empirical results of studies based on the past experience of a large number of countries. Developing countries, both individually and as a group, could learn from past experiences and focus on effective perusal of sustainable development at a much earlier stage of economic development than has been the case in the past. They also need to pay much more attention to the environment at local and regional levels.

The capabilities and resources of environmental activists (and even of individual governments) are increasingly stretched to the limit due to the progressive globalization of economic activities and the associated intense competition for attracting investments. Moreover, global arrangements are not replacing the lessened capabilities of the national governments to sustain the quality of their environment. Disagreements also exist regarding the extent to which market mechanisms can be used to bring about policies and practices consistent with sustainable development.

Economists in general are far more optimistic about the market's ability (Stigson, 1996) than are those who emphasize people and values (Rich, 1996). Markets could assist in achieving targets and objectives of sustainable development but certainly not their determination.

Given the prevalent controversies regarding the meaning and policy relevance of sustainable development, it is perhaps wise to depend on a few simple guidelines, such as preventing any significant loss of the natural capital. At the same time we need to improve our knowledge, technology and databases. It is also necessary to go through the arduous and slow task of global consensus building.

15.4 The United Nations Conference on Environment and Development (UNCED)

The conference at Rio de Janeiro on environment and development (3–14 June 1992) was another remarkable event like Stockholm. Unlike Stockholm, however, 178 governments actively participated at the conference and most heads of state and governments deemed that they should be present. The global nature of environmental problems as detailed in Chapter 14 and the contents of the Brundtland Report (especially the concept of linking environment and development together on a sustainable basis) necessitated the meeting.

A number of international agreements have come into force in the twenty years between Stockholm and Rio de Janeiro. Examples include treaties and conventions on World Heritage (1972), Endangered Species (1973) formerly known as the Convention on International Trade in Endangered Species of Wild Flora and Fauna (CITES), the Antarctic (1980), the Law of the Sea (1982), the Montreal Protocol on CFCs (1987), and the movement of Hazardous Waste (1989). A number of agreements regarding aid and trade practices between the developed and developing countries could be added to the list. Besides the Brundtland Commission, two other UN commissions have submitted their reports: the Brandt Commission on North–South issues and the Palme Commission on security and disarmament. The recommendations of the Brundtland Commission were followed by a UNEP report, *Environmental Perspectives to the Year 2000 and Beyond* and the IUCN publication, *World Conservation Strategy*. The time seemed appropriate to call a global meeting to translate the recommendations into action and usher in the age of sustainable development. In December 1989, the General Assembly of the United Nations convened the United Nations Conference on Environment and Development (UN Resolution 44/228), and the Brazilian government offered to host the meeting.

Apart from the necessary acronym of UNCED, the conference is also known as the Rio Conference or the Earth Summit. It was a huge affair with 178 participating governments; a parallel Global Forum attended by more than 500 different groups; a large-scale ECOTECH forum and conference; many associated special events and exhibitions; and more than 30 000 visitors to the city (Grubb *et al.*, 1993). It enjoyed huge publicity and the participatory governments affirmed five agreements

(Box 15.1). Controversies, however, exist regarding the outcome of UNCED. It has been described as an outstanding success and also as a dismal failure (Chatterjee and Finger, 1994). Perhaps the rational view lies somewhere in between. It raised more expectations (at least from the developing countries and environmentally conscious people) than it delivered, but it did deliver a few agreements and sowed the seeds of future environmental arrangements. Perhaps it was too optimistic to expect a miraculous solution of the world's environmental problems from a conference attended by so many states with diverse interests, knowledge and enthusiasm.

The preparation for UNCED required, as for Stockholm, a set of preparatory meetings to determine conference procedures; and to negotiate on the agreements and to cover the differences. The first of these meetings (PrepCom I) was held in Nairobi; PrepCom II and III were held in Geneva, and the final meeting, PrepCom IV, in New York. A number of differences between the developed and developing countries (which collectively called themselves Group of 77 or G77) and China needed to be resolved before UNCED started. Differences existed even within the developed countries and countries of G77, especially regarding priorities, the importance to be given to development, and the source of funding for environmental management. In the event, UNCED became more than a single conference; it acted as a catalyst for a set of other meetings and environment-related deliberations. Most of the governments submitted country reports on the national environment at UNCED. Although the tone and quality of these reports varied, the production of such reports provided documentation on an unprecedented scale. A number of these reports involved non-governmental personnel, and the act of preparing one involved environmental stock-taking at the national level. The NGOs had a strengthened role to play and they became part of the official delegation for certain governments at the Earth Summit. A number of international conferences were held piggyback with UNCED. A parallel NGO forum was held at Rio during the conference, encouraged by the conference secretariat. Tremendous media interest and coverage spread the news of UNCED across the world. The agreements and differences at UNCED shaped five documents, along with the recognition of several other important environmental issues (Box 15.1).

Box 15.1

Agreements at UNCED

Framework Convention on Climate Change

It is a legally binding treaty aimed at lowering the emission of CO_2 and other greenhouse gases. The treaty was expected to come into operation immediately; the developed countries should lead, and the developing countries should be compensated for costs incurred for taking necessary measures for controlling climate change. Governments should submit reports on their activities and meet regularly to review progress.

Convention on Biological Diversity

It is another legally binding treaty aimed at halting the destruction of biological diversity, protection of species, and also establishing standards of conduct in the sharing of research information, profits and technology in the area of genetic research. It was agreed that states have sovereign rights over their biological resources but fair and equitable sharing should take place.

Rio Declaration

The declaration is a set of 27 non-binding principles which emphasize the link between environment and development.

Agenda 21

The agenda for the twenty-first century is a document of over 500 pages that contains a non-binding action programme for sustainable development which was adapted by consensus by more than 170 countries. It covers the different aspects of environment, resources, population and development. A participatory and community-based approach to problem-solving is preferred.

Forest Principles

Forest Principles is a shortened version of a more comprehensive plan for forest protection and management which was originally discussed but failed to secure world-wide affirmation. It also recognizes the sovereign rights of states to utilize forest resources.

Apart from these five agreements, other environmental issues and management arrangements were discussed, including (1) desertification and droughts at the insistence of a number of African nations; (2) institutional arrangements for environmental management including the agreement to create a high-level commission on sustainable development, financial resources and mechanisms; and (3) the role of women related to environment and development.

UNCED did not lead to firm global commitments at the scale which many expected. It was certainly not the final global conference on environment; rather a large step in the right direction. The final agreement has been considered by some as an achievement given the discussions at the PrepCom meetings and the refusal of the United States under the Bush administration to sign the Biodiversity Convention. The United States has ratified the Climate Change Convention but has been lukewarm about its implementation (Chapter 17).

Certain very positive results came out of Rio. Tremendous enthusiasm and publicity were generated regarding environment and development. The concept of sustainable development was globally recognized. The role of unofficial organizations and common people were recognized. Various post-Rio arrangements were set up to supervise the global environmental scene as discussed earlier. However, a number of shortcomings remained, to the disappointment of various environmental groups and member countries of G77. Two of the shortcomings are disturbing. First, the financial arrangements for governing the environment of the world fell short of the target. The UNCED secretariat in Agenda 21 estimated the need for $600 billion a year, most of which had to come from the developed countries. A minuscule portion of this was forthcoming. Second, a number of countries felt that the convention did not give enough importance to the environmental problems of the developing countries (Grubb *et al.*, 1993). The real test of UNCED will, of course, be the extent to which the governments ratify and follow up the commitments they have made in Rio, especially regarding the climate change and biodiversity conventions. Finally, the role of the United Nations was strengthened as the legitimate leading organization for advancing sustainable development (Grubb *et al.*, 1993), and also for supervising global environmental agreements and conditions.

15.5 The Post-Rio Institutional Arrangements

Later in 1992, the United Nations General Assembly in its 47th session deliberated on UNCED. UNCED and its recommendations were approved by the General Assembly which passed seven resolutions in support, including one which strongly endorsed the outcomes from Rio. The Assembly also agreed to review periodically the progress towards implementation of the UNCED arrangements. The new arrangements essentially accepted Agenda 21 and the concept of sustainable development.

On 12 February 1993, the United Nations Economic and Social Council (ECOSOC) established a high level monitoring agency, the United Nations Commission on Sustainable Development (CSD) to encourage and oversee progress in implementing Agenda 21 at various levels: national, regional and global. CSD reports to ECOSOC, and through it to the General Assembly, which is the supreme world body for intergovernmental policy-making (World Resources Institute, 1994). CSD has 53 seats distributed between the following units: Africa, Asia, Eastern Europe, Latin America and the Caribbean, Western Europe, and North America. It is a forum for reviewing sustainable development issues, and again for supervising and co-ordinating environmental issues across economic and social sectors and countries. Contact with NGOs is actively encouraged. The first meeting of the CSD took place in June 1993, a year after UNCED.

Several other agencies were also set up to supervise and co-ordinate Agenda 21 issues. The United Nations Administration Committee on Co-ordination brings together major UN organizations and multilateral financial agencies. A 21-member high-level advisory group consisting of people prominent in the areas of environment and development was set up in July 1993, following an Agenda 21 recommendation, to advise intergovernmental bodies, the Secretary General, and the United Nations. This was also an Agenda 21 recommendation. Agenda 21 also recommended strengthening UNEP and the UNDP. The UNDP is building technical and organizational capacity at various levels to meet the demands of Agenda 21. This plan is known as Capacity 21 and its main objective is to assist developing countries in improving their environmental management capabilities so that they can incorporate the concept of sustainable development into their national programmes. Other arrangements arising out of Agenda 21 include setting up a convention to control desertification, a global conference on the sustainable development of small island states, and a conference on fish stocks.

UNCED also acted as a catalyst at other levels. It encouraged the national governments to strengthen their environmental programmes and NGOs to continue in their bid for a better world. Maurice Strong and other prominent individuals in the area of environment and development established an organization, the Earth Council, in Costa Rica in late 1992. The Earth Council is expected to independently evaluate, in co-ordination with other NGOs, the progress towards meeting the objectives set out in Agenda 21.

The results of all this lie in the future. Meanwhile controversies continue regarding the post-Rio achievements and the co-operation of the national governments. The biggest problem, however, is financing the UNCED proposals.

Box 15.2
Financing Sustainable Development

Accepting sustainable management as an objective has been easier to achieve than that of the contentious issue of financing it. The available financial estimates vary in quality, and should be treated only as indicative and order of magnitude estimates. They do, however, provide the starting point for analysing financial requirements for sustainable estimates. This lack of robustness of the estimates also makes firm international agreements difficult to achieve.

Financing sustainable development involves three interrelated issues:

- the level or total amount of funding needed, depending on the scope of the overall programmes and targets to be achieved in specific areas;
- terms and conditions involved in the sharing of costs between countries, specifically between the developed (high income) and the developing (low and middle income) countries;
- countrywise mobilization of the requisite internal resources.

Funding necessary to provide sustainable developments to the developing countries has been estimated in three different ways as discussed below.

Estimate 1
The two conventions of UNCED (Framework Convention on Climatic Change and Convention on Biological Diversity) established the principle that the developing countries should be refunded the *full agreed incremental costs* associated with measures taken under them, with the Global Environment Facility (GEF, discussed in detail in Section 16.2) as the interim funding mechanism. However, as environmental protection and development are inextricably linked, the term *full agreed incremental cost* has proved to be difficult to define and operationalize. The two conventions did not mention any specific cost.

The GEF was endowed with an initial funding of $1.3 billion for a three-year period: July 1991–June 1994. It concentrates on four key areas: global warming, biological diversity, pollution of international waters, and stratospheric ozone depletion. Since March 1994, the GEF has become a permanent financial mechanism.

The GEF is managed through a collaborative tripartite arrangement involving the World Bank, UNEP and the UNDP. In practice, the World Bank, where the facilities are located, has been the dominant partner. Up to 80 percent of the GEF funds can be attached to the Bank's sector loans. The GEF grants up to early 1996, totalled over $1 billion to fund 113 projects in 63 developing countries, with approximately 45 percent of the total commitments funding biodiversity projects and a further 40 percent funding projects dealing with climate change (Sharma, 1996). A group of 30 wealthy donor countries are already committed to tripling the GEF core fund to $4 billion. The actual GEF funding of around $1 billion is regarded as too small, especially in comparison to the official development assistance ($51.7 billion in 1993) and private financial flows ($157.7 billion in 1993) to make a significant difference (World Bank, 1995a). The supporters of the GEF argue that although the amount involved is small, it can act as a catalyst for global sustainability. For example, by providing only the incremental cost for financing more environmentally friendly technology, the GEF can optimize its leverage from the limited resources, and have a much larger impact that its own resources would suggest.

Estimate 2
The UNCED Secretariat estimated the average annual (over the 1993–2000 period) cost of implementing Agenda 21 programmes for the developing countries. The estimates were based on incomplete supporting documents and not reviewed by the national governments. The estimated annual cost was just below $600 billion, with $130 billion (22 percent) provided through grant or concessional requirements (Grubb *et al.*, 1993). This amount is put in perspective when the GDP of all low- and middle-income countries is noted: $5276 billion in 1994 (World Bank, 1996b). The $470 billion ($600 – $130 billion) cost to be borne by the developing countries is 8.9 percent of their 1994 GDP.

The $130 billion concessional element to be funded by the high-income countries is 0.65 percent of their 1994 GDP of $20 120 billion. However, the total official long-term net resource flows to developing countries in 1993 was only $51.7 billion (World Bank, 1995a), or 40 percent of the estimated $130 billion. The private net flows of 1993 to the developing countries were $157.7 billion. Thus, the importance of official flows is now much less than that of private flows.

Even if the Newly Industrialized Countries (NICs: Taiwan, South Korea, Singapore) begin making overdue contributions to net official flows of assistance to the developing countries, the requisite amount is unlikely to be realized. The fiscal stringency being faced by the high-income countries, along with the ideological shift towards the private sector as opposed to the public sector solutions, make it unlikely that the estimated concessional resource flows would be forthcoming.

The UNCED Secretariat estimates were carried out according to the Chapter/Programme areas of Agenda 21. The largest sums were for the following areas: human settlements ($215 billion); freshwater ($55 billion); human health ($51 billion); agriculture ($32 billion); deforestation ($31 billion); poverty ($30 billion); biotechnology ($20 billion); education, public awareness, and training ($15 billion). It should be stressed that Agenda 21 takes a broader view of the environment and therefore poverty reduction, trade liberalization, health care, shelter, etc., are included as part of the environmental management effort.

Estimate 3
The 1992 World Development Report of the World Bank (1992) was devoted to development and the environment. The report estimated that in AD 2000, the *additional* investment in selected environmental programmes would amount to $75 billion, or about 1.4 percent

of the GDP of the developing countries for that year. The major items of *additional* investments are soil conservation and afforestation ($15–20 billion); reducing emissions, affluents, and wastes from industry ($10–15 billion), controlling main pollutants from vehicles ($10 billion); and family planning ($7 billion). The Report also recommended that such investments be made on a self-financing basis to the extent consistent with equity and other objectives.

These three examples of financial estimates suggest that clarity in this area is much below the desired level, in spite of its crucial importance. The methods used to arrive at these estimates not always being explicit, their robustness cannot be determined. It is also not clear whether in kind services are included in the estimates, and what is the level of efficiency assumed in carrying out environmental projects.

While increasing recognition of the regional, and in many cases global, nature of the environmental degradation has happened, supranational institutions with adequate fiscal powers for dealing with environmental problems do not exist. The world is organized around the nation states which are unwilling to cede fiscal powers to supranational bodies. Consequently, transaction costs of reaching an agreement, particularly relating to sharing of costs, remain quite high while the resources mobilized remain inadequate. An innovative global environmental fund could be derived from a very small tax on global trade and capital flows and on foreign exchange markets (whose daily turnover in mid-1996 was $1200 billion). This would ensure that those benefiting from international trading and financial arrangements bear some of the costs of providing environment-related international public goods. Such a tax on foreign exchange transactions is known as the Tobin Tax after the Nobel prize winning proposer James Tobin. He, however, proposed it in the 1970s as a way of deterring short-term currency speculation. A proposal to use it for wider purposes, although not for an international environment-related fund, may be found in Haq, Kaul and Grunberg (1996). The main difficulties are not technical but political.

16

Global Governance for Environment

16.1 Global Governance for Environment

A 'soft' hierarchical structure is emerging for planning and overseeing environmental issues at the global level and also for linking them with development. This arrangement operates in two ways.

First, the establishment of international organizations, conferences and agreements such as the Montreal Protocol, CSD, Agenda 21, and UNCED, leads to periodic reviews of the state of the world's environment at global and national levels, and also to arrangements for its improvement which often requires a distribution of resources, technology and goodwill. We live in an imperfect world and such arrangements suffer from existing inequity and lack of trust. One may recall the necessity for having four PrepCom meetings before UNCED and also the continuing shortage of environmental funds. There are occasional success stories (such as controlling the CFCs) where it has been possible to work out an acceptable arrangement. With time and understanding of individual difficulties and expectations, a more efficient global environmental management may come into existence, especially if linkages between good environment and development are universally recognized.

Second, environmental degradation often occurs with a development project such as power plant siting, construction of dams and reservoirs, or planning a new town. In a developed country, the economic and technical sources for such projects are mostly available at the national level and the project is reviewed by the national environmental agency prior to implementation. In most cases an EIA, as discussed in Chapter 12, is required. The developing countries also have such arrangements, but given their limited resources, an external source for both funding and technical assistance is frequently necessary. The UNDP and the World Bank are two organizations which usually provide such assistance, and both organizations are currently very much aware of the prospect of project-related environmental degradation. For example, in the past the World Bank has financed projects that resulted in destruction of the tropical rain forests or threatening established human settlements, activities which had generated wide and unfavourable publicity. The Bank organized a set of structural changes in 1987 to strongly screen development projects for their impact on the environment (World Bank, 1993). The assisted projects these days are usually reviewed environmentally by the major donor organizations, which are often international bodies, and governed by Agenda 21 and similar agreements.

Three case studies illustrate how environmental governance works at this level and its strengths and shortcomings. We discuss the Global Environment Facility and the environmental management organizations of the UNDP and the World Bank. Development work of course is carried out by other international agencies but the UNDP and the World Bank are involved with a very large number of projects of various sizes, some of which could conceivably cause serious environmental degradation unless redesigned before implementation. In both organizations, techniques now exist to discover this possibility at the outset, and to either take preventive measures or abandon the proposed project. These techniques have only recently come into operation, and several years in the future, it will be worthwhile to look back and examine their role in reducing the environmental impacts of development projects.

16.2 Global Environment Facility

The Global Environment Facility (GEF) has been described in one of the documents of the UNEP as a financial mechanism for providing grants and concessional funds to developing countries for specific types of projects that would enhance the protection of global environment. It grew out of a proposal put forward by France and Germany as members of the Developing Committee (a joint World Bank–International Monetary Fund ministerial advisory group) in September 1989. On 28 October 1991, the GEF was formally established (UNDP *et al.*, 1994) as a pilot programme for three years (1991–94). The GEF is jointly run by three implementing agencies: UNEP, UNDP and the World Bank. The World Bank, where GEF is located, has been the dominant partner. The responsibilities are as follows:

- The UNEP is expected to contribute its research experience, and select areas suitable for GEF projects. Its involvement in strategic planning is used to merge the GEF with the existing global and national environmental issues and to ensure that the GEF policy framework is consistent with the operating conventions and legal instruments. It provides scientific and technical guidelines with the help of the Scientific and Technical Advisory Panel (STAP) of GEF.
- The UNDP is designated to link development with environment, and draw on its past experience to organize technical assistance, institution building, and training. It is also expected to co-ordinate pre-investment studies and assistance so that these are consistent with national development strategies and action plans. Through its world-wide network of offices, it helps to select projects that should be funded.

- The World Bank functions as the Trust Fund Administrator, and organizes project identification, evaluation and supervision with both UNEP and UNDP. The Bank is responsible for investment projects but not technical assistance. The chair of the Global Environment Facility is held by the World Bank.

This tripartite arrangement was expected to bring together the strengths of the three agencies in different areas. The pilot phase of the GEF was funded at a cost of approximately $1.2 billion to focus on four areas:

- reduction of global warming
- protection of biological diversity
- protection of international waters
- reduction of ozone layer depletion

In addition, a small proportion of the funding could be used for supporting projects under the Small Grants Programme or for preparatory arrangements for various environmental projects.

Funding for the pilot phase of the GEF (UNDP *et al.*, 1994) came from three sources:

- a core fund (Global Environmental Trust fund, GET) of about $800 million was contributed or pledged by 28 countries, including 12 developing ones;
- cofinancing arrangements totalling more than $300 million were subscribed by seven developed countries;
- about $200 million came from the Multilateral Fund for Montreal Protocol (MFMP) for phasing out ozone-depleting gases.

The funds were originally planned on easy terms as extra financing in addition to regular development assistance to countries with a 1989 per capita GNP below $4000 and a UNDP programme in place. The GEF is run by an Assembly, a Council and a Secretariat, with a Scientific and Technical Advisory Panel (STAP) providing technical advice. Apart from its primary responsibility of being a funding agency, the GEF was also expected to be (1) a programme catalyst for focusing attention on global environmental concerns and mobilizing resources from various sources, thereby integrating national and international efforts; and (2) a research and development facility (UNDP *et al.*, 1994). Most of the projects in the pilot phase and the associated funding went to the focal areas of biodiversity and global warming. The geographical distribution of projects supported is reasonably even, although six large countries have turned out to be the major recipients of GEF assistance during the pilot phase: China ($52 million), the Philippines ($50 million), India ($40.5 million), Brazil ($37.7 million), Mexico ($35 million), and Poland ($29.5 million). It was, however, decided to fix a limit on the size of the individual projects in order to have a widely distributed pattern of assistance out of the Global Environment Facility. For example, the World Bank set a limit of $30 and $10 million for regular World Bank projects and free-standing ones respectively.

The following list provides a sample of the types of projects that have been supported by the GEF:

- global development of a programme and approaches for a regional strategy for the reduction of greenhouse gases
- wind power in Costa Rica
- non-conventional energy project in India
- a dynamic farmer-based approach to the conservation of African plant genetic resources in Ethiopia
- institutional support for the protection of East African biodiversity
- coastal wetlands management project in Ghana
- Bhutan trust fund for environmental conservation
- environmental management and protection of the Black Sea
- pollution control and biodiversity in the Gulf of Guinea large marine ecosystem
- ship waste disposal in China
- ozone-depleting substances reduction in the Czech and Slovak Federal Republics
- monitoring and research network for ozone depletion and greenhouse gases in the Southern Cone (Argentina, Brazil, Chile, Paraguay and Uruguay)

The GEF has not always met the high expectations it was established with. It was expected that the GEF would cover areas not usually covered by the existing instruments of assistance. The new programmes and projects needed to be 'innovative, experimental and replicable, besides aiming at leveraging the "modest" GEF resources by the mobilization of additional financial support from other sources' (UNDP *et al.*, 1994). Some of the expectations of the donor and the developing countries matched. These include innovative approaches to environmental problems, local participation and transparency, support in institutional strengthening and policy formulation, and resource mobilization for the developing countries. The major difference between the two groups apparently lies in the use of the GEF for ameliorating global environmental deterioration as the developing countries in general perceive this deterioration to be caused by the developed world's exploitation of global commons.

The developing countries would rather concentrate on issues such as poverty alleviation, sustainable environmental development, and strengthening local resources. The NGOs perceived the GEF as a measure that would reform the destructive development policies associated with the lending practices of the multilateral development banks (UNDP *et al.*, 1994). There is also the perception that as the GEF is essentially a financing institution, the obligation and duty to act tend to shift with the allotment of funding to the developing countries, whereas there is no similar obligation on the donor countries to fulfil their share of environmental recovery. Transparency and democratization of the GEF also should be reexamined (Khor and Chee, 1992).

An independent evaluation of the pilot phase of the GEF was organized by the three implementing agencies and published in 1994 (UNDP *et al.*, 1994). The GEF was extended in March 1994 when 73 participating governments agreed to restructure the facility and replenish the core fund with more than $2 billion for another three-year period. However, in this context it is profitable to examine some of the conclusions reached by the independent evaluators of the pilot phase. Their report pointed out the following shortcomings:

- The role of the GEF in global environmental protection was not well defined or articulated.
- The selection of focal areas and guidelines for resource allocation was not apparent.
- The global strategies for the focal areas were not well developed and thereby the resource allotment had been arbitrary.
- The GEF had encouraged a narrow project approach and weakened the sense of mutual responsibility rather than building a global viewpoint and taking into consideration national policies, programme strategies, and institutional capacities.
- The GEF was planned to encourage innovation in its areas of support but in reality the word innovation had been rather liberally interpreted.
- Some of the principles behind the formulation of the GEF are exceedingly difficult to implement in the field.
- Instead of building a substantial portfolio of projects, the GEF should have been blended much more with national and regional policy framework.

Over-emphasis on incremental costs may also lead to projects that are not sustainable.

Originally the GEF was expected to be a three-year experiment (UNDP *et al.*, 1994), and was designed to provide modest amounts of financial resources to assist programmes and projects that affect the global environment. This was supposed to be achieved by strengthening local capacities and technologies. This is not an easy task, especially shifting to local issues while maintaining an international focus, and apparently at the end of the allotted span of three years the objectives have not been fully reached. It is hoped that the restructured GEF, drawing on the experience of the pilot phase, would do better. It may even be part of the permanent funding mechanism for the Framework Convention on Climate Change and the Convention on Biological Diversity, if the permanent bodies of these two conventions so desire. The membership also is expected to rise from the total of 94 states towards the end of the pilot phase in April 1994. Any UN member country can now become a GEF participant by depositing a notification of participation with the GEF Secretariat. The country eligibility for funding has been extended to all countries that can borrow from the World Bank or which have a UNDP programme in place. The three implementing agencies have responsibilities similar to the pilot phase and so the administration structure has been maintained. There is a continuing commitment to work with NGOs and community groups.

The relationship of the GEF with the World Bank has been viewed with apprehension (Mikesell and Williams, 1992; Rich, 1994) because of the history of the World Bank as a Bretton Woods institution, which is top-down, secretive, averse to providing information, not welcoming of participation of those most directly affected by its projects and policies, and overly influenced by its main shareholders from the high-income economies, particularly the United States. The GEF is also regarded as imposing additional conditionalities on the borrowers from the World Bank, and focusing insufficiently on linking environment with development in the low- and middle income countries and more on the narrower environmental concerns of the high-income economies.

The GEF, however, represents an additional source, albeit relatively small, of concessionary financing. The World Bank is also becoming more sensitive to criticism. For example, the GEF now includes a Small Grants Programme (SGP) which enables environmental NGOs from the low-income countries to receive grants ranging between \$50 000 and \$200 000. The administrative and monitoring costs of the GEF are also minimized by using the three existing agencies. It also should be noted that the GEF is currently the only institution that aims specifically to fund the incremental costs of environmental projects where domestic costs are greater than domestic benefits, but global benefits are greater than domestic costs.

Certain improvements to the present GEF, however, may be suggested. The Bretton Woods weighted voting mechanism which, for example, gives the United States 20 percent of the voting power and Egypt only 0.5 percent, could be reconsidered. The GEF has already modified this system by requiring that all decisions of its governing council are made by a 'double weighted majority', i.e. an affirmative vote representing both 60 percent of the total number of participants and 60 percent of the majority of the total contributors. Sharma (1996) has suggested consideration of a system modelled after the Montreal Protocol Fund, which provides an equal number of seats to both developed and developing countries and operates by consensus. Such an arrangement, however, is unlikely to attract significant resources to the GEF from the donor countries. Innovative use of the debt-for-nature concept can be considered. Part of the total debt of the developing countries (currently running at about \$1100 billion) can be forgiven for concrete verified measures adopted by the debtor countries to help achieve global environmental goals.

The Global Environment Facility illustrates how an attempt to enhance the quality of the global environment by assisting regional and national projects failed to make a considerable impact during the pilot phase because of loosely structured objectives and quick implementation. It is conceivable, of course, that because of this experience the GEF subsequently would work as a much tighter, better-focused and well-integrated unit, thereby making a greater contribution towards a better global environment. Attempts at globally co-ordinated environmental management procedures have achieved mixed success. The success rate is high where it has been possible to make lifestyle adjustments across the entire planet. The final version of the Montreal Protocol is an example.

16.3 United Nations Development Programme (UNDP)

The UNDP organizes and administers most of the technical assistance provided by the UN system to individual countries. It assists the developing countries (particularly the least developed countries) in their social and economic development by providing systematic and continuing assistance tailored to national development objectives. Although the UNDP is centrally administered, each country is served by a country office that operates on a five-year country programme designed for (1) local needs and (2) the amount of UNDP assistance available for the period. Individual projects are approved of at appropriate stages. Projects are jointly executed by the country governments, UN agencies and organizations, and regional economic institutions including development banks (donor agencies). The planning and implementation of development projects necessarily require environmental consideration which is carried out within the UNDP. Under the present system, each project proposal should be reviewed for possible negative impacts on the environment. The UNDP is also involved in building country-level capacity for sustainable development as formulated in Agenda 21: information gathering, project evaluation and project monitoring. In 1992, the Governing Council of the UNDP urged the UNDP Administrator 'to ensure that environmental concerns are taken into account in all United Nations Development Programme programmes and projects to the maximum extent possible' (31st meeting of the UNDP Governing Council, 26 May 1992). The summary of the annual report of the Administrator for 1992 states:

> Faced with the mounting recent evidence of accelerating environmental degradation and the serious obstacles it presents to poverty eradication, UNDP has intensified its efforts to integrate environmental concerns into the mainstream of its development activities. Since the Earth Summit, UNDP has also moved quickly to launch a major programme in support of Agenda 21. This programme includes, among other components, reviewing UNDP's own capacities for supporting sustainable development activities; reviewing the fifth cycle country programmes and preparing for a more detailed review to be carried out during the cycle to ensure that they fully support a sustainable development strategy; and launching a major capacity-building programme – capacity 21 to support recipient countries in their implementation of Agenda 21.
> (Annual Report of the Administrator, UNDP, for 1992, 15 May 1993)

The UNDP has put in place three major environment-related activities:

- environmental overviews for country programmes (EOP);
- environmental overviews of specific projects/programmes and a formulation of management strategies (EOPMS);
- Capacity 21.

The UNDP programmes for specific countries are planned on the basis of five-year units. At the end of each five-year period, a new series of programmes are designed for each specific country, which are now called Country Co-operative Frameworks (CCF). It has been recommended that the basic geographic conditions and environmental signatures of a country are considered and incorporated within the programme when a new country programme is set up. This is the environmental overview for the country programme (EOP) whose objective is to highlight the critical environmental conditions so that any development project or programme may utilize the existing strengths such as technical facilities, and at the same time, avoid accelerating possible major environmental degradation such as soil erosion. An EOP will include the following:

- a brief description of the natural environment of the country;
- identification of the main environmental issues;
- an assessment of the relationship between economic development and the environment;
- an evaluation of the ability of the country to achieve appropriate environmental management and sustainable development;
- the identification of likely environmental impacts associated with the implementation of the UNDP country programme;
- a comparison of the alternatives to the proposed country programme.

This is not supposed to be a large document and is usually restricted to less than 10 pages. But even a summary document like this is remarkably effective in highlighting the strengths and weaknesses of a country regarding its general environmental conditions and environmental management procedures. It can be interpreted as a document which helps to prevent putting in place a programme that may cause drastic environmental degradation.

A technique for carrying out an environmental overview of a specific project (such as concerning a land development scheme) and the related management strategy (EOPMS) exists in the UNDP. Its objective is to ensure that no approved project results in environmental degradation of a serious nature. It goes further by exploring whether the proposed project can be redesigned to improve the physical, social or economic conditions of the project area as a piggyback development on a project designed mainly for some other purpose. This is also a brief document which is expected to be attached to a project proposal. The different sections of an EOPMS are as follows:

- a brief description of the local environment of the area of the project
- the main environmental issues in the project area
- the economics and the environment in the project area
- environmental management in the project area
- major natural and socio-economic impacts and opportunities associated with the project implementation
- alternatives for project design
- identification of environmental objectives of the chosen alternatives
- identification of conflicts of interest
- formulation of an operational strategy
- monitoring of the EOPMS

This brief document evaluates the project proposal for negative environmental effects, examines the possibility of finding any opportunity to improve the environmental conditions with minor modifications, considers any viable alternatives, identifies the expected problems and conflicts, and organizes periodic supervision of project implementation. Both these documents (EOP and EOPMS) are exercises which identify potential environmental problems and areas of conflict of interest without going through complicated and long-term assessment procedures.

Capacity 21 was fully launched by the UNDP in June 1993. The idea behind Capacity 21 is to ensure that the national countries are able to act according to Agenda 21 and follow the principle of sustainable development. The UNDP has described the broad objectives of Capacity 21 as

- to assist the integration of principles of sustainable development into national development policies
- to assist the involvement of all 'stakeholders' (including local communities, NGOs, the private sector and local assemblies) in developing planning and sustainable environmental management
- to create a body of expertise in sustainable development and capacity building that will be of continued material value to developing countries, UNDP, specialised agencies, NGOs and contributing countries.

(UNDP (1994) Capacity 21, management report on the first year of operation: 2)

Capacity 21 is also an inter-agency effort involving UNEP, WHO and UNESCO in specific countries. A management committee oversees it within the UNDP, and it is supported by funds from a number of donor countries. The basic purpose of the programme is to achieve development, increase wealth and eradicate poverty without destroying the environmental base. Currently, UNDP uses the term *sustainable human development*, putting emphasis on improving the quality of life.

16.4 The World Bank and the Environment

The International Bank for Reconstruction and Development (IBRD), also known as the World Bank, is one of the two economic institutions set up at the Bretton Woods Conference at the end of the Second World War. The other institution is the International Monetary Fund (IMF). The World Bank initially concentrated on the reconstruction of Europe, and subsequently on development, essentially through projects, programmes and sector loans. The total capital of the Bank was $176.4 billion on 30 June 1995, of which only $10.9 billion was paid in, the rest being subject to call. The Bank is owned by the governments of the 178 countries that have subscribed to its capital. The voting power of the countries is based on their subscription and not uniform. The countries with the largest voting power on 30 June 1995 were the United States (16.98 percent), Japan (6.24 percent), Germany (4.82 percent), and the United Kingdom (4.62 percent). Some attempts have been made in recent years to widen the participation of the developing countries in Bank management and, to a lesser extent, in decision-making.

The activities of the Bank are supplemented by those of three affiliated organizations:

1. the International Finance Corporation (IFC), established in 1956 to promote private sector investment;

2. the International Development Association (IDA), established in 1960 to provide concessional loans to low-income countries;
3. the Multilateral Investment Guarantee Agency (MIGA), established in 1988 and designed to encourage flow of dircct foreign investment by providing guarantees against non-commercial risks.

During 1991–95, the annual loan commitment of the Bank and the IDA averaged $15.9 billion and $5.1 billion respectively. In recent years, the share of loans by the World Bank and other multilateral institutions has been decreasing, while that of the private sector has increased sharply.

The linking of Bank-assisted projects with environmental degradation, primarily in the 1970s, prompted structural and policy changes. In 1970, the Bank established the post of Environmental Advisor. In May 1984, the Operational Manual Statement 2.36, *Environmental Aspects of Bank Work*, was issued, requiring the introduction of environmental considerations at the time of project identification and preparation. This was to be followed by modifications, if necessary, at later stages of the project. The speed of the integration of environmental concerns into the normal working procedure of the Bank is partly due to the realization that environmental degradation is incompatible with promoting development (World Bank, 1991a). It was further stipulated that projects with severe environmental impacts would not be financed without proper mitigatory measures. However, incidences of environmental degradation caused by bank-financed development projects continued, with some of these developing into well-publicized cases and confrontations with local NGOs. Projects that contributed to the destruction of tropical rain forest with associated loss of biodiversity and resettlement of indigenous people are clear examples (World Bank, 1993).

In May 1987, a series of structural changes were implemented to strengthen the Bank's environmental policies, procedures and resources. These included the creation of an Environment Department along with Environmental Divisions in each of the four regional technical departments. Currently, the Environmental Department includes the Land, Water and Natural Habitats Division (ENVLW); the Pollution and Environmental Economics Division (ENVPE); and the Social Policy and Resettlement Division (ENVSP). October 1989 saw the introduction of the Operational Directive on Environmental Assessment (EA) which required an environmental assessment for all projects with possible environmental connotations. All prospective loans needed to be screened and classified into categories depending on the nature and magnitude of probable environmental impact. The original four categories (A–D) were later reduced to three (A–C). The determination of the screening category depends on a combination of factors: project location, sensitivity of environmental issues, nature of impact, and magnitude of impact. For example, the location of a project near sensitive and valuable ecosystems, archaeological sites, cultural and social institutions, a high density of population, water courses, etc., may give it an A rating, which is the highest potential class for adverse environmental impact.

- *Category A.* A full EA is required for projects that fall into Category A as such projects are expected to have 'adverse environmental effects that may be sensitive, irreversible and diverse'. Such projects may involve large-scale pollution discharge directly into soil, water or air; large-scale physical disturbances; replacement of an existing forest; modification of hydrological cycle components; storage of hazardous material; or displacement of people.

- *Category B*. A full EA is not required but a limited environmental analysis is necessary. The impact is seen as less significant and not irreversible. Examples include rehabilitation, agro-industries, aquaculture, small-scale irrigation and drainage, and tourism.
- *Category C*. Category C includes projects that do not require an EA or other environmental analysis. Examples include general projects dealing with education, family planning, health matters, and human resource development.

The Bank now requires that affected groups and local NGOs be informed and consulted as part of the environmental assessment procedure of a development project. Environmental assessment, as a standard procedure for bank-financed projects, is expected to be a flexible instrument, making environmental considerations an integral part of project preparation, and allowing environmental issues to be addressed during both project formulation and implementation. This avoids both delays and expenses arising out of unforeseen environmental degradation as a result of project implementation. The general procedure for EIA preparation has been discussed in Chapter 12. The Bank uses the following different stages of environmental assessment:

- screening
- scoping and terms of reference (TOR) development
- preparing the EA report
- EA review and project appraisal
- project implementation

The responsibility for carrying out the necessary environmental assessment remained with the borrower but the Bank decided to opt for a follow-up role to ensure that proper assessment has been carried out. The operational directive was revised in 1991 and the Environmental Assessment Sourcebooks published in the Bank's technical papers series (World Bank, 1991b, c, d) began to be used throughout the environmental assessment process. The assessment sourcebooks were updated in 1993. A growing number of recent publications show the Bank's continuing concern with development-related environmental impacts. Such studies are published both in the technical and environmental series of the World Bank papers. Examples may include papers written for internal circulation (World Bank, 1995b); technical papers dealing with the effect of dams on local environment (Dixon, Talbot and Le Moigne, 1989); proper methodology for watershed development (Doolittle and Magrath, 1990); and publications arising from conferences on rain forest conservation (Cleaver *et al*., 1992). Written primarily for the Bank personnel, they are useful commentaries and handbooks, and above all, are indicators of the Bank's progressive integration of environment with development. A vice-presidency for environmentally sustainable development was established in

1993. The regional agencies, such as the Asian Development Bank (ADB), have also started to focus on environmental issues. It is, however, difficult (and requires a certain quantum of time) to properly assess possible environmental impacts, which may build up a conflict between attempts at environmental protection and pressure to rapidly disburse financial assistance.

16.5 National and Local Governance of the Environment

National governments usually have either a Ministry of Environment (MoE) or a comparable institution to oversee matters related to the environment of the country and also to liaise with global organizations. This, however, is a modern addition to the administration, and in many countries, certain aspects of environmental protection or management are shared with other ministries. These ministries had been charged with such responsibilities for a long time before the MoEs were created. For example, ministries of irrigation and power or of health have looked after very important aspects of our environment for a long time. Apart from job-sharing, the other common limitation of the environment ministries is the lack of the clout of 'important' ministries such as finance or industry. As a result, ministries of environment in most countries are not always able on their own to halt development projects with negative environmental impacts. Often a well-organized campaign by the NGOs is needed as a catalyst for this to happen. One of the reasons for the early success in making Singapore an eminently livable city was the placement of the newly-formed Ministry of Environment in the office of the Prime Minister.

One of the early examples of a federal organization that is entrusted with protection and improvement of the physical environment is the Environmental Protection Agency (EPA) of the United States. EPA began in 1970, following the centralization of environment-related units of the federal government into a single agency in recognition of the seriousness of national environmental degradation. Various programmes were transferred to it from the Department of Health, Education and Welfare, the Department of Interior, the Department of Agriculture, Food and Drug Administration, and the Atomic Energy Commission. The EPA, in the general area of environment, has the following responsibilities:

- establishing and enforcing standards
- enforcing environment-related legislation
- monitoring of pollution
- conducting research into problem areas and disseminating information
- assisting state and local governments in environment-related issues

In other countries, MoEs are expected to carry out similar functions. The efficiency of an MoE clearly depends on a number of factors:

- its relative strength in the national government
- the quality and dedication of the ministerial staff

- the available technical and scientific capacity
- the inclination (or lack of it) of the government to enhance the quality of the ambient environment
- the quality and coverage of the legislative acts which are used to combat matters of environmental degradation or to demand implementation of assessment procedures prior to project approval

In large countries a hierarchical arrangement of governance requires environmental supervision at provincial and local levels also.

Clearly, the efficiency and capacity for environmental supervision will vary from country to country, and especially among the developing countries. It is also intuitively obvious that the quality of the national environment is best preserved by a dedicated and efficient national body and not solely by global consensus. The motivation and capability of national ministries of environment are therefore crucial. So are the built-in procedures for incorporating environmental concerns into economic and social development planning and budgeting, and also interagency cooperation on cross-sectoral environmental issues, and the knowledge of the local environment.

Most countries provided a country environmental report for UNCED in 1992. Apart from providing a country environmental profile, these reports include important information regarding the capability of a country to monitor, protect and improve the national environment. International donor agencies such as the UNDP and the World Bank also periodically review country environmental capacities and such documents contain basic information for integrating environmental and development issues. The EOP document of UNDP, described earlier in this chapter, fulfils this function. Environmental information now exists in summary form for many countries.

Besides building the technical capacity for environmental evaluation, the legislative capacity for environmental protection, and the administrative capacity for environmental monitoring, each country should have a clear understanding of the overall nature of its environment, and also be in possession of a general strategy or plan which integrates environment enhancement with economic development. This can be achieved by (1) accounting for national environmental assets (Repetto, 1992), and (2) preparing a national environmental strategy.

The cost of environmental degradation is difficult to compute in monetary terms, but where it has been done, the figures are extremely disturbing. In Bangkok, for example, the annual health costs of air pollution concerning particulate matter and lead have been estimated to be between 0.6 and 2.8 percent of Thailand's GDP. In Ghana, annual degradation from the present pattern of crop and livestock production costs the country more than 1 percent of its GDP (World Bank, 1995b). Such losses, to a large extent, are avoidable and could be prevented by incorporating environmental concerns into development planning.

A number of environmental strategies and action plans are currently being formulated at the national level. The national governments draft these plans with assistance from NGOs and international and bilateral organizations. These plans are based on environmental profiles and sectoral and economic analyses. The documents produced need to change with time, with changing national conditions and with a better understanding of the national environment and its development-related alterations. The national environmental strategies come in various types (World Bank, 1995c):

- *National Conservation Strategy (NCC)*. Promoted by World Conservation Union (IUCN), this type of document is intended to provide a comprehensive analysis of country conditions for integrating environmental conditions into the development process.
- *National Environmental Action Plan (NEAP)*. This identifies major environmental concerns and problems and formulates necessary ameliorating policies and actions. This is intended to be a continuing process integrating environmental management with the national decision-making which is favoured by the World Bank and other donors. By 1995, 57 International Development Association (IDA) members from the main low-income countries receiving aid completed their NEAPs. Another 16 NEAPs were at least at advanced stages of completion.
- *National Sustainable Development Strategy (NSD)*. This genre of cyclic and participatory processes, connected with Agenda 21, includes a series of national strategies which are designed to integrate economic, ecological and social objectives.

Whatever the favoured format and whatever their level of success, national plans or strategies which attempt to integrate environmental enhancement with economic and social development make two very important contributions. First, they at least provide documents which highlight the problems and concern areas, and to a variable extent, destroy environmental ignorance and the lack of concern. Second, they highlight the fact that environmental strategic implementation is an on-going process that needs to be continuously monitored and kept in operation. National environmental strategies are new exercises, and whether these will successfully incorporate environmental concerns into national decision-making remains to be seen. The instruments, however, may certainly be taken as national commitments towards a better environment, and furthermore as vehicles for continuous communication between the national governments, the country NGOs, and the international agencies. Viable linkages between the local, state and national governments in environmental management are also essential.

16.6 The Non-governmental Organizations (NGOs)

By the early 1990s, non-governmental organizations had developed into a diverse and growing set of institutions operating at various levels. Globally they numbered in thousands, and the total numbers of members and associates of individual organizations showed a remarkable increase. This recent proliferation is indicated by the age of the NGOs. Only about a half of all the NGOs from the developing countries are more than 10 years old. They still carry out the primary function of the environmental NGOs by publicizing instances of environmental degradation, although most modern NGOs will also be competent enough to suggest an alternative to the proposed project which they consider to be environmentally unfriendly. It has probably become easier to function, as the NGOs have a network of their own and are recognized as an integral part of environmental management by UN institutions, the World Bank, and a number of national governments. The Friends of the Earth (FoE) network which covers USA, most western European countries, Malaysia and South Africa is a good example. The funding situation is also better as they may receive financial assistance from a number of sources. NGOs have a variable geographical distribution, being scarce in certain regions such as China or North Africa due to a combination of political, economic and cultural reasons. In certain countries (e.g. India, the Philippines and Brazil), well-organized groups operate as autonomous NGOs, whereas in other parts of the world support from international NGO networks is crucial for their success.

Although NGOs, including large environmental NGOs, have been perceived as pressure groups (Willetts, 1982), their proliferation has created problems of identity. The interests of the NGOs now span a wide range of issues: economic, environmental, political and social. NGOs can in many cases depend on network support and do not have to battle environmental degradation or face powerful vested interests individually. Even then, in the developing countries, a number of NGOs remain community-based, comparatively small, and at times fragile. Survival depends on the courage and persistence of a small number of individuals. In certain countries they are considered as organizations hostile to the national government. In extreme cases, members operate at the risk of their lives (World Resources Institute, 1992). In 1989, news of the murder of Chico Mendes, the leader of the rubber tappers of the Brazilian Amazon who successfully led a local campaign for preserving the forest, reverberated round the world.

The vast range of structure, objectives and scale that the NGOs currently span is perhaps best indicated by the attempt by the World Resources Institute (1992) to identify the common types:

- Grassroots NGOs are community organizations in both urban and rural areas of the developing world concentrating on local issues such as deforestation or better housing.
- Service NGOs are units that support grassroots groups, often as part of a national or regional federation.
- Thematic NGOs operate at regional, national and global scales with a clearly defined objective such as environment or disaster relief; often in support of grassroots and service groups.
- Regional, national and international networks and coalitions operate on a much bigger scale, are well-publicized and connected, and deal with larger environmental issues.

From the 1980s onwards, the strength of the NGO movement was enhanced by the establishment of direct contact with international organizations. The NGO–World Bank Committee was founded in 1982 to provide a forum for policy discussion between senior bank officials and NGO leaders from different parts of the world. In recent years, subsequent meetings have discussed popular participation in development-related decision-making. A selection of NGOs are currently kept informed of bank-related projects with a possible environmental impact, or of those projects where the potential for possible NGO involvement in the project exists (World Bank, 1991a). In the 1980s, a group of US environmental NGOs campaigned successfully against Brazil's Polonoreste road-building and colonization project. Another group of NGOs has created the Development Bank Assessment Network which urges the Bank to examine project-impact on the poor and on the environment in particular. The World Bank has progressively encouraged the involvement of NGOs in projects financed by the Bank. Over time, the profile of the NGO partners of the Bank has shifted from a majority of international NGOs to a majority of grassroots groups, the people who are directly affected by development projects (World Resources Institute, 1992). Both the Stockholm and the Rio de Janeiro conferences have contributed to the linkage between the UN institutions and the NGOs. Hundreds of NGOs have a consultative status with the Economic and Social Council (ECOSOC) of the United Nations. This gives the NGOs the right to attend ECOSOC meetings; to submit statements; to testify at meetings; and to some extent, propose items for consideration. Certain UN agencies work in collaboration with the NGOs. Small grants ($25 000 per country) are made to grassroots NGOs in support of innovative projects through UNDP's 'Partners in Development Programme'. The Sustainable Development Network Programme (SDN) of the UNDP identifies and links organizations (including NGOs) that could contribute to sustainable development. The UNEP has organized the Environmental Liaison Centre in Nairobi, the African NGOs Environmental Network, and the Green Belt Movement, all of which require NGO networking. The United Nations Development Fund for Women (UNIFEM) is another UN institution that relates to NGOs in supporting the women's movement. UNIFEM provides funding for supporting poor women in the developing world in their attempt towards building a better life in an enhanced environment. For example, UNIFEM has been involved in a sericulture project for village women near Udaipur, India; in introducing biodigesters as replacement for firewood gathering to the Yemeni women at Al-Habeel; and in training the women in building low-cost water pumps in Tempoal County, Mexico.

The success of the NGOs varies from place to place and project to project. The *chipko* movement of India, which prevented the local government from destroying valuable forest on steep Himalayan slopes, is one of the best-known examples of success (Weber, 1988; Guha, 1989). Here, outside contractors with tree-felling contracts from the forest department were prevented from forest destruction by the local villagers by the simple and expeditious act of hugging the trees when they were about to be felled. This technique was first used in 1974 by a group of village women and children, and was subsequently repeated by a number of local villagers. The regional forest was ultimately saved by a central government ban on tree felling in the area. This was a situation where a number of factors came together to bring success:

- There had been a long tradition of social and environmental work in the Uttarkhand Himalayas which was led by disciples of Mahatma Gandhi.

- A tradition of non-violent movements existed in the country.
- A tightly organized movement growing up under the able leadership of Sundarlal Bahuguna and Chandi Prasad Bhatt united the villagers and got the scientific community of India involved.
- A political network existed from the village grassroots units to the country's capital.
- The climactic act of hugging the trees by local inhabitants produced an emotional impact far beyond the region.
- Lastly, the scientific knowledge of the local people was of a higher order than that of the forest department, the villagers associating the destruction of the forest with increasing local mass movements, soil erosion and floods, whereas the forestry department concentrated on selling the timber without working out the consequences.

Bandyopadhyay (1995) identified the capacity for thinking globally and acting locally as one of the basic strengths of the movement. The success of NGO environmental movements often depends on a combination of such factors. In fact, *chipko* has almost become a generic name for movements for forest protection, and this technique for preventing tree felling has been tried out elsewhere; for example, in South India, where the movement is called *appiko*. In 1996, trees were embraced in a similar fashion in southern England.

A number of instances can be given, albeit with varying success, where non-governmental organizations have forced modification or redesigning of development projects or existing practices with negative environmental impact (Ghai and Vivian, 1995). Enhancement of the environment is often achieved via the introduction of simple techniques provided with NGO help, such as biogas plants, basic and low-cost water-retention devices on slopes, or providing information on indigenous trees (World Resources Institute, 1992). In most countries a development project will not pass unexamined by at least one NGO, and the NGOs are adopting more and more the role of environmental guardians. A number of NGOs now include economists, earth scientists and technicians among their members, and are capable of carrying out environmental assessments of acceptable quality. NGOs now have a large number of environment-related publications to their credit, publications which go beyond raising public consciousness or network building, such as that of the Third World Network (1992). These publications evaluate the state of the national or regional environment and existing organizations (Agarwal and Narain, 1986; Samad, Watanabe and Kim, 1995). Large global networks exist, such as the US-based Pesticide Action Network (PAN) which is an international coalition of 300 organizations in 50 countries. Such coalitions provide support and information. The availability of these two attributes will undoubtedly be at a greater and faster pace in the near future via networking with computers.

A new development has been the contributions from dedicated individuals at universities and research institutions who are involved in the careful monitoring and evaluation of specific environmental factors. Such work does not fall within the ambits of mainstream NGOs, but the value of the work is enormous and the reliability of the findings leads rapidly to their applicability. A considerable amount of data is now available and is passed around the world via various media, from printed pages to electronic mail. It is becoming increasingly difficult to commit an act of environmental degradation out of ignorance; almost all such acts are deliberate. It should be added, however, that our knowledge of the environment, especially the tropical environment, is still quite limited, and continuation of research concerning both the nature of the tropical environment and its degradation is absolutely vital for its proper management.

Although non-governmental organizations of the developing world are now numerous and well established, they tend to suffer from financial limitations, structural weaknesses, and, in some countries, a lack of trust between NGOs and the government. The relationships between governments and NGOs remain complicated. Successful management of specific environmental projects can be attributed to joint participation of both the government and NGOs. On the other hand, in certain instances NGOs are firmly entrenched in opposition to governmental development projects and little dialogue takes place. NGOs, however, have become an important and oft-recognized actor on the environmental scene.

16.7 The Changing Industrial Outlook and Environmental Management

A striking amount of environmental degradation, ranging from polluting local drainage systems to destroying the global ozone layer, has been attributed to industry. In the past, the common reaction has been to prevent the inappropriate disposal of industrial wastes. Industry is also required to comply with a set of regulations imposed by governmental agencies, which require minimizing of waste, substituting for toxic material, and controlling of harmful emissions. In recent years, industry has shown increasingly innovative attempts to reduce environmental degradation by controlling harmful pollutants by various means, and a new term, 'industrial ecology', has come into operation (Graedel and Allenby, 1995). As Graedel and Allenby describe it, industrial ecology 'requires familiarity with industrial activities, environmental processes, and societal interaction, a combination of specialities that is rare'. Industrial ecology involves redesigning industrial plants and maintaining a strict account of material in use so that environmentally harmful effects are minimized inside the plant, rather than being dealt with at the emission end. If this is the trend for the future, it is a positive sign as nothing controls industry as well as industry itself.

Industrial laws have existed for a long time, usually drawn up and implemented after a specific type of damage has occurred. Some of these problems could have been anticipated by the industry itself given the knowledge available at that point in time, or when an industrial plant already in operation is replicated at a second location. Some of the instances of serious environmental damage were, however, not realized at the time of industrial establishment. The classic example of this is the production of CFCs, which were originally designed as an inflammable and non-toxic material. Traditional environmental regulation of industries is focused on individual phenomenon, and the controlling legislation imposed by a central authority is very specific.

Compliance is usually measured at the production and emission end. Such regulations, however, do produce results. A new type of regulation is emerging in the developed countries. This extends beyond policing of pollution at the end of the production line, and is focused more on operational techniques in the industry. The industry is expected to control itself in order to achieve incentives (usually financial) given by the state for appropriate environmental friendliness. The new approach calls for technological sophistication, collaboration between the interested actors (industry, government and environmentalists), and avoidance of past adversarial attitudes towards environmental compliance.

This vision for the future involves the industry in (1) integration of environment and technology, (2) incorporating environmental costs into economic decision-making, and (3) recognition of its societal responsibility (Graedel and Allenby, 1995). At present, this is not common but as this prediction of future industrial behaviour comes from the industry itself, it is possible that industrial establishments will play a significant role in environmental management in the future, and authoritative regulations would not be the only way to reduce environmental damage from industrial sources. The new approach and techniques will, of course, need a global diffusion to prevent a dichotomy between the progressive industry of the developed world and the old-style manufacturing techniques restricted to the developing countries. The role of the industrial establishments in sustainable development has been described by Schmidheiny *et al.* (1992) in the book *Changing Course*, which also includes a set of case studies where business enterprises have been successful in reducing the industry-related environmental impacts. Case studies of industrial best practices are also available from the website of UNEP.

Industrial ecology requires that the products of manufacturing cause minimal damage to the outside world. The technique for achieving this objective is known as life-cycle assessment (LCA). This implies careful stewardship of materials, services, processes, products and technologies over their entire life-span. LCA methodology is complex and this is not the appropriate place for a detailed presentation. Basically it is carried out in three successive stages:

1. *Inventory analysis*. This establishes quantitatively the types and levels of energy and material inputs to an industrial system and the effect on the environment of the outputs; in other words, which outputs are valuable and which are potential liabilities.
2. *Impact analysis*. This measures the effect of the outputs on the external environment.
3. *Improvement analysis*. This examines the existing industrial procedure and determines the need and opportunity for reducing environmental impact by modifying the methodology.

LCA was originally developed for simple industrial products (e.g. drinking cups), which has limited its use in more complex production processes. The use of LCA needs to be satisfactorily demonstrated for complex industrial products such as aeroplanes (Graedel and Allenby, 1995). However, it is a step in the right direction, and it is expected that tighter auditing throughout the life-cycle of a product would minimize harmful process residues. The examination of the various types of industry has already raised certain conflicts with conventional wisdom (is recycling automatically environment-friendly?) and discovered new ground rules (what makes recycling economically desirable?). As Graedel and Allenby (1995) stated, 'Many of the world's largest corporations are coming to the realisation that business activities and environmental concerns are not antithetical but rather are essential components of excellent organisational approaches'. It has been shown that the ecological considerations of companies provide an opportunity to be rewarded with new economic instruments such as taxes and tradable permits, to acquire bank loans and insurance easily, to avoid conflicts with stringent governmental regulations, and to maintain a good profile with the customers and employees. Such stimuli will undoubtedly grow even stronger in future. Two other vital acts of encouragement are

- the marks of approval given by country governments for industrial products; examples include the Green Seal or Energy Star (USA), Blauer Engel (Germany), Ecomark (Japan), Green Label (Singapore), etc.;
- the International Standards Association sign of approval ISO 9000, which ensures reasonable standards of quality in manufacturing and is much coveted by the industrial world. ISO 9000 also guarantees a detailed and quality auditing system throughout the manufacturing procedure. ISO 9000 is currently being replaced by a new system, ISO 14000.

This type of competitive labelling increases the turning out of environmentally responsible products. As the environmental regulations will not be less demanding in the future, industry could be expected to be environmentally concerned at least because of its own interest, and devise new technologies which are part of industrial ecology which rejects the concept of wastes. The future may produce a new style of industrial engineers who will emphasize the use of less material, more substitution by environmentally friendly material, and more efficient use of energy. This of course is still very much in the future but an attitudinal change is taking place in the industrial world. There is certainly a less adversarial approach to environmental issues and regulations. Only time will tell whether we are going to benefit from an environmentally enlightened industry or not. The other nagging question is whether the enlightened industry will be truly global in the future, and whether industrial pollution, either small-scale or as large as the killer gas of Bhopal, will be much reduced even in the developing countries.

16.8 Trade and Environment

Trade and environment are linked as follows (UNCTAD, 1994):

- Multilateral trade obligations, such as those under the World Trade Organization (WTO), may come into conflict with domestic environmental protection arrangements consistent with national priorities and preferences.

- Trade agreements should permit transborder and global environmental concerns to be addressed in a manner that allows for equitable distribution of responsibilities between countries at different levels of income and development.

The high and growing level of world trade and the globalization of economic activities have increased the need to tackle the link between trade and environment (Anderson, 1996). The value of the world output at market prices was $29 852 billion in 1995, while the world export of goods and services (by definition, world imports) was $6201 billion or 20.8 percent of the world output. In recent years, the volume of world trade has been growing at a much faster rate than the world output. The world output grew at an average annual rate of 3.3 percent during the period 1978–87, and the volume of trade at 4.0 percent. During the 1990s, the gap between the two rates widened. In 1995, the volume of trade at 8.7 percent was nearly 2.5 times the increase in world output of 3.5 percent (International Monetary Fund, 1996).

The preamble to the Agreement establishing the WTO recognizes the need to protect and preserve the environment. A number of international agreements on environmental issues also deal with trade. Such agreements include the Montreal Protocol on Ozone Depleting Substances, the Basle Convention on Hazardous Substances, and the Convention on International Trade in Endangered Species (CITES). It is essential that these provisions and the WTO rules and practices regarding environment be consistent with each other. One possible approach would be to use the WTO dispute resolution mechanism to enforce the trade-related aspects of the international environmental agreements.

WTO established a committee on Trade and Environment (CTE) in 1994. The committee is expected to

- analyse the trend towards unilateral action in addressing extra-jurisdictional environmental problems
- find a balance between multilateral solutions to environmental problems and ensure that the WTO members are not unduly affected by either disguised protectionism or the imposition of unacceptable environmental values and standards through trade policy actions

The CTE is also expected to examine the implementation of the polluter pays principle in an economically efficient manner.

Specific WTO rules and provisions related to trade and environment are as listed below.

- *Agreement on Technical Barriers to Trade (TBT).* This agreement governs the use of product standards and brings technical regulations adopted for environmental objectives more explicitly within its scope. Countries are encouraged to follow international standards, and are required to notify their standards or regulations when they are likely to have significant trade effects. The technical regulations should not be more restrictive than necessary. Participation of the developing countries is necessary for these regulations to be effective in protecting the environment and promoting trade. Any perception that these could be used by the developed countries in a protective manner could create a backlash against multilateral trading arrangements and environmental agreements.

- *Agreement on the Application of Sanitary and Phytosanitary Measures (SPS).* This agreement deals with measures applied to protect human or animal life or health within the territory of the importing country from risks arising from additives, contaminants, toxins, or disease-causing organisms in food, beverages or feedstuffs, and to prevent the establishment or spread of pests. Special provisions exist for developing countries, including technical assistance.
- *Agreement on Subsidies and Countervailing Measures (SCM).* This deals with subsidies that a country may provide to its exporters and the countervailing duties that an importing country may levy. Subsidies may be allowed when designed to help businesses enact higher environmental standards.
- *Agreement on Trade Related Intellectual Property Rights (TRIPS).* Under such conditions, countries may exclude from patentability inventions which may cause serious harm to the environment. The protection of indigenous community rights and biodiversity is organized by a combination of patent and *sui-generis* arrangements. The developing countries have a particular interest in preserving biodiversity and the traditional use of medicinal plants.

Important and complex linkages occur between trade and environment. This requires more research, both to identify the impact of trade on environment and in order to translate the polluter pays principle into actual measures. This is unlikely to be achieved soon.

16.9 Conclusions

Environmental management has come a long way since the Stockholm Conference and at all three scales, local, national and global. Environmental awareness has permeated political boundaries and economic levels of existence. A number of agreements, regulations and procedures are now in operation to safeguard our environment. The global unification to prevent ozone depletion is a benchmark example of what can be achieved in a very short time if the economic and political will is there to back up good science. The potential for a better future exists but it certainly has not been achieved.

We live in an imperfect world, and future management of the global environment undoubtedly will fall short of ideal, but at least the following problem areas should be immediately addressed:

- global availability of financing and technology for environmental management
- research in areas of environmental science, degradation and management
- eradication of extreme poverty

- integration of correct environmental management procedures among organizations and governments at various levels (international, country, state and local), and a strong commitment to these procedures
- communication with local people
- contribution from an environmentally enlightened industry

All these have been partially achieved but fall far short of the required. The success of the future environmental management of our planet depends on a greater achievement of these objectives. Otherwise, in the middle of a distressed world, we will perpetually remember the words of the Brundtland Commission, 'The Earth is one but the world is not'. The alternative is much better.

Exercise

1. Select a large-scale project that has generated considerable controversy because of its possible environmental impact. Your selection should be a project you are personally familiar with or one about which a significant amount of literature exists. Prepare two tables. The first table should list the benefits and the negative impacts of the project. Use your discretion, do not include unimportant observations.

   In the second table compile in brief the opinions of the environmental actors such as national, state and local governments (whichever is applicable), donor agencies, developers, industrialists, NGOs, and ordinary people who would be affected.

   Write your carefully considered opinion regarding the project. Should it go ahead or be modified or be abandoned? Now state who among the environmental actors listed above are guilty of exaggeration and in what direction.

2. List all the national environmental reports that are available for a country you are familiar with. Who compiled these reports? If you have reports from both governmental sources and the NGOs, do these substantially agree or disagree? In which areas?

17
The Main Issues

17.1 Interrelationships and Interdependencies

The first part of this book introduced basic concepts in physical geography, global population distribution and environmental economics, including a discussion on benefit–cost analysis. Such an introduction is useful for comprehending and managing environmental issues which tend to be multifaceted and multidisciplinary in nature. The second section discussed specific environmental topics such as vegetation, land, water, etc., and the techniques for environmental impact assessment. Environment at the global level was the subject matter of the third part.

Irrespective of the coverage, all sections indicate the interrelationships and interdependencies of the environmental attributes. Environmental studies and environmental management are both fascinating and demanding because of such relationships. For example, agricultural development in an area may involve providing the appropriate mix of the following:

- land rights
- water
- fertilizer
- seeds
- pesticides
- power
- financial credit
- marketing arrangements

Changing any of these from the existing conditions may lead to a specific environmental problem. An environmental manager may have to deal with a multiplicity of such problems. Some of these problems may even create unexpected synergistic effects. Such impacts and interdependencies of these factors should therefore be examined at the planning stage. If the project is very large, it may even contribute to bigger environmental problems such as global warming or large-scale water or air pollution. The effects of this agricultural development on the socio-economic conditions of the country also have to be examined. Environmental management is a wonderfully complicated and challenging subject.

17.2 Common Problems of Environmental Management

Complications also extend to the basic principles of environmental management. Wijkman (1996) describes sustainable development as a common denominator for global understanding and action of people and environment. Sustainable development is also the basic principle of Agenda 21 which was accepted at the UNCED conference in Rio de Janeiro in 1992. But certainly no consensus exists regarding sustainable development, as shown earlier in this book. Some of the specific areas where difficulties have arisen are as follows:

- more emphasis on development than on environment
- a shortage of theoretical knowledge and necessary environmental models
- in many cases, insufficient emphasis on poverty elimination
- the national governments (of both developed and developing countries) failing to meet their commitments made at the UNCED meeting; in fact certain countries have interpreted the support for continuous growth as 'business as usual' with a few environmental adjustments

In order to make development both people-oriented and environmentally sound, UNDP currently uses the term *human sustainable development*. Such an approach was also supported at the 1995 World Summit for Social Development in Copenhagen (Wijkman, 1996).

Sustainable management could be difficult to achieve in real life. Forest management, based on carefully regulated timber production, theoretically reconciles the economic interests of the producer with the objectives of the environmentalists. Rice, Gullison and Reid (1997) found sustainability of tropical forests in Bolivia surprisingly problematical in their study of the mahogany trees of the Chimanes Permanent Timber Production Forest. Mahogany trees grow scattered in this area, only following sizeable natural disturbances which create openings. Extraction of mahogany is therefore a low-impact practice, but unsustainable for mahogany, unless large openings are made artificially in the forest. Furthermore, sustainable development does not provide the timber companies with any economic incentive. In many developing countries the option of cutting down all valuable trees and investing the resulting proceeds provides a much higher income than long-term sustainable forestry. The former provides an annual return of 14 percent in Peru, 17 percent in Bolivia, and 24 percent in Guyana. In comparison, the delayed timber harvest as required by sustainable development would provide only 5 percent annual return. Such a comparison explains why less than one-eighth of one percent of all tropical production forest was managed on a sustainable basis in the late 1980s. Rice, Gullison and Reid (1997) are of the opinion that the management of tropical forests for sustainable production has a low probability of becoming a widespread phenomenon in the near future due to contrary economic incentives, limited government control, and very little local political support. This is a reality that should be recognized.

Even if we are concerned only with protecting the environment, we are hampered by our incomplete knowledge of it. This lack of knowledge occurs at all levels, from global to local. For example, neither global warming nor stratospheric ozone depletion was an issue at the Stockholm meeting only a quarter century ago. Our knowledge of the biodiversity of the tropical rain forests is limited. We do not know enough about the nature and behaviour of rivers to correctly anticipate all the possible changes following the implementation of a development project.

The lack of data and only a partial understanding of the physical environment are particularly relevant for the developing countries. Determining the impact of a project often requires baseline data of good quality and long duration, such as river discharge figures. Lack of such baseline information limits the quality of project-related environmental impact assessments. Even simple data sets have limited geographical coverage. For example, rainfall records are seldom available for the middle of a large forest or from inhospitable mountainous areas. It is crucial that such databases (and of good quality) are built to cover the developing countries, and that technical and, if necessary, financial assistance is provided for it. Even in the developed countries the requisite information occasionally has to be synthesized or a surrogate used. The first step in environmental management is academic research or a routine collection of good quality information by a government agency. Contributions by the NGOs also help to fill in the information gap, although so far only a small number of organizations have attempted this task (Agarwal and Narain, 1986; Agarwal and Chak, 1991).

The crucial problem perhaps is the quality of growth (Wijkman, 1996). In order to control waste generation, it is necessary to have a changed approach with redesigned production technology so that far smaller amounts of energy and material throughputs are needed. In the developing countries growth is necessary for alleviating poverty but such growth should be channelled according to the concept of sustainable development. Growth should not lead either to serious resource depletion or to large-scale waste generation. The holistic nature of the earth requires that knowledge, techniques and financial resources required for this should be available across the planet but that is seldom the case.

It is possible to list the current common and important problems of environmental management:

- environmental management is complicated and multifaceted
- many of the management objectives are rather loosely defined
- the background data are not always available, and when available, they are often of limited quality
- the relationship between environment and development is not always linear
- a change in mindset, particularly of the policy-makers and the élite, is imperative
- environmental management suffers from a paucity of funds, especially in the developing countries

- a more efficient system of co-operation between the developed and the developing countries is required
- poverty and inequity hinder proper environmental management and sustainable development

Attempts have been made in the past, particularly over the last three decades, to resolve such problems and shortcomings although with limited success. The level of this success is reviewed in the following section.

17.3 Success in Environmental Management

The determination of the level of success in environmental management (at various scales, from local to global) varies with evaluators, often depending on their political beliefs. It is certainly true, however, that our knowledge, will, funding, techniques and organizations regarding the environment have increased a great deal over the last three decades, certainly from the time of the Stockholm Conference. It is equally true that the level of such achievements should have been much higher, and could have been much higher, if the economic conditions of the developing countries were better, and if stronger political will and co-operation (at international, national and local levels) between governments and social organizations were in existence.

Box 17.1

Burning of Forests and Production of Haze in Southeast Asia

Traditionally, many farmers of Southeast Asia (including sedentary ones) take advantage of the dry conditions before the arrival of the monsoon to clear the land for cultivation. This is done by burning the secondary vegetation and leaving the ash on the ground before the rains appear. The vegetation that is fired is usually not part of primary rain forest as such a wet forest is extremely difficult to burn.

The scale of the burning of vegetation, however, has increased tremendously in recent years due to various causes. The number of farmers who follow this practice has increased, often due to deliberate government policy to clear land and resettle farmers from populated areas. Destruction of the rain forest, both for timber and plantation expansion, has also become widespread. This forest destruction is carried out by large companies with contacts in high places, the political elite. If the land is not used for plantation, the settlers may move in after timber extraction and burn the residual biomass in order to clear the ground. Usually environmental degradation is limited to the standard list associated with rain forest destruction. Near the Equator where this happens, rains are frequent and the duration and extent of burning limited. This type of agricultural practice and resource extraction is extremely common on two large islands, Sumatra and Borneo. Here the burning starts around August–September and is over by November when the rains appear from the northeast.

However, in a particularly dry year the rains are delayed, and the smoke from the fire persists and turns into large-scale and transboundary air pollution. The El Niño years are particularly dry in Southeast Asia and the monsoon rainfall is both delayed and reduced in amount. Such incidents have happened three times in the last two decades: 1982–83, 1991–94, and 1997–98.

El Niño was particularly strong in 1982–83 and a number of fires started on Borneo, affecting both the Indonesian part of Kalimantan and the Malaysian State of Sabah. A large volume of dry biomass was available in the Kalimantan rain forest, parts of which had earlier been cleared for settling farmers from a different island. The fires were visible via remote sensing satellites, burnt for weeks, and where the underlying peat was ignited the situation was beyond control. A large part of the forest was destroyed in a piecemeal fashion. The smoke from the burning drifted over to West Malaysia and Singapore as a short-lived haze phenomenon.

The early 1990s saw an El Niño that persisted for several years. This time smoke from the uncontrollable fires of southern Sumatra drifted over the cities of southern Malaysia and Singapore which were shrouded in haze for days (Figure 17.1). Conditions were much worse in the Malaysian capital of Kuala Lumpur than in Singapore because of the local physiography which prevented haze dissipation. The situation in 1994 was worse, with the PSI index reading 'unhealthful' for days. This led to discussions among the countries concerned and an office to deal with environmental matters was set up between the member countries of ASEAN (a group of Southeast Asian countries). The fires, however, were too large to be put out and hazy conditions persisted for months before the monsoon finally arrived and put the fires out. Living conditions near the fires must have been horrendous.

In 1997, in exactly the same fashion, destruction of the forests of Indonesia and biomass burning combined with a large El Niño. The haze of 1997 was the worst. In Kuching (Sabah, East Malaysia), the PSI index rose to 839. Readings above 200 were common in Singapore and in the cities of Malaysia and Western Indonesia. Streets were deserted, airports were temporarily closed, and in particularly affected areas life almost came to a standstill. Visibility was extremely limited, raising the fear of collisions between ships in the busy shipping channels of the narrow Malacca Straits. People fell ill and incidences of death have even been reported. Again, the fires were out of control and only the monsoon rains were able to put them out. In an El Niño year, the rains are delayed, which means that millions of people will continue to live in terrible conditions for months. The total cost of this haze was estimated to be more than \$60 million for Singapore (The *Straits Times Weekly Edition*, 21 March 1998).

This periodic suffering from haze illustrates anthropogenic environmental degradation which becomes uncontrollable in bad years. This is also a case of transboundary air pollution which necessitates regional co-operation. The co-operation, however, has been cosmetic in nature, involving surprising endeavours such as attempting to seed clouds in the middle of a dry period, spraying water from tops of tall towers, and making authoritative statements. The problem would not arise if the farmers were supported by an agricultural service, if people remembered the effect of the past El Niños, and if in spite of their excellent connections timber companies were not allowed a free hand during the El Niño years. The human suffering in parts of Southeast Asia is therefore due to a combination of lack of knowledge about local physical conditions on a regional scale, the shortage of political will, and the greed of the powerful cohorts that exploit the natural resources. At times, knowledge and unpleasant steps are necessary to resolve an environmental problem. No firm step seems to have been taken but President Suharto of Indonesia apologised to other countries of the region. This is an example of an environmental disaster for which the solution is known but is not implemented. A similar environmental disaster is likely to reoccur with the next strong El Niño, unless the current methods of regional co-operation are vastly improved and firm steps are taken to prevent biomass burning.

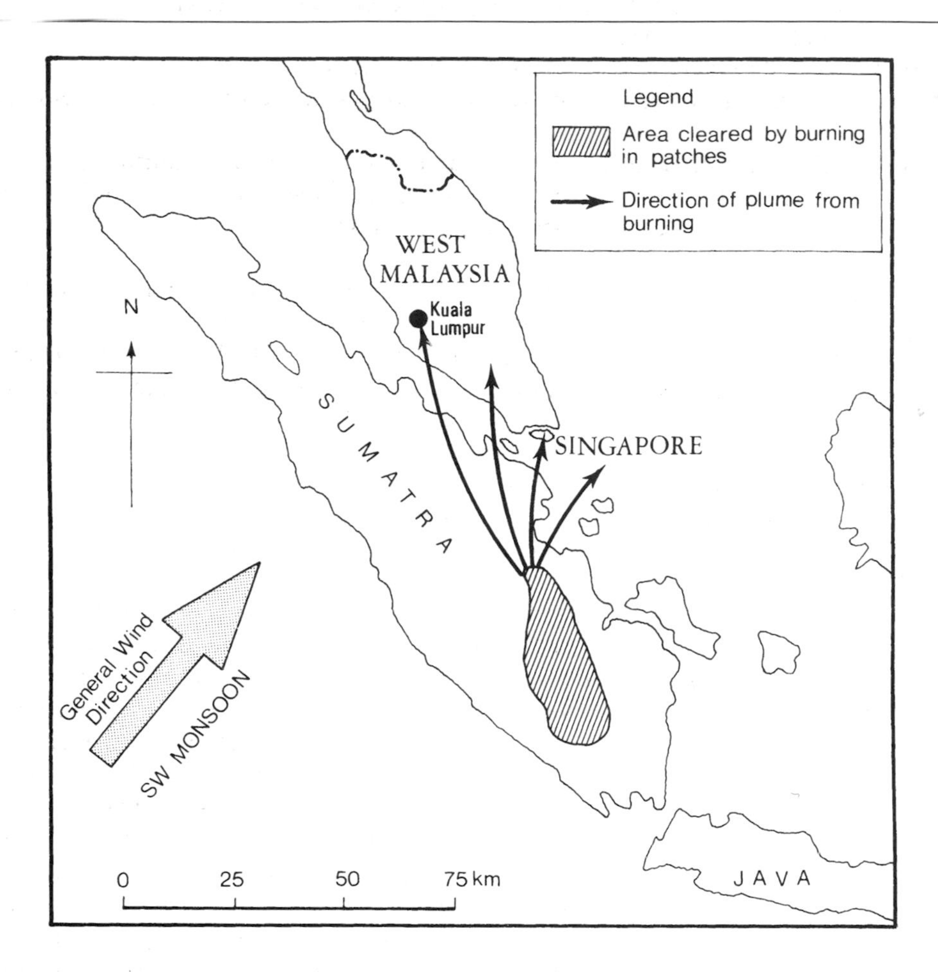

Figure 17.1 *A generalized diagram showing the source and path of pollution from smoke plumes, Sumatra to Kuala Lumpur and Singapore*

This book has described some of the general successes, both in environmental knowledge and in environmental management. Important achievements in these areas include the following:

- a general awareness of environmental issues across the world;
- the development of environmental science and technology; new findings ranging from stratospheric ozone depletion and global warming to problem solving at local levels such as the use of biodigesters or social forestry;
- general acceptance of the interrelationship between environment and development;

- increased environmental awareness built into projects financed by multilateral funding institutions such as the World Bank;
- acceptance of environmental impact assessment for projects;
- the development of agencies of governance exclusively for environmental management at various levels;
- partial acceptance of the greening of national accounts;
- partial provision for funding of environment-related projects in the developing world;
- the development of environment-related non-governmental organizations;
- involvement of industry and business in environmental management, e.g. the World Business Council for Sustainable Development.

The last development is a recent one but it has far reaching capabilities in terms of pollution control and resource depletion. The World Business Council for Sustainable Development (WBCSD) (Schmidheiny, 1992), which currently includes more than 600 business leaders, emphasises the following practice along with four partner organizations:

- projects related to sustainable development
- training programmes which use less resources
- policy framework reforms, in collaboration with governments

The Council realizes that the industry of the twenty-first century will be very different, with eco-efficiency (involving less pollution and low resource depletion) as normal practice. Such an approach is becoming more and more common.

A recent publication of the National Academy of Engineering of the United States (Schulze, 1996) discusses the shifting of the expectations that engineers have to fulfil with cultural evolution of the societies involved. The growing concern about the environment has brought a new constraint to engineers which the book refers to as ecological constraint, and which needs to be added to others such as budgetary constraint and design constraint. Ecological constraint requires closer co-operation between the engineers and the ecologists. The word 'ecology' is apparently used in a broader context, involving all environmental topics. Acceptance of this viewpoint, however, raises operational difficulties regarding the nature of ecological constraints, their relative importance, the uncertainty of environmental consequences of engineering projects, and the difficulty of finding proper solutions (Schulze, Frosch and Risser, 1996). Even the term 'ecological engineering' has come into use. This indicates project design and implementation with as little alteration as possible to the ambient conditions. For example, the building of a dam should involve procedures that cause little disturbance to the physical conditions of the river, so that the river may continue to be the same type of ecological habitat as it was before the dam was built. This is a far cry from the project impacts discussed in Chapter 7. To provide another example, oil exploration in an Ecuadorian rain forest has been carried out with as little disturbance as possible (Lindstedt-Siva *et al.*, 1996). This involved

- building no new roads; instead using existing roads, helicopters, and approaching on foot
- minimizing the scale of disturbance, including vegetation removal and use of heavy equipment
- consultation with local communities
- restoration of disturbed sites

Such environmental consciousness among the engineers and business leaders is not as widespread as it should be, but a change is certainly taking place. This is diametrically different from the past confrontations between project designers and ecologists and local inhabitants, or the acrimonious discussions that took place in the early days of global warming. What is perhaps needed most is a change in the mindset of the governmental bureaucracy and business leaders of the developing world so that the environment is given much more importance officially than it enjoys at present (Box 17.1), and so that the elite in particular feel responsible for environmental sustenance. A similar change in mindset is required of the consumers. They should be encouraged to take into account environmental consequences of their consumptive decisions. This is necessary as businesses do respond to consumer demands.

One interesting development has been the strong role played by the Indian judiciary in recent years. On a number of occasions industries and municipalities have been ordered to control emissions and install treatment plants. The active NGOs and news media regularly publicize the cost of environmental degradation. The pressure from certain importers from the developed countries and the local availability of pollution-control equipment and services have helped. Although incidences of water and air pollution in India are still very high, such a state of environment is no longer tolerated as normal.

To the United Nations, 1997 is the year of evaluation regarding the procedures for better environmental management accepted at the Rio Conference five years ago. The evaluation was attempted at a meeting at the United Nations in New York in June 1997 but the result was disappointing. In comparison with Rio, this meeting (called Rio Plus Five) was a gloomy muted affair with very little commitment from the United States. Some progress was reported, mainly in the areas of population control and CFC control, but not on most major issues such as deforestation or the emission of greenhouse gases. Very little was accomplished regarding funds or technology transfer to developing countries. Very little evidence of political will to halt global environmental degradation was displayed. Emission of greenhouse gases was discussed in December 1997 at an international meeting (COP 3) in Kyoto and a new climate treaty is scheduled for signing. Before Kyoto, the European Union proposed significant cuts in greenhouse gas emissions, limiting the developed countries to 92.5 percent and 85 percent of their 1990 level by the years 2005 and 2010 respectively. However, in a speech in October 1997, President Clinton proposed delaying the cuts, phasing down the US emission to the 1990 level only by 2012. Japan and Australia also voiced their unwillingness to reduce greenhouse gas emissions on a crisis footing. Such decisions have undoubtedly weakened the agreement at Rio and provided very little cause for optimism as the greenhouse gas emissions of a number of countries are on the increase.

Prolonged negotiations took place in Kyoto at various levels. The United States was not willing to accept the level of reduction of greenhouse gases proposed by the European Union. The developing countries saw global warming as related to emissions from the developed countries over years and expected bigger efforts from them. Australia was concerned about the future of its coal export. It was suggested that as different countries started controlling their emissions at different times and from different levels, this should also be taken into consideration in determining the reduction rate. Strong anti-control lobbying also took place from organizations such as the energy-related industry.

The agreement, which was reached after 11 days of discussion, called on the attending 159 countries to reduce their emission of greenhouse gases by 5.2 percent of the 1990 level by the year 2012. However, countries were given individual targets and a few nations were even allowed to increase their emission levels. The reduction targets for the European Union, the United States and Japan were 8, 7 and 6 percent of their 1990 figures, respectively. In all, reductions for 38 developed countries were considered mandatory, whereas for the developing countries controls were voluntary. The agreement in Kyoto now has to be ratified by individual countries for it to be effective.

17.4 The Future Trends

The environmental scene in the developing world is progressively changing. The Physical Quality of Life Index (PQLI) of M. D. Morris is calculated from life expectancy at one year of age, rates of literacy, and infant mortality. Over the years, the PQLI indicates considerable improvements in the quality of life in the developing countries, in spite of a very large increase in population (Doyle, 1996).

In this context, it is worthwhile examining the future prospects. However, as past events have shown, estimates beyond a limited number of years can be difficult and erroneous because of new environmental problems and new technical resources.

Based on current conditions and ongoing trends, it seems very likely that in the immediate future the developing countries will need to concentrate on improving the environmental conditions in certain areas. Such areas include the following:

- urban environments, especially large cities
- coastal environments, especially deltas, mangroves and coral reefs
- deforested hillslopes
- large-scale river projects
- water to meet various demands
- problems of pollution of various types

It is difficult to add to this list with confidence. The uncertainty regarding the rate of global warming and its effects also adds to this difficulty. Areas which might also turn out to be vulnerable include arid and semi-arid regions, tropical rain forests, currently degraded lands, and, following global warming in the next century, small and low islands.

At present, however, the countries of the developing world need to improve their knowledge of the local environment; national, technical, legal and managerial resources; and procedures for managing national problems against the background concept of sustainable development. Most developing countries already have some resource base but there is a wide range in ability and awareness. The balancing of economic development with environmental management could bring growth, equity and sustainable resources to the developing countries. Their success of course depends to a large extent on global co-operation, such as the transfer of technologies and resources. Such a transfer, however, will only reach optimality if there is a desire in the developing countries themselves to use such technologies, and also to be able to modify them to suit local conditions better.

There should, however, be a healthy balance between borrowing techniques and finding new ones. The role of innovation, in the resource-starved developing world, is extremely important. Certain examples have been discussed in the book, e.g. the urban planning of Curitiba and the use of wetlands in Calcutta for the processing of wastewater. The ability to innovate depends to a great extent on local knowledge and skills. As discussed at the beginning of this book, an environmental manager should not only have a multidisciplinary background covering earth sciences, ecology, economics and engineering, but also needs to be a specialist in any of these or other relevant disciplines. Environment and its proper management are important for all countries, irrespective of national incomes or lifestyles. We still have a long road to travel. The World Bank, taking stock five years after the Rio Earth Summit, concluded that in spite of some remarkable successes, overall progress has fallen short of expectation (World Bank, 1997). This involved both the physical environment and the quality of life of about three billion people who still survive on less than $2 per day.

Exercise: Shrimp and Paddy in a Tropical Delta

This exercise requires role playing as a member of a team. You have to take the role of one of the environmental actors listed below. The objective of this participatory exercise is to provide you with an understanding of how people perceive environmental issues from their own viewpoints and how decisions are taken on such issues. Your participation should lead to sensible decisions being taken. The environmental problem discussed is a common problem for tropical deltas.

Traditionally rice has been grown on the edge of a tropical delta. It is transplanted rice and takes several months to grow. The young rice plants are transferred to the fields immediately after the arrival of the wet monsoon. Over the last few years, some of the local people, especially those who have land with access to water, have started to build ponds for fish and shrimps. In this area, embankments constructed and maintained by the government protect the land from flooding by rivers during the rainy season. People with access to water have built smaller embankments to create ponds where saline water is introduced by small channels which are then blocked. Fish and shrimps are grown in these ponds during the non-paddy season which is a dry period. Before paddy is transplanted, the ponds are drained and flushed with rainwater. The farmers thus supplement their income from paddy by harvesting fish and shrimps at the end of the dry period.

The cultivation of shrimps has become very profitable, and the export of the headless shrimps to developed countries is bringing in a considerable amount of foreign exchange into this developing country. This has led to a proliferation of ponds, the destruction of riverbank vegetation, and an extension of the shrimp-growing period. Paddy has taken a back seat. Waterfront land has become prime property and powerful landowners are trying to evict small farmers. A type of absentee lease-holder has sprung up. They rent the land from the farmers for shrimp cultivation but do not live in the countryside. They protect their investment by hiring thugs and keeping the local officials happy.

The national government is seeking financial and technical support from international banks and organizations to build shrimp hatcheries, infrastructure, export facilities, and institutions for offering soft-option loans to the local farmers. However, the first attempt resulted in most benefits going to the big landowners and the absentee lease-holders who continue to pay a fixed rent in spite of rising profits. Also, a general tendency to grow more shrimps and neglect paddy has developed. The international donors and the national government want to revise the project so that most of the benefits go to the local farming population and the degradation of the environment is prevented. A meeting is called to identify the needs and interests of various groups so that the project design and implementation can be improved. The problem lies in the conflict between various interest groups and, to a lesser extent, among the various government organizations. There are two donor agencies, and one of the two is more interested in the financial viability of the project than in everything else.

Environmental actors attending the meeting

- representatives of the following ministries:
  - finance
  - agriculture
  - fisheries and livestock
  - irrigation and flood control
  - environment
  - forest
- representatives from the local offices of the two international donor agencies
- the local NGO supporting the small farmers
- a representative of the large landowners
- a lawyer for the absentee lease-holders

Evaluate the role of each environmental actor. Play the role of the actor given to you. You are free to invent situations, interests, etc. Certain roles are environmentally friendly, others are not. See where the discussion leads you. This game is usually a sobering experience.

References

Agarwal, A. and Chak, A. (Eds) (1991) *Floods, Flood Plains and Environmental Myths: State of India's Environment, A Citizens' Report 3*, Centre for Science and Environment, New Delhi.

Agarwal, A, and Narain, S. (Eds) (1986) *The State of India's Environment 1984–85: The Second Citizens' Report*, Centre for Science and Development, New Delhi.

Ahmad, Y.J. and Sammy, G.K. (1985) *Guidelines to Environmental Impact Assessment*, Hodder and Stoughton, London.

Aitken, A.P. (1981) Aspects of erosion and sediment transport in Java, Indonesia, in *Proceedings, The South-East Asian Regional Symposium on Problems of Soil Erosion and Sedimentation* (T. Tingsanchali and H. Eggers, Eds), Asian Institute of Technology, Bangkok, 81–91.

Allen, B.J. (1993) The problems of upland land management, in *South-East Asia's Environmental Future: The Search for Sustainability* (H. Brookfield and Y. Byron, Eds), United Nations University Press, Tokyo, 225–237.

American Society of Planning Officials (1968) *Trees in the City*, Planning Advisory Service Report 236, Chicago.

Anderson, J.M. and Spencer, T. (1991) *Carbon, Nutrient and Water Balances of Tropical Rain Forest Ecosystems Subject to Disturbance: Management Implications and Research Proposals*, MAB Digest 7, UNESCO, Paris.

Anderson, K. (1996) Environmental standards and international trade, in *Annual World Bank Conference on Development Economics* (M. Bruno and B. Pleskovic, Eds), The World Bank, Washington, DC, 317–342.

Andrews, R.N.L. (1988) Environmental impact assessment and risk assessment: learning from each other, in *Environmental Impact Assessment: Theory and Practice* (Wathern, P., Ed.), Routledge, London, 85–97.

Anon. (1994) *United Nations Handbook 1994*, New Zealand Ministry of Foreign Affairs and Trade, Wellington.

Anon. (1995) Ozone recovery? *Earth in Space*, **8**(2), 3.

Bahuguna, V.K., Luthra, V. and Rathor, B.M.S. (1994) Collective forest managament in India, *Ambio*, **23**, 269–273.

Bandyopadhyay, J. (1995) From environmental conflicts to sustainable mountain transformation: ecological action in the Garhwal Himalaya, in *Grassroots Environmental Action: People's Participation in Sustainable Development* (D. Ghai and J.M. Vivian, Eds), Routledge, London, 259–278.

Barber, C.V., Johnson, N.C. and Hafild, E. (1994) *Breaking the Logjam: Obstacles to Forest Policy Reform in Indonesia and the United States*, World Resources Institute, Washington, DC.

Barron, E.J. (1995) Researchers Assess Projections of Climate Change, *Earth in Space*, **8**(1), 4–5.

Berner, E.K. and Berner, R.A. (1996) *Global Environment: Water, Air and Geochemical Cycles*, Prentice Hall, Upper Saddle River, NJ.

Biswas, A.K. (1978) Water development and environment, in *Water Pollution Control in Developing Countries, Proceedings of the International Conference, Bangkok, 21–25 February* (B.N. Lohani and N.C. Thanh, Eds), Vol. 2, Asian Institute of Technology, Bangkok.

Biswas, A.K. and Agarwala, S.B.C. (1992) *Environmental Impact Assessment for Developing Countries*, Butterworth-Heinemann, Oxford.

Biswas, A.B. and Saha, A.K. (1985) Environmental hazards of the recession of piezometric surface of groundwater under Calcutta, *Proceedings, Indian National Science Academy*, **51A**, 610–621.

Björk, M., Mohammed, S.M., Björklund, M. and Semesi, A. (1995) Coralline algae, important coral-reef builders threatened by pollution, *Ambio*, **24**, 502–505.

Blackwell, J.N., Goodwillie, R.N. and Webb, R. (1991) *Environment and Development in Africa: Selected Case Studies*, The World Bank, Washington, DC.

Bloem, A.M. and Weisman, E. (1996) National accounts and the environment, in *Macroeconomics and the Environment* (V.P. Gandhi, Ed.), International Monetary Fund, Washington, DC, 136–149.

Brown, S. and Lugo, A.E. (1990) Tropical secondary forests, *Journal of Tropical Ecology*, **6**, 1–32.

Bruijnzeel, L.A. with C.N. Bremer (1989) *Highland–Lowland Interactions in the Ganges Brahmaputra River Basin: A Review of Published Literature*, ICIMOD Occasional Paper 11, International Centre for Integrated Mountain Development, Kathmandu.

Brundtland, G.H. (1996) Our Common Future revisited, *The Brown Journal of World Affairs*, **3**(2), 173–175.

Bryson, R.A. and Ross, J.E. (1972) The climate of the city, in *Urbanization and Environment: The Physical Geography of the City* (T.R. Detwyler and M.G. Marcus, Eds), Duxbury Press, Belmont, 51–68.

Buckholz, R.A. (1993) *Principles of Environmental Management: the Greening of Business*, Prentice-Hall, Englewood Cliffs.

Buckley, C.B. (1902) *An Anecdotal History of Old Times in Singapore, 1819–1867,* Fraser and Neave, Singapore.

Canter, L.W. (1996) *Environmental Impact Assessment*, McGraw-Hill, New York.

Carlowicz, M. (1995) Global warming picture remains clouded, *Eos, Transactions, American Geophysical Union*, **76**(45), 452.

Carson, R. (1962) *Silent Spring*, Houghton Mifflin, Boston.

Chakraborti, D., Van Vaeck, L. and Van Espen, P. (1988) Calcutta pollutants: part II. Polynuclear aromatic hydrocarbon and some metal concentration on air particulates during winter 1984, *International Journal of Environmental Analytical Chemistry*, **32**, 109–120.

Chakraborti, D., Das, D., Chatterjee, A., Zhao, J. and Jiang, S.G. (1992) Direct determination of some heavy metals in urban air particulates by electrothermal atomic absorption spectrometry using Zeeman background correction after simple acid decomposition. Part IV, application to Calcutta air particulates, *Environmental Technology*, **13**, 95–100.

Chansang, H. (1988) Coastal tin mining and marine pollution in Thailand, *Ambio*, **17**, 223–228.

Chatterjee, P. and Finger, M. (1994) *The Earth Brokers*, Routledge, London.

Chau Kwai-cheong (1995) The Three Gorges project of China: resettlement prospects and problems, *Ambio*, **24**, 98–102.

Christiansson, C. (1988) Degradation and rehabilitation of agropastoral land – perspectives on environmental change in semiarid Tanzania, *Ambio*, **17**, 144–152.

Clark, R.B. (1992) *Marine Pollution*, Clarendon Press, Oxford.

Cleaver, K., Munasinghe, M., Dyson, M., Egli, N., Peuker, A. and Wencélius, F. (Eds) (1992) *Conservation of West and Central African Rainforests*, World Bank Environment Paper No. 1, Washington, DC.

Coase, R.H. (1960) The problem of social cost, *Journal of Law and Economics*, **3**, 1–44.

Colchester, M. (1993) Colonizing the rainforests: the agents and causes of deforestation, in *The Struggle for Land and the Fate of the Forests* (M. Colchester and L. Lohmann, Eds), The World Rainforest Movement, Penang, 1–15.

Colchester, M. and Lohmann, L. (1990) *The Tropical Forestry Action Plan: What Progress?* World Rainforest Movement, Penang.

Collins, N.M., Sayer, J.A. and Whitmore, T.C. (1991) *The Conservation Atlas of Tropical Forests: Asia and the Pacific*, Macmillan, London.

Coltrinari, L. (1996) Natural and anthropogenic interactions in the Brazilian tropics, in *Geoindicators: Assessing Rapid Environmental Changes in Earth Systems* (A.R. Berger and W.J. Iams, Eds), A. A. Balkema, Rotterdam, 295–310.

Common, M. (1995) *Sustainability and Policy*, Cambridge University Press, Cambridge.

Commoner, B. (1971) *The Closing Circle: Nature, Man and Technology*, Knopf, New York.

Corlett, R. (1992) The changing urban vegetation, in *Physical Adjustments in a Changing Landscape: The Singapore Story* (A. Gupta and J. Pitts, Eds), Singapore University Press, Singapore, 190–214.

Crosby, A. (1986) *Ecological Imperialism: The Biological Expansion of Europe*, Cambridge University Press, Cambridge.

Darby, H.C. (1956) The clearing of the woodland in Europe, in *Man's Role in Changing the Face of the Earth* (W.L. Thomas, Ed.), University of Chicago Press, Chicago, 183–216.

Das, D., Chatterjee, A., Samanta, G. and Chakraborti, D. (1992) Preliminary estimation of tetraalkyllead compounds (TAL) in Calcutta city air, *Chemical Environmental Research*, **1**, 279–287.

Dasgupta, P. (1995) The population problem: theory and evidence, *Journal of Economic Literature*, **33**, 1879–1902.

DasGupta, S.P. (Ed.) (1984) *The Ganga Basin*, Part II, Central Board for the Prevention and Control of Water Pollution, India, New Delhi.

Detwyler, T.R. (1972) Vegetation of the city, in *Urbanization and Environment: The Physical Geography of the City* (T.R. Detwyler and M.G. Marcus, Eds), Duxbury Press, Belmont, 229–259.

Dixon, J.A., Talbot, L.M. and Le Moigne, G.J.-M. (1989) *Dams and the Environment: Considerations in World Bank Projects*, World Bank Technical Paper 110, The World Bank, Washington, DC.

D'Monte, D. (1985) *Temples or Tombs? Industry Versus Environment, Three Controversies*, Centre for Science and Environment, New Delhi.

Doolittle, J.B. and Magrath, W.B. (Eds) (1990) *Watershed Development in Asia: strategies and technologies*, World Bank Technological Paper 127, Washington, DC.

Douglas, I. (1978) The impact of urbanisation on fluvial geomorphology in the humid tropics, *Geo-Eco-Trop*, **2**, 229–242.

Doyle, R. (1996) The changing quality of life, *Scientific American*, **275** (1), 18.

Dunne, T. and Leopold, L.B. (1978) *Water in Environmental Planning*, W.H. Freeman, San Francisco.

Durning, A.T. (1993) *Saving the Forests: What Will it Take?* World Watch Paper 117, Worldwatch Institute, Washington, DC.

Eckholm, E.P. (1976) *Losing Ground: Environmental Stress and World Food Prospects*, W.W. Norton, New York

Ehrlich, P.R. (1968) *The Population Bomb*, Ballantine Books, New York.

Ehrlich, P.R. and Ehrlich, A.H. (1970) *Population, Resources, Environment*, W.H. Freeman, San Francisco.

ESCAP Secretariat (1988) Geological information for planning in Bangkok, Thailand, in *Geology and Urban Development: Atlas of Urban Geology* (UN-ESCAP, Ed.), Vol. 1, United Nations, Economic and Social Commission for Asia and the Pacific, Bangkok, 24–60.

Faeth, P. (Ed.) (1993) *Agricultural Policy and Sustainability: Case Studies from India, Chile, the Philippines and the United States*, World Resources Institute, Washington, DC.

Fair, G., Geyer, J. and Okun, D. (1968) *Water and Wastewater Engineering*, John Wiley, New York.
Falkenmark, M. (1986) Fresh water – time for a modified approach, *Ambio*, **15**, 192–200.
FAO (1993) *Forest Resources Assessment, 1990: Tropical Countries*, FAO Forestry Paper 112, Food and Agricultural Organization of the United Nations, Rome.
Fearnside, P.M. (1986) Spatial concentration of deforestation in the Brazilian Amazon, *Ambio*, **15**, 74–81.
Frankel, E. (1995) *Ocean Environmental Management: A Primer on the Role of the Oceans and How to Maintain Their Contributions to Life on Earth*, Prentice Hall, Englewood Cliffs.
Gadgil, M. and Guha, R. (1992) *This Fissured Land: An Ecological History of India*, Oxford University Press, Oxford.
Gandhi, V.P. (1996) Macroeconomics and the environment: summary and conclusions, in *Macroeconomics and the Environment* (V.P. Gandhi, Ed.), International Monetary Fund, Washington, DC, 1–21.
Geertz, C. (1963) *Agricultural Involution: The Processes of Ecological Changes in Indonesia*, University of California Press, Berkeley.
GESAMP (1990) *The State of the Marine Environment*, Blackwell Scientific, Oxford.
Ghai, D. and Vivian, J.M. (1995) *Grassroots Environmental Action: People's Participation in Sustainable Development*, Routledge, London.
Ghassemi, F., Jakeman, A.J. and Nix, H.A. (1995) *Salinisation of Land and Water Resources: Human Causes, Extent, Management and Case Studies*, Centre for Resources and Environmental Studies, Australian National University, Canberra.
Ghosh, D. (1995) Integrated wetlands system (IWS) for wastewater treatment and recycling for the poorer parts of the world with ample sunshine, Basic Manual, Calcutta, cyclostyled.
Gilpin. A., (1995) *Environmental Impact Assessment (EIA): Cutting Edge for the Twenty-First Century*, Cambridge University Press, Cambridge.
Goldsmith, E. and Hildyard, N. (1984) *The Social and Environmental Effects of Large Dams*, Sierra Club Books, San Francisco.
Goodstein, E.S. (1995) *Economics and the Environment*, Prentice-Hall, Englewood Cliffs.
Gore, A. (1992) *Earth in the Balance: Ecology and the Human Spirit*, Houghton Mifflin, Boston.
Graedel, T.E. and Allenby, B.R. (1995) *Industrial Ecology*, Prentice-Hall, Englewood Cliffs.
Gramlich, E.M. (1990) *A Guide to Benefit–Cost Analysis*, Prentice-Hall, Englewood Cliffs.
Griffiths, J.F. (1972) Eastern Africa, in *Climates of Africa: World Survey of Climatology* (J.F. Griffiths, Ed.), Vol. 10, Elsevier, Amsterdam, 313–347.
Groombridge, B. (1992) *Global Biodiversity: Status of the Earth's Living Resources, A Report Compiled by the World Conservation Monitoring Centre*, Chapman and Hall, London.
Grossman, G.M. and Kruegar, A.B. (1995) Economic growth and the environment, *Quarterly Journal of Economics*, May 1995, 353–377.
Grubb, M., Koch, M., Thomson, K., Munson, A. and Sullivan, F. (1993) *The 'Earth Summit' Agreements: A Guide and Assessment*, Earthscan, London.
Gugler, J. (Ed.) (1996) *The Urban Transformation of the Developing World*, Oxford University Press, Oxford.
Guha, R. (1989) *The Unquiet Woods: Ecological Change and Peasant Resistance in the Himalayas*, Oxford University Press, New Delhi.
Guha Mazumder, D.N., Das Gupta, J., Chakraborty, A.K., Chatterjee, A., Das, D. and Chakraborti, D. (1992) Environmental pollution and chronic arsenicosis in South Calcutta, *Bulletin, World Health Organization*, **70**, 481–485.
Gupta, A. (1992) Floods and sediment production in Singapore, in *Physical Adjustments in a Changing Landscape: The Singapore Story* (A. Gupta and J. Pitts, Eds), Singapore University Press, Singapore, 301–326.
Gupta, A. (1996) Erosion and sediment yield in Southeast Asia: a regional perspective, in *Erosion and Sediment Yield: Global and Regional Perspectives* (D.E. Walling and B.W. Webb, Eds), International Association of Hydrological Sciences Publication No. 236, Wallingford, UK, 215–222.

Gupta, A. (1998) *Ecology and Development in the Third World*, Routledge, London.

Gupta, A. and Ahmad, R. (1998) Geomorphology and the urban tropics, in *Engineering Geomorphology: Changing the Face of the Earth* (J.R. Giardino, R.A. Marston and M. Morisawa, Eds), Elsevier, Amsterdam, in press.

Gupta, A. and Krishnan, P. (1994) Spatial distribution of sediment discharge to the coastal waters of South and Southeast Asia, in *Variability in Stream Erosion and Sediment Transport* (L.J. Olive, R.J. Loughran and J.A. Kesby, Eds), International Association of Hydrological Sciences Publication No. 224, Wallingford, UK, 457–463.

Guzmán, H.M. and Jarvis, K.E. (1996) Vanadium century record from Caribbean reef corals: a tracer of oil pollution in Panama, *Ambio*, **25**, 523–526.

Haas, P. (1996) Is 'sustainable development' politically sustainable? *The Brown Journal of World Affairs*, **3**(2), 239–247.

Hamill, P. and Toon, O.B. (1991) Polar stratospheric clouds and the ozone hole, *Physics Today*, **44**(12), 34–42.

Hammerton, D. (1972) The Nile River – a case history, in *River Ecology and Man* (R.T. Oglesby, C.A. Carlson and J.A. McCann, Eds), Academic Press, New York.

Hansen, J. and Lebedeff, S. (1987) Global trends of measured surface air temperature, *Journal of Geophysical Research*, **92**, 13 345–13 372.

Hansen, J., Fung, I., Lacis, A., Rind, D., Lebedeff, S., Ruedy, R. and Russell, G. (1988) Global climate changes as forecast by Goddard Institute for Space Studies three-dimensional model, *Journal of Geophysical Research*, **93**(D8), 9341–9364.

Haq, M. Ul., Kaul, I. and Grunberg, I. (Eds) (1996) *The Tobin Tax: Coping with financial volatility*, Oxford University Press, Oxford.

Hardin, G. (1968) The Tragedy of the Commons, *Science*, **162**, 1243–1248.

Harvey, D. (1993) The nature of environment: the dialectics of social and environmental change, *The Socialist Register*, 1–51.

Hastenrath, S. and Kruss, P. (1992) The dramatic retreat of Mount Kenya's glaciers between 1963 and 1987: greenhouse forcing, *Annals of Glaciology*, **16**, 127–133.

Hecht, S. (1993) Land speculation and pasture-led deforestation in Brazil, in *The Struggle for Land and the Fate of the Forests* (M. Colchester and L. Lohmann, Eds), The World Rainforest Movement, Penang, 164–178.

Hellden, U. (1991) Desertification – time for an assessment? *Ambio*, **20**, 372–383.

Helm, D. and Pearce, D. (1990) The assessment: economic policy towards the environment, *Oxford Review of Economic Policy*, **6**(1), 1–16.

Henderson-Sellers, A. (1994) Numerical modelling of global climates, in *The Changing Global Environment* (N. Roberts, Ed.), Blackwell, Cambridge, MA, 99–124.

Holdgate, M. (1996) The ecological significance of biological diversity, *Ambio*, **25**, 409–416.

Holdridge, L.R. (1967) *Life Zone Ecology*, Tropical Science Center, San José, Costa Rica.

Hsu, M. (1996) Introduction, *The Brown Journal of World Affairs*, **3**(2), 167–172.

Htun, N. (1988) The EIA process in Asia and the Pacific region, in *Environmental Impact Assessment: Theory and Practice* (Wathern, P., Ed.), Routledge, London, 225–238.

Hudson, N. (1971) *Soil Conservation*, Cornell University Press, Ithaca.

Ibrahim, F.N. (1988) Causes of the famine among the rural population of the Sahelian zone of the Sudan, *GeoJournal*, **17**, 133–141.

International Energy Agency (1993) *Cars and Climate Change*, OECD, Paris.

International Monetary Fund (1996) *World Economic Outlook*, IMF, Washington, DC, May 1996.

Ives, J.D. and Messerli, B. (1989) *The Himalayan Dilemma: Reconciling Development and Conservation*, Routledge, London.

Jackson, P. (1983) The tragedy of our tropical rainforests, *Ambio*, **12**, 252–254.

Jenkins, G. and Lamech, R. (1994) *Green Taxes and Incentive Policies*, ICS Press, San Francisco.

Johansson, P.O. (1990) Valuing environmental damage, *Oxford Review of Economic Policy*, **6**(1), 34–50.
Johnson, N. and Cabarle, B. (1993) *Surviving the Cut: Natural Forest Management in the Humid Tropics*, World Resources Institute, Washington, DC.
Jones, P.D. and Wrigley, T.M.L. (1990) Global warming trends, *Scientific American*, **263**(2), 66–73.
Kardell, L., Steen, E. and Fabião, A. (1986) Eucalyptus in Portugal – a threat or a promise, *Ambio*, **15**, 6–13.
Kendall, H.W. and Pimentel, D. (1994) Constraints on the expansion of the global food supply, *Ambio*, **23**, 198–205.
Khantaprab, C. and Boonop, N. (1988) Urban geology of Bangkok metropolis: a preliminary assessment, in *Geology and Urban Development: Atlas of Urban Geology*, (UN-ESCAP, Ed.) Vol. 1, UN-ESCAP, Bangkok, 107–135.
Khor, M. and Chee, Y.L. (1992) The Global Environment Facility: democratisation and transparency principles, *Earth Summit Briefings*, Third World Network, Penang, 44–45.
Kishk, M.A. (1986) Land degradation in the Nile Valley, *Ambio*, **15**, 226–230.
Kühlmann, D.H.H. (1988) The sensitivity of coral reefs to environmental pollution, *Ambio*, **17**, 13–21.
Landsberg, H.E. (1956) The climate of towns, in *Man's Role in Changing the Face of the Earth* (W.L. Thomas, Jr., Ed.), University of Chicago Press, Chicago, 584–600.
Landsberg, H. (1981) *The Urban Climate*, Academic Press, New York.
Lavigne, F. and Thouret, J.-C. (1995) Rain triggered lahars in Boyong river after the 22 November 1994 eruption of Merapi volcano (Central Java, Indonesia), paper presented at the International Association of Geomorphologists Southeast Asia Conference, Singapore, June 1995.
Laxmi, V., Parikh, J. and Parikh, K.S. (1997) Environment, in *India Development Report 1997* (K.S. Parikh, Ed.), Oxford University Press, New Delhi, 95–106.
Lean, J. and Warrilow, D.A. (1989) Simulation of the regional climatic impact of Amazon deforestation, *Nature*, **342**, 411–413.
Leopold, A. (1966) *A Sand County Almanac*, Ballantine Books, New York.
Leopold, L.B. (1968) Hydrology for urban land planning, *US Geological Survey Circular 554*, Washington, DC.
Leopold, L.B. (1969) Quantitative comparison of some esthetic factors among rivers, *US Geological Survey Circular 620*.
Leopold, L.B. and Marchand, M.O. (1968) On the quantitative inventory of the riverscape, *Water Resources Research*, **4**, 709–717.
Leopold, L.B., Clark, F.E., Hanshaw, B.B. and Balsey, J.R. (1971) A procedure for evaluating environmental impact, *US Geological Survey Circular 645*.
Levi, B.G. (1988) Ozone depletion at the poles: the hole story emerges, *Physics Today*, **41**(7), 17–21.
Lin, D.Y. (1956) China, in *A World Geography of Forest Resources* (S. Haden-Guest, J.K. Wright and E.M. Teclaff, Eds), American Geographical Society and Ronald Press, New York, 529–550.
Lindstedt-Siva, J., Soileau, L.C., IV, Chamberlain, D.W. and Wouch, M.L. (1996) Engineering for development in environmentally sensitive areas: oil explorations in a rain forest, in *Engineering within Ecological Constraints* (P.C. Schulze, Ed.), National Academy of Engineering, National Academy Press, Washington, DC, 141–161.
Ludwig, D., Hilborn, R. and Walters, C. (1993) Uncertainty, resource exploitation, and conservation: lessons from history, *Science*, **260**, 17 and 36.
Lundin, C.G. and Lindén, O. (1993) Coastal ecosystems: attempts to manage a threatened resource, *Ambio*, **22**, 468–473.
Lutz, W. (1996) Epilogue: dilemmas in population stabilization, in *The Future Population of the World* (W. Lutz, Ed.), Earthscan, London, 429–435.
Malin, J.C. (1948) *The Grassland of North America: Prolegomena to Its History*, James C. Malin, Lawrence, Kansas.

Markandya, A. (1997) Economic instruments: accelerating the move from concepts to practical application, in *Finance for Sustainable Development: The Road Ahead,* United Nations, New York, 221–252.

Marlow, M.L. (1995) *Public Finance*, Harcourt Brace, Orlando, FL.

Marsh, J.P. (1898) *The Earth as Modified by Human Action. A Last Revision of 'Man and Nature'*, Scribner, New York.

Martin, P.S. and Klein, R.G. (Eds) (1984) *Quaternary Extinctions: A Prehistoric Evolution*, University of Arizona Press, Tucson.

Masters, G.M. (1991) *Introduction to Environmental Engineering and Science*, Prentice-Hall, Englewood Cliffs.

Mazurski, K.R. (1990) Industrial pollution: the threat to Polish forests, *Ambio*, **19**, 70–74.

McCauley, D.S. (1991) Watershed management in Indonesia: the case of Java's densely populated upper watersheds, in *Watershed Resources Management: Studies from Asia and the Pacific* (K.W. Easter, J.A. Dixon and M.M. Hufschmidt, Eds), Institute of Southeast Asian Studies, Singapore, 177–190.

McCormick, J. (1989) *The Global Environmental Movement: Reclaiming Paradise*, Belhaven Press, London.

McGee, T.G. (1971) *The Urbanization Process in the Third World: Explanation in Search of a Theory*, Bell, London.

McManus, J.W. (1988) Coral reefs of the ASEAN region: status and management, *Ambio*, **17**, 189–193.

Meadows, D.H., Meadows, D.L., Randers, J. and Behrens, W.W. III (1972) *The Limits to Growth: A Report for the Club of Rome's Project on the Predicament of Mankind*, Signet, New York.

Mee, L.D. (1992) The Black Sea in crisis: a need for concerted international action, *Ambio*, **21**, 278–286.

Mikesell, R.F. and Williams, L. (1992) *International Banks and the Environment: From Growth to Sustainability – An Unfinished Agenda*, Sierra Club, San Francisco.

Mikhalevsky, P.N., Baggeroer, A.B., Gavrilov, A. and Slavinsky, M. (1995) Experiment tests use of acoustics to monitor temperature and ice in Arctic Ocean, *Eos, Transactions, American Geophysical Union*, **76**, 265, 268–269.

Miller, A.S. and Mintzer, I.M. (1986) *The Sky is the Limit: Strategies for Protecting the Ozone Layer*, Research Report 3, World Resources Institute, Washington, DC.

Miller, G.T., Jr. (1993) *Environmental Science: Sustaining the Earth*, Wadsworth, Belmont.

Mitchell, J.F.B. (1989) The 'greenhouse' effect and climate change, *Reviews of Geophysics*, **27**(1), 115–139.

Moffat, D. and Lindén, O. (1995) Perception and reality: assessing priorities for sustainable development in the Niger River Delta, *Ambio*, **24**, 527–538.

Mohan, R. (1996) Urbanization in India: patterns and emerging policy issues, in *The Urban Transformation of the Developing World* (J. Gugler, Ed.), Oxford University Press, Oxford, 93–131.

Molina, M.J. and Rowland, F.S. (1974) Stratospheric sink for chlorofluoro-methanes: chlorine atom catalysed destruction of ozone, *Nature*, **249**, 810–812.

Munasinghe, M. (1992) *Water Supply and Environmental Management: Developing World Applications*, Westview Press, Boulder, CO.

Munasinghe, M. and Lutz, E. (1993) Environmental economics and valuation in development decisionmaking, in *Environmental Economics and Natural Resource Management in Developing Countries* (M. Munasinghe, Ed.), The World Bank, Washington, DC, 17–71.

Myers, N. (1986) Economics and ecology in the international arena: the phenomenon of linked linkages, *Ambio*, **15**, 296–300.

Nelson, R. (1990) *Dryland Management: The Desertification Problem*, World Bank Technical Paper 116, The World Bank, Washington, DC.

Ngoile, M.A.K. and Horrill, C.J. (1993) Coastal ecosystems, productivity and ecosystem protection: coastal ecosystem management, *Ambio*, **22**, 461–467.

Nutalaya, P., Yong, R.N., Chumnankit, T. and Buapeng, S. (1996) Land subsidence in Bangkok during 1978–1988, in *Sea-Level Rise and Coastal Subsidence* (J.D. Milliman and B.U. Haq, Eds), Kluwer Academic, Dordrecht, 105–130.
Odum, W.E. (1976) *Ecological Guidelines for Tropical Coastal Development*, International Union for Conservation of Nature and Natural Resources, Morges, Switzerland.
OECD (Organization for Economic Co-operation and Development) (1996) *Implementation Strategies for Environmental Taxes*, OECD, Paris.
Oerlemans, J. (1994) Quantifying global warming from the retreat of glaciers, *Science*, **264**, 243–245.
Oldeman, L.R., Hakkeling, R.T.A. and Sombroek, W.G. (1991) *World Map of the Status of Human-Induced Soil Degradation: An Explanatory Note*, International Soil Reference and Information Centre, Wageningen.
Olsson, L. (1993) On the cause of famine – drought, desertification and market failure in the Sudan, *Ambio*, **22**, 395–403.
Pagiola, S. (1995) *Environmental and Natural Resource Degradation in Intensive Agriculture in Bangladesh*, The World Bank, Washington, DC.
Parikh, K. (1995) Sustainable development and the role of tax policy, *Asian Development Review*, **13**(1), 127–166.
Pearce, D., Barbier, E. and Markandya, A. (1990) *Sustainable Development: Economics and Environment in the Third World*, Earthscan, London.
Peet, J. (1992) *Energy and the Ecological Economics of Sustainability*, Island Press, Washington, DC.
Phang, S.Y., Wong, W.K. and Chia, N.C. (1996) Singapore's experience with car quotas: issues and policy processes, *Transport Policy*, **3**(4), 145–153.
Pitts, J. (1992) Slope stability in Singapore, in *Physical Adjustments in a Changing Landscape: The Singapore Story* (A. Gupta and J. Pitts, Eds), Singapore University Press, Singapore, 259–300.
Pretty, J.N., Thompson, J. and Kiara, J.K. (1995) Agricultural regeneration in Kenya: the catchment approach to soil and water conservation, *Ambio*, **24**, 7–15.
Prinn, R.G. (1994) The interactive atmosphere: global atmospheric–biospheric chemistry, *Ambio*, **23**, 50–61.
Putnam, P.C. (1953) *Energy in the Future*, Van Nostrand, New York.
Rabinovitch, J. and Leitman, J. (1996) Urban planning in Curitiba, *Scientific American*, **274**(3), 26–33.
Rajasuriya, A., De Silva, M.W.R.N. and Öhman, M.C. (1995) Coral reefs of Sri Lanka: human disturbance and management issues, *Ambio*, **24**, 428–437.
Ramanathan, V. (1988) The greenhouse theory of climate change: a test by an inadvertent global experiment, *Science*, **240**, 293–299.
Randrianarijaona, P. (1983) The erosion of Madagascar, *Ambio*, **12**, 308–311.
Rapp, A., Murray-Rust, D.H., Christiansson, C. and Berry, L. (1972) Soil erosion and sedimentation in four catchments near Dodoma, Tanzania, *Geografiska Annaler*, **54A**, 255–318.
Reid, W.V., Laird, S.A., Meyer, C.A., Gámez, R., Sitttenfeld, A., Janzen, D.H., Gollin, M.A. and Juma, C. (1993) *Biodiversity Prospecting: Using Genetic Resources for Sustainable Development*, World Resources Institute, Washington, DC.
Renberg, I., Korsman, T. and Anderson, N.J. (1993) A temporal perspective of lake acidification in Sweden, *Ambio*, **22**, 264–271.
Repetto, R. (1988) *The Forest for the Trees? Government Policies and the Misuse of Forest Resources*, World Resources Institute, Washington, DC.
Repetto, R. (1990) Deforestation in the tropics, *Scientific American*, **262**(4), 18–24.
Repetto, R. (1992) Accounting for environmental assets, *Scientific American*, **266**(6), 64–70.
ReVelle, C. and ReVelle, P. (1974) *Sourcebook on the Environment: The Scientific Perspective*, Houghton Mifflin, Boston.
Rice, R.E., Gullison, R.E. and Reid, J.W. (1997) Can sustainable management save tropical forests? *Scientific American*, **276**(4), 34–39.

Rich, B. (1994) *Mortgaging the Earth: The World Bank, Environmental Impoverishment and the Crisis of Development*, Beacon Press, Boston.
Rich, B. (1996) Sustainable development: a broader perspective (an interview by M. Hsu), *The Brown Journal of World Affairs*, **3**(2), 305–316.
Richardson, S.D. (1956) *Forestry in Communist China*, Johns Hopkins University Press, Baltimore.
Rohde, H. (1989) Acidification in a global perspective, *Ambio*, **18**, 155–160.
Rosemarin, A. (1990) Some background on CFCs, *Ambio*, **19**, 280.
Rosencranz, A. and Milligan, R. (1990) CFC abatement: the needs of developing countries, *Ambio*, **19**, 312–316.
Rowland, F.S. (1990) Stratospheric ozone depletion by chlorofluorocarbons, *Ambio*, **19**, 281–292.
Ruangpanit, N. (1985) Percent crown cover related to water and soil losses in mountainous forest in Thailand, in *Soil Erosion and Conservation* (S.A. El-Swaify, W.C. Moldenhauer and A. Lo, Eds), Soil Conservation Society of America, 462–471.
Sainath, P. (1996) *Everybody Loves a Good Drought*, Penguin Books, New Delhi.
Salati, E. (1987) The forest and the hydrologic cycle, in *The Geophysiology of Amazonia: Vegetation and Climate Interactions* (R.E. Dickinson, Ed.), John Wiley, New York, 273–296.
Samad, S.A., Watanabe, T. and Kim, S.-J. (1995) *People's Initiatives for Sustainable Development: Lessons of Experience*, Asian and Pacific Development Centre, Kuala Lumpur.
Sayer, J.A. (1992) Development assistance strategies to conserve Africa's rainforests, in *Conservation of West and Central African Rainforests* (K. Cleaver, M. Munasinghe, M. Dyson, N. Egli, A. Peuker and F. Wencélius, Eds), Environment Paper 1, the World Bank, Washington, DC.
Sayer, J.A., Harcourt, C.S. and Collins, N.M. (1992) *The Conservation Atlas of Tropical Forests: Africa*, Macmillan, London.
Schimel, D.S. (1995) Terrestrial ecosystems and the carbon cycle, *Global Change Biology*, **1**, 77–91.
Schmandt, J. and Clarkson, J. (Eds) (1992) *The Regions and Global Warming: Impacts and Response Strategies*, Oxford University Press, New York.
Schmidheiny, S. with the Business Council for Sustainable Development (1992) *Changing Course: A Global Business Perspective on Development and the Environment*, MIT Press, Cambridge, MA.
School of Environmental Studies, Jadavpur University (1994) *Arsenic in Groundwater in Six Districts of West Bengal: The Biggest Arsenic Calamity in the World*, unpublished preliminary report, Jadavpur University, Calcutta.
Schubert, C. (1992) The glaciers of the Sierra Nevada de Merida (Venezuela): a photographic comparison of the recent deglaciation, *Erdkunde*, **46**, 58.
Schulze, P.C. (Ed.) (1996) *Engineering within Ecological Constraints*, National Academy of Engineering, National Academy Press, Washington, DC.
Schulze, P.C., Frosch, R.A. and Risser, P.G. (1996) Overview and perspectives, in *Engineering within Ecological Constraints* (P.C. Schulze, Ed.), National Academy of Engineering, National Academy Press, Washington, DC, 1–10.
Schumacher, E.F. (1973) *Small is Beautiful*, Harper and Row, New York.
Sen, A. (1981) *Poverty and Famines: An Essay on Entitlement and Deprivation*, Oxford University Press, Oxford.
Serageldin, I. (1996) Sustainability as opportunity and the problem of social capital, *The Brown Journal of World Affairs*, **3**(2), 187–203.
Setzer, A.W. and Pereira, M.C. (1991) Amazonia biomass burnings in 1987 and an estimate of their tropospheric emissions, *Ambio*, **20**, 19–22.
Shalaby, A.M. (1988) High Aswan Dam and environmental impact, *Transactions, 16th Congress on Large Dams*, San Francisco, ICOLD, Paris.

Sham Sani (1987) *Urbanization and the Atmospheric Environment in the Low Tropics: Experiences from the Kelang Valley Region, Malaysia*, Penerbit Universiti Kebangsaan Malaysia, Bangi.
Sharifa Mastura, S.A. (1987) Conflict at Port Dickson, Malaysia, in *International Geomorphology 1986* (V. Gardner, Ed.), John Wiley, Chichester, 319–335.
Sharma, S. (1996) The World Bank and the Global Environment Facility: challenges and prospects for sustainable development, *Brown Journal of World Affairs*, **3**(2), 275–287.
Shepherd, J.B. (1971) A study of earthquake risk in Jamaica and its influence on physical development planning, Ministry of Finance and Planning, Kingston, unpublished report.
Shiva, V., Bandyopadhyay, J. and Jayal, N.D. (1985) Afforestation in India: problems and strategies, *Ambio*, **14**, 329–333.
Shome, P. (1995) *Global Taxes*, National Institute of Public Finance and Policy, New Delhi, Working Paper No. 5.
Shukla, J., Nobre, C. and Sellers, P. (1990) Amazon deforestation and climate change, *Science*, **247**, 1322–1325.
Singh, D.R. and Gupta, P.N. (1982) Assessment of siltation in Tehri Reservoir, Proceedings, International Symposium on Hydrological Aspects of Mountainous Watersheds, Vol. VII, University of Roorkee, Roorkee, Mangalik Prakashan, Saharanpur, India, pp. 60–66.
Smil, V. (1984) *The Bad Earth: Environmental Degradation in China*, Zed Press, London.
Smith, D.B. and van der Wansem, M. (1995) *Strengthening EIA Capacity in Asia: Environmental Impact Assessment in the Philippines, Indonesia, and Sri Lanka*, World Resources Institute, Washington, DC.
Solow, R.M. (1996) *Human Development Report 1996, UNDP*, Oxford University Press, Oxford.
Sperling, D. (1996) The case for electric vehicles, *Scientific American*, **275**(5), 36–41.
Ståhl, M. (1993) Land degradation in East Africa, *Ambio*, **22**, 505–508.
Steer, A. and Lutz, E. (1994) Measuring environmentally sustainable development in *Making Development Sustainable: From Concepts to Action* (I. Serageldin and A. Steer, Eds) ESD Occasional Paper 2, The World Bank, Washington, DC, 17–20.
Steinbeck, J. (1939) *The Grapes of Wrath*, Heinemann, London.
Stiglitz, J.E. (1988) *Economics of the Public Sector*, W.W. Norton, New York.
Stigson, B. (1996) Eco-efficiency as the business norm for the 21st century: the challenge to industry and government, *The Brown Journal of World Affairs*, **3**(2), 289–297.
Stoker, H.S. and Seager, S.L. (1976) *Environmental Chemistry: Air and Water Pollution*, Scott, Freeman and Co., Glenview.
Strahler, A.N. (1975) *Physical Geography*, John Wiley, New York.
Strahler, A.H. and Strahler, A.N. (1992) *Modern Physical Geography*, John Wiley & Sons, New York.
Tegart, W.J.McG., Sheldon, G.W. and Griffiths, D.C. (Eds) (1990) *Climate Change: The IPCC Impacts Assessment*, Australian Government Printing Service, Canberra.
Theophrastus (1916) *Enquiry into Plants and Minor Works on Odours and Water Signs with an English translation by Sir Arthur Hort*, Heinemann, London.
Third World Network (1992) *Earth Summit Briefings*, Third World Network, Penang.
Thompson, L.G., Mosley-Thompson, E., Davis, M.E., Lin, P.N., Yao, T., Dyurgerov. M. and Dai, J. (1993) Recent warming: ice core evidence from tropical ice cores with emphasis on Central Asia, *Global Planetary Changes*, **7**, 145.
Tiffen, M., Mortimore, M. and Gichuki, F. (1994) *More People, Less Space: Environmental Recovery in Kenya*, John Wiley, Chichester.
Trexler, M.C. and Haugen, C. (1994) *Keeping it Green: Tropical Forestry Opportunities for Mitigating Climate Change,* World Resources Institute, Washington, DC.
Turner, R.K., Pearce, D. and Bateman, I. (1994) *Environmental Economics*, Prentice-Hall, Englewood Cliffs.
Udall, S.L. (1963) *The Quiet Crisis*, Holt, Rinehart and Winston, New York.

United Nations (1993a) *Handbook of National Accounting: Integrated Environmental Accounting, Studies in Methods*, Series F, 61, Department of Economic and Social Accounting and Policy Analysis, United Nations.
United Nations (1993b) *World Urbanization Prospects: The 1992 Revision*, United Nations Publications ST/ESA/SER.A/136, New York.
United Nations Conference on Trade and Development (1994) *The Outcome of the Uruguay Round: An Initial Assessment*, United Nations, New York.
United Nations, Department for Development Support and Management Services, Energy Branch (1994) *Trends in Environmental Impact Assessment of Energy Projects*, United Nations Publications ST/TCD/EB/3, New York.
United Nations (1994) *Population, Environment and Development: Proceedings of the United Nations Expert Group Meeting on Population, Environment and Development*, United Nations Headquarters, 20–24 January 1992, UN Publication ST/ESA/SER.A/129, United Nations, New York.
United Nations (1995) *World Urbanization Prospects: The 1994 Revision, Estimates and Projections of Urban and Rural Populations and of Urban Agglomerations*, United Nations Publications ST/ESA/SER.A/150, United Nations, New York.
United Nations (forthcoming) *World Urbanization Prospects: The 1996 Revision*, United Nations Publications ST/ESA/SER.A/170, Population Division of the Department of Economic and Social Affairs, United Nations Secretariat, United Nations, New York.
United Nations Development Programme (1995) *Human Development Report*, Oxford University Press, Oxford.
United Nations Development Programme, United Nations Environment Programme and the World Bank (1994) *Global Environment Facility: Independent Evaluation of the Pilot Phase*, The World Bank, Washington, DC.
Van der Leun, J.C., Tang, X. and Tevini, M. (1995) Environmental effects of ozone depletion: 1994 assessment, *Ambio*, **24**, 138.
Viles, H. and Spencer, T. (1995) *Coastal Problems: Geomorphology, Ecology and Society at the Coast*, Edward Arnold, London.
Ward, B. (1979) *Progress for a Small Planet*, Penguin Books, Harmondsworth.
Ward, B. and Dubos, R. (1972) *Only One Earth: The Care and Maintenance of a Small Planet*, Penguin Books, Harmondsworth.
Wathern, P. (Ed.) (1988) *Environmental Impact Assessment: Theory and Practice*, Routledge, London.
Weber, T. (1988) *Hugging the Trees: The Story of the Chipko Movement*, Penguin Books, Harmondsworth.
Whipple, C.G. (1996) Can nuclear waste be stored safely at Yucca Mountain? *Scientific American*, **274**(6), 56–64.
White, G.F. (1988) The environmental effects of the high dam at Aswan, *Environment*, **30**, 7.
Whitmore, T.C. (1991) *An Introduction to Tropical Rain Forests*, Clarendon Press, Oxford.
Wijkman, A. (1996) Stumbling blocks on the road to sustainable development, *The Brown Journal of World Affairs*, **3**(2), 177–186.
Wilkinson, C.R., Chou, L.M., Gomez, E., Ridzwan, A.R., Soekarno, S. and Sudra, S. (1993) Status of coral reefs in Southeast Asia: threats and responses, in *Colloquium on Global Aspects of Corals: Health, Hazard and History* (Compiler: R. Ginsburg), University of Miami, Miami.
Willetts, P. (1982) *Pressure Groups in the Global System: The Transnational Relations of Issue-oriented Non-governmental Organizations*, Frances Pinter, London.
Williams, G.P. and Wolman, M.G. (1984) Downstream effects of dams on alluvial rivers, *US Geological Survey Professional Paper 1286*.
Wolman, A. (1965) The metabolism of cities, *Scientific American*, **213**, 179–188.
Wolman, M.G. (1967) A cycle of sedimentation and erosion in urban river channels, *Geografiska Annaler*, **49A**, 385–395.
World Bank (1991a) *The World Bank and the Environment: A Progress Report, Fiscal 1991*, World Bank, Washington, DC.

World Bank, Environment Department (1991b) *Environment Assessment Sourcebook, Vol. I: Policies, Procedures, and Cross-sectional Issues*, World Bank Technical Paper 139, Washington, DC.
World Bank, Environment Department (1991c) *Environmental Assessment Sourcebook, Vol. II: Sectoral Guidelines*, World Bank Technical Paper 140, Washington, DC.
World Bank, Environment Department (1991d) *Environmental Assessment Sourcebook, Vol. III: Guidelines for Environmental Assessment of Energy and Industry Projects*, World Bank Technical Paper 154, Washington, DC.
World Bank (1992) *World Development Report 1992: Development and the Environment*, Oxford University Press, Oxford.
World Bank, Environmental Department (1993) *The World Bank and Environmental Assessment: An Overview*, Environmental Assessment Sourcebook Update, 1, World Bank, Washington, DC.
World Bank (1994) *Averting the Old Age Crisis*, Oxford University Press, Oxford.
World Bank (1995a) *Global Economic Prospects and the Developing Countries*, The World Bank, Washington, DC.
World Bank, Environment Department (1995b) *Environment Assessment: Challenges and Good Practice*, The World Bank, Washington, DC, Paper No. 018.
World Bank, Environment Department (1995c) *National Environmental Strategies: Learning from Experience*, World Bank, Washington, DC.
World Bank (1996a) *The World Bank Atlas 1996*, The World Bank, Washington, DC.
World Bank (1996b) *From Plan to Market: World Development Report 1996,* Oxford University Press, Oxford.
World Bank (1997) *Advancing Sustainable Development: The World Bank and Agenda 21*, Environmentally Sustainable Development Studies and Monographs Series 19, Washington, DC.
World Commission on Environment and Development (1988) *Our Common Future*, Oxford University Press, Oxford.
World Health Organization and United Nations Environmental Programme (1992) *Urban Air Pollution in Megacities of the World*, WHO, UNEP, Blackwell, Oxford.
World Rainforest Movement (1990) *The Battle for Sarawak's Forests*, World Rainforest Movement and Sahabat Alam Malaysia, Penang.
World Resources Institute (1988) *World Resources 1988–89*, Basic Books, New York.
World Resources Institute (1990) *World Resources 1990–91*, Oxford University Press, New York.
World Resources Institute (1992) *World Resources 1992–93*, Oxford University Press, New York.
World Resources Institute (1994) *World Resources 1994–95*, Oxford University Press, New York.
World Resources Institute (1996) *World Resources 1996–97: The Urban Environment*, Oxford University Press, New York.
Zorpette, G. (1996) Hanford's nuclear wasteland, *Scientific American*, **274**(5), 72–81.

Index

Note: page references in **bold** refer to figures